Nikolay I. Kolev

Multiphase Flow Dynamics 3

Nikolay I. Kolev

Multiphase Flow Dynamics 3
Turbulence, Gas Absorption and Release, Diesel Fuel Properties

1st Edition

With 55 Figures

 Springer

Nikolay Ivanov Kolev, PhD., DrSc.
Möhrendorferstr. 7
91074 Herzogenaurach
Germany
Nikolay.Kolev@herzovision.de

Library of Congress Control Number: 2007929113

ISBN 978-3-540-71442-2   Springer Berlin Heidelberg New York

This work is subject to copyright. All rights are reserved, whether the whole or part of the material is concerned, specifically the rights of translation, reprinting, reuse of illustrations, recitation, broadcasting, reproduction on microfilm or in other ways, and storage in data banks. Duplication of this publication or parts thereof is permitted only under the provisions of the German Copyright Law of September 9, 1965, in its current version, and permission for use must always be obtained from Springer-Verlag. Violations are liable to prosecution under German Copyright Law.

Springer is a part of Springer Science+Business Media
springer.com

© Springer-Verlag Berlin Heidelberg 2002, 2005 and 2007

The use of general descriptive names, registered names, trademarks, etc. in this publication does not imply, even in the absence of a specific statement, that such names are exempt from the relevant protective laws and regulations and therefore free for general use.

Production: Integra Software Services Pvt. Ltd., Pondicherry, India
Cover design: Erich Kirchner, Heidelberg

Printed on acid-free paper   SPIN: 11928270   45/3100/Integra   5 4 3 2 1 0

To Iva, Rali and Sonja with love!

(*Somewhere in Bulgaria, 2005, painted by Nikolay Ivanov Kolev*)

(*My first physics teacher, my father Ivan Gutev, 2000, painted by Nikolay Ivanov Kolev*)

Родина

Не си нещо повече дълбоко в мене сещам,
отколкото земя в граници обърната,
по тръпнещата болка, която често чувствам,
с косурите ти спорейки и с вяра невърната.

Намирам те в изгарящия порив
да те видя ковяща мощни технологии
и новото пол синия ти покрив,
да помита фразеологии и демагогии.

Намирам те в пулсиращата мисъл
от своя интелект нещо да ти дам,
че мускули и лакти животът е отписал
от средствата градящи прогреса тъй желан.

В онези хора те намирам,
които с възрожденски дух горят
и с пламъка си бъдното трасират,
единствена полза те за теб да извлекат.

Те често си остават неразбрани,
понякога ги смазва простотия,
но макар и след години, лекувайки тез рани,
превръщат делото им в светиня.

1983 Sofia

Summary

Volume 3 is devoted to selected subjects in the multiphase fluid dynamics that are very important for practical applications but could not find place in the first two volumes of this work.

The state of the art of the turbulence modeling in multiphase flows is presented. As introduction, some basics of the single-phase boundary layer theory including some important scales and flow oscillation characteristics in pipes and rod bundles are presented. Then the scales characterizing the dispersed flow systems are presented. The description of the turbulence is provided at different levels of complexity: simple algebraic models for eddy viscosity, algebraic models based on the *Boussinesq* hypothesis, modification of the boundary layer share due to modification of the bulk turbulence, modification of the boundary layer share due to nucleate boiling. Then the role of the following forces on the matematical description of turbulent flows is discussed: the lift force, the lubrication force in the wall boundary layer, and the dispersion force. A pragmatic generalization of the *k-eps* models for continuous velocity field is proposed containing flows in large volumes and flows in porous structures. Its large eddy simulation variant is also presented. Method of how to derive source and sinks terms for multiphase k-eps models is presented. A set of 13 single- and two-phase benchmarks for verification of k-eps models in system computer codes are provided and reproduced with the IVA computer code as an example of the application of the theory. This methodology is intended to help other engineers and scientists to introduce this technology step-by-step to their own engineering practice.

In many practical applications gases are dissolved in liquids under given conditions, released under other conditions and therefore affecting technical processes for good or for bad. There is almost no systematic description of this subject in the literature. That is why I decided to collect in Volume 3 useful information on the solubility of oxygen, nitrogen, hydrogen and carbon dioxide in water valid within large interval of pressures and temperatures, provide appropriate mathematical approximation functions and validate them. In addition methods for computation of the diffusion coefficients are described. With this information solution and dissolution dynamics in multiphase fluid flows can be analyzed. For this purpose the non-equilibrium absorption and release on bubble-, droplet- and film-surfaces under different conditions is mathematically described.

VIII Summary

In order to allow the application of the theory from all the three volumes also to processes in combustion engines a systematic set of internally consistent state equations for diesel fuel gas and liquid valid in broad range of changing pressure and temperature are provided also in Volume 3.

Erlangen, October 2006   Nikolay Ivanov Kolev

Table of contents

1 Some basics of the single-phase boundary layer theory 1
    1.1 Flow over plates, velocity profiles, share forces, heat transfer 1
        1.1.1 Laminar flow over the one site of a plane 1
        1.1.2 Turbulent flow parallel to plane 2
    1.2 Steady state flow in pipes with circular cross sections 4
        1.2.1 Hydraulic smooth wall surface 6
        1.2.2 Transition region 14
        1.2.3 Complete rough region 14
        1.2.4 Heat transfer to fluid in a pipe 15
    1.3 Transient flow in pipes with circular cross sections 21
    Nomenclature 23
    References 26

2 Introduction to turbulence of multi-phase flows 29
    2.1 Basic ideas 29
    2.2 Isotropy 40
    2.3 Scales, eddy viscosity 41
        2.3.1 Small scale turbulent motion 41
        2.3.2 Large scale turbulent motion, *Kolmogorov-Pandtl* expression 42
    2.4 k-eps framework 44
    Nomenclature 48
    References 53

3 Sources for fine resolution outside the boundary layer 55
    3.1 Bulk sources 55
        3.1.1 Deformation of the velocity field 55
        3.1.2 Blowing and suction 55
        3.1.3 Buoyancy driven turbulence generation 56
        3.1.4 Turbulence generated in particle traces 57
    3.2 Turbulence generation due to nucleate boiling 61
    3.3 Treatment of the boundary layer for non boiling flows 62
    3.4 Initial conditions 65
    Nomenclature 66
    References 73

4 Source terms for *k-eps* models in porous structures ... 75
- 4.1 Single phase flow ... 75
  - 4.1.1 Steady developed generation due to wall friction ... 75
  - 4.1.2 Heat transfer at the wall for steady developed flow ... 79
  - 4.1.3 Heat transfer at the wall for non developed or transient flow ... 80
  - 4.1.4 Singularities ... 81
- 4.2 Multi-phase flow ... 81
  - 4.2.1 Steady developed generation due to wall friction ... 81
  - 4.2.2 Heat transfer at the wall for forced convection without boiling ... 83
  - 4.2.3 Continuum-continuum interaction ... 84
  - 4.2.4 Singularities ... 85
  - 4.2.5 Droplets deposition at walls for steady developed flow ... 87
  - 4.2.6 Droplets deposition at walls for transient flow ... 87
- Nomenclature ... 88
- References ... 91

5 Influence of the interfacial forces on the turbulence structure ... 93
- 5.1 Drag forces ... 93
- 5.2 The role of the lift force in turbulent flows ... 93
- 5.3 Lubrication force in the wall boundary layer ... 98
- 5.4 The role of the dispersion force in turbulent flows ... 99
  - 5.4.1 Dispersed phase in laminar continuum ... 99
  - 5.4.2 Dispersed phase in turbulent continuum ... 100
- Nomenclature ... 104
- References ... 106

6 Particle–eddy interactions ... 109
- 6.1 Three popular modeling techniques ... 109
- 6.2 Particle–eddy interaction without collisions ... 110
  - 6.2.1 Response coefficient for single particle ... 110
  - 6.2.2 Responds coefficient for clouds of particles ... 112
  - 6.2.3 Particle–eddy interaction time without collisions ... 112
- 6.3 Particle–eddy interaction with collisions ... 113
- Nomenclature ... 114
- References ... 116

7 Two group *k-eps* models ... 119
- 7.1 Single phase flow ... 119
- 7.2 Two-phase flow ... 120
- Nomenclature ... 121
- References ... 123

8 Set of benchmarks for verification of *k-eps* models in system computer codes ... 125
- 8.1 Introduction ... 125
- 8.2 Single phase cases ... 126

    8.3 Two-phase cases 137
    Conclusions...139
    Nomenclature..140
    References..142

9 Simple algebraic models for eddy viscosity in bubbly flow.......................145
    9.1 Single phase flow in rod bundles 145
        9.1.1   Pulsations normal to the wall 146
        9.1.2   Pulsation through the gap 147
        9.1.3   Pulsation parallel to the wall 150
    9.2 Two phase flow...150
        9.2.1   Simple algebraic models..150
        9.2.2   Local algebraic models in the framework
                of the *Boussinesq*'s hypothesis..154
        9.2.3   Modification of the boundary layer share due to modification
                of the bulk turbulence..162
    Nomenclature..163
    References..169

10 Large eddy simulations 173
    10.1 Phenomenology 173
    10.2 Filtering – brief introduction..173
    10.3 The extension of the *Amsden* et al. LES model to porous structures 177
    Nomenclature..182
    References..184

11 Solubility of O_2, N_2, H_2 and CO_2 in water 185
    11.1 Introduction..185
    11.2 Oxygen in water...193
    11.3 Nitrogen water 199
    11.3 Hydrogen water..203
    11.4 Carbon dioxide–water..206
    11.5 Diffusion coefficients 209
    11.6 Equilibrium solution and dissolution 211
    Nomenclature..212
    References..214

12 Transient solution and dissolution of gasses in liquid flows 215
    12.1 Bubbles 216
        12.1.1 Existence of micro-bubbles in water..219
        12.1.2 Heterogeneous nucleation at walls..221
        12.1.3 Steady diffusion mass transfer of the solvent
                across bubble interface...224
        12.1.4 Initial bubble growth in wall boundary layer...............................228
        12.1.3 Transient diffusion mass transfer of the solvent across
                the bubble interface...229

12.2 Droplets 239
    12.2.1 Steady state gas site diffusion 239
    12.2.2 Transient diffusion inside the droplet..................................... 243
12.3 Films 246
    12.3.1 Geometrical film-gas characteristics 246
    12.3.2 Liquid side mass transfer due to molecular diffusion............. 248
    12.3.3 Liquid side mass transfer due to turbulence diffusion............ 249
Nomenclature.. 257
References 263

13 Thermodynamic and transport properties of diesel fuel 269
13.1 Introduction 269
13.2 Constituents of diesel fuel 271
13.3 Averaged boiling point at atmospheric pressure 273
13.4 Reference liquid density point... 274
13.5 Critical temperature, critical pressure.. 275
13.6 Molar weight, gas constant.. 275
13.7 Saturation line 276
13.8 Latent heat of evaporation 279
13.9 The liquid density.. 280
    13.9.1 The volumetric thermal expansion coefficient 281
    13.9.2 Isothermal coefficient of compressibility................................ 283
13.10 Liquid velocity of sound.. 284
13.11 The liquid specific heat at constant pressure 285
13.12 Specific liquid enthalpy 288
13.13 Specific liquid entropy 290
13.14 Liquid surface tension 292
13.15 Thermal conductivity of liquid diesel fuel 292
13.16 Cinematic viscosity of liquid diesel fuel 294
13.17 Density as a function of temperature and pressure
     for diesel fuel vapor.. 295
13.18 Specific capacity at constant pressure for diesel vapor 296
13.19 Specific enthalpy for diesel fuel vapor 298
13.20 Specific entropy for diesel fuel vapor.. 299
13.21 Thermal conductivity of diesel fuel vapor... 300
13.22 Cinematic viscosity of diesel fuel vapor.. 301
References 301
Appendix 13.1 Dynamic viscosity and density for saturated
    n-octane vapor 302

Index 305

1 Some basics of the single-phase boundary layer theory

Hundreds of very useful constitutive relations describing interactions in multi-phase flows are based on the achievements of the single-phase boundary layer theory. That is why it is important to recall at least some of them, before stepping to more complex interactions in the multi-phase flow theory. My favorite book to begin with learning the main ideas of the single-phase boundary layer theory is the famous monograph by *Schlichting* (1982). This chapter gives only the basics which helps in understanding the following chapters of this book.

1.1 Flow over plates, velocity profiles, share forces, heat transfer

Consider continuum flow parallel to a plate along the x-axis having velocity far from the surface equal to u_∞. The share force acting on the surface per unit flow volume is then

$$f_w = \frac{F_w \tau_w}{V_{flow}}, \qquad (1.1)$$

where the wall share stress is usually expressed as

$$\tau_w = c_w(x) \frac{1}{2} \rho u_\infty^2. \qquad (1.2)$$

Here the friction coefficient c_w is obtained from the solution of the mass and momentum conservation at the surface.

1.1.1 Laminar flow over one site of a plane

For laminar flow over one site of a plane, the solution of the momentum equation delivers the local share stress as a function of the main flow velocity and of the distance from the beginning of the plate as follows

$$c_w(x) = \frac{\tau_w(x)}{\frac{1}{2}\rho u_\infty^2} = \frac{0.332}{\left(\frac{u_\infty x}{\nu}\right)^{1/2}}, \qquad (1.3)$$

Schlichting (1982) Eq. (7.32) p.140. The averaged drag coefficient over Δx is then

$$\overline{c_{w,\Delta x}} = \frac{F_w/(\Delta y \Delta x)}{\frac{1}{2}\rho u_\infty^2} = \frac{1.328}{\left(\frac{u_\infty \Delta x}{\nu}\right)^{1/2}}, \quad \mathrm{Re}_{\Delta x} < 5\times 10^5, \qquad (1.4)$$

Eq. (7.34) *Schlichting* (1982) p. 141. The corresponding heat transfer coefficients h are reported to be

$$Nu_x = \frac{hx}{\lambda} = \frac{1}{\sqrt{\pi}}\left(\frac{u_\infty x}{\nu}\right)^{1/2}\mathrm{Pr}^{1/2} \quad \text{for } \mathrm{Pr}\to 0 \text{ for liquid metals}, \qquad (1.5)$$

$$Nu_x = \frac{hx}{\lambda} = 0.332\left(\frac{u_\infty x}{\nu}\right)^{1/2}\mathrm{Pr}^{1/3} \quad \text{for } 0.6<\mathrm{Pr}<10, \qquad (1.6)$$

$$Nu_x = \frac{hx}{\lambda} = 0.339\left(\frac{u_\infty x}{\nu}\right)^{1/2}\mathrm{Pr}^{1/3} \quad \text{for } \mathrm{Pr}\to\infty, \qquad (1.7)$$

Schlichting (1982) p. 303. Averaging over Δx results in

$$\overline{Nu_{\Delta x}} = 2 Nu_{\Delta x}. \qquad (1.8)$$

1.1.2 Turbulent flow parallel to plane

For turbulent flow over one site of a plane the solution of the momentum equation gives the local share stress as a function of the main flow velocity and of the distance from the beginning of the plate as follows

$$c_w(x) = \frac{\tau_w(x)}{\frac{1}{2}\rho u_\infty^2} = \frac{0.0296}{\left(\frac{u_\infty x}{\nu}\right)^{1/5}}, \qquad (1.9)$$

Eq. (21.12) *Schlichting* (1982) p. 653. This equation is obtained assuming the validity of the so-called 1/7-th velocity profile,

$$\frac{u(x,y)}{u_{max}} = \left(\frac{y}{\delta(x)}\right)^{1/7} \qquad (1.10)$$

with boundary layer thickness varying with the distance from the beginning of the plate in accordance with

$$\delta(x) = 0.37 x \left(\frac{u_\infty x}{v}\right)^{1/5}. \qquad (1.11)$$

At the distance from the wall

$$y = \delta_{99\%} \approx 5\sqrt{\frac{u_\infty x}{v}} \qquad (1.12)$$

the velocity reaches 99% of the flow mean velocity. $\delta_{99\%}$ is called *displacement thickness*. The averaged *steady state* drag coefficient over Δx is then

$$\overline{c_{w,\Delta x}} = \frac{F_w/(\Delta y \Delta x)}{\frac{1}{2}\rho u_\infty^2} = \frac{0.074}{\text{Re}_{\Delta x}^{1/5}}, \; 5 \times 10^5 < \text{Re}_{\Delta x} < 10^7, \qquad (1.13)$$

Eq. (21.11) *Schlichting* (1982) p. 652. Here $\text{Re}_{\Delta x} = u_\infty \Delta x / v$. The corresponding steady state local and averaged heat transfer coefficient h are reported to be

$$Nu_x = \frac{hx}{\lambda} = 0.0296 \left(\frac{u_\infty x}{v}\right)^{0.8} \text{Pr}^{1/3}, \qquad (1.14)$$

$$\overline{Nu_{\Delta x}} = \frac{h \Delta x}{\lambda} = 0.037 \left(\frac{u_\infty \Delta x}{v}\right)^{0.8} \text{Pr}^{1/3}, \qquad (1.15)$$

respectively. The influence of the wall properties in the last equation is proposed by *Knudsen* and *Katz* (1958) to be taken into account by computing the properties at the following effective temperature

$$T_{eff} = T + \frac{0.1\text{Pr} + 40}{\text{Pr} + 72}(T_w - T).$$

The only information known to me for the influence of the unsteadiness of the far field velocity is those by *Sidorov* (1959)

$$\overline{c_{w,\Delta x}} = \frac{0.0263}{\left\{ \mathrm{Re}_{\Delta x} \left[1 - \left(1 - \frac{0.78}{\mathrm{Re}_{\Delta x}^{1/14}} \right)^{-1} \frac{1}{u_\infty^2} \frac{du_\infty}{d\tau} \right] \right\}^{1/7}} \quad (1.16)$$

1.2 Steady state flow in pipes with circular cross sections

Consider continuum flow along the *x*-axis of a circular pipe having velocity cross section averaged velocity equal to $\overline{w}$. The share force acting on the surface per unit flow volume is then

$$f_w = \frac{F_w \tau_w}{V_{flow}}, \quad (1.17)$$

where the wall share stress is usually expressed as

$$\tau_w = c_w \frac{1}{2} \rho \overline{w}^2. \quad (1.18)$$

Here the friction coefficient c_w, called *Fanning* factor in Anglo-Saxon literature, is obtained from the solution of the developed steady state mass and momentum conservation in the pipe. Replacing the wall surface to pipe volume ratio with $4/D_h$ we have for wall friction force per unit volume of the flow

$$f_w = \frac{F_w}{V_{flow}} c_w \frac{1}{2} \rho \overline{w}^2 = \frac{4}{D_h} c_w \frac{1}{2} \rho \overline{w}^2 = \frac{\lambda_{fr}}{D_h} \frac{1}{2} \rho \overline{w}^2. \quad (1.19)$$

Here

$$\lambda_{fr} = 4 c_w \quad (1.20)$$

called friction coefficient is usually used in Europe. Note the factor 4 between the *Fanning* factor and the friction coefficient and

$$\tau_w = \frac{\lambda_{fr}}{8} \rho \overline{w}^2. \quad (1.21)$$

Note that for a steady developed single flow the momentum equation reads $\frac{1}{r}\frac{d}{dr}(r\tau) - \frac{dp}{dx} = 0$. With $\frac{dp}{dx} = \frac{dp_w}{dx}$ and therefore $\frac{1}{r}\frac{d}{dr}(r\tau) - \frac{dp_w}{dx} = 0$ we have $d(r\tau) = \frac{dp_w}{dx} r dr$ or after integrating

$$\tau(r) = \frac{dp_w}{dx}\frac{r}{2}, \tag{1.22}$$

which gives at the wall the relation between the wall share stress and the pressure gradient due to friction

$$\tau_w = \frac{dp_w}{dx}\frac{R}{2}. \tag{1.23}$$

Usually for describing turbulent flows in pipe the following dimensionless variables are used: The friction velocity

$$w^* = \sqrt{\frac{\tau_w}{\rho}} = \bar{w}\sqrt{\frac{\lambda_{fr}}{8}}, \tag{1.24}$$

the dimensionless cross section averaged velocity

$$w^+ = w/w^*, \tag{1.25}$$

and the dimensionless distance from the wall

$$y^+ = yw^*/\nu, \tag{1.26}$$

where y is the distance from the wall. Note that yw^*/ν is in fact the definition of a boundary layer *Reynolds* number. With this transformation the measured mean velocity distribution near the wall is not strongly dependent on the *Reynolds* number as shown in Fig. 1.1. *Hammond* (1985) approximated this dependency by a continuous function of the type $y^+ = y^+(u^+)$ which have to be inverted iteratively if one needs $u^+ = u^+(y^+)$.

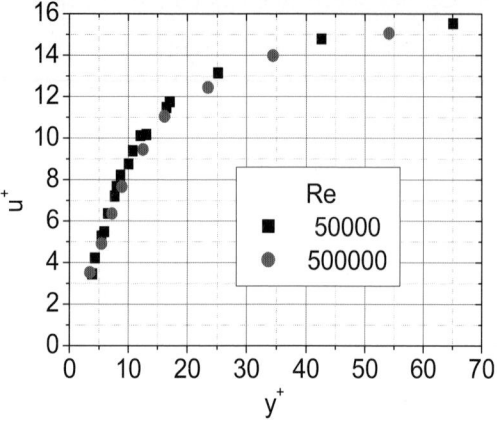

Fig. 1.1. Mean velocity distribution near the wall, *Laufer* (1953)

The penetration of the wall roughness k into the boundary layer dictates different solutions of practical interests. Usually the dimensionless roughness of the surface $k^+ = kw^*/\nu$ is compared to the characteristic dimensionless sizes of the boundary layer to define the validity region of specific solution of the momentum equation.

1.2.1 Hydraulic smooth wall surface

Hydraulic smooth surfaces are defined if

$$0 \leq k^+ \leq 5. \tag{1.27}$$

1.2.1.1 The Blasius solution

Historically *Blasius* obtained in 1911 the following equation,

$$\lambda_{fr} = 0.3164/\mathrm{Re}^{1/4}, \tag{1.28}$$

where $\mathrm{Re} = \overline{w} D_h / \nu$, validated with his data and the data of other authors for $\mathrm{Re} < 10^5$. Later it was found that the velocity profile that can be assumed to obtain this friction coefficient has the form

$$w(y) = \frac{(2n+1)(n+1)}{2n^2} \overline{w} \left(\frac{y}{R}\right)^{1/n} = w_{\max} \left(\frac{y}{R}\right)^{1/n}. \tag{1.29}$$

It is known from the *Nikuradze* measurements that the exponent is a function of *Reynolds* number: for $\text{Re} \leq 1.1 \times 10^5$, $n = 7$

$$w(y) = w_{max}\left(\frac{y}{R}\right)^{1/7} = \frac{60}{49}\overline{w}\left(\frac{y}{R}\right)^{1/7} \qquad (1.30)$$

or

$$w^+(y) = 8.74(y^+)^{1/7}. \qquad (1.31)$$

and for $\text{Re} \leq 3.2 \times 10^6$, $n = 10$.

1.2.1.2 The Collins et al. solution

More sophisticated than the *Blasius* profile which depends on the *Reynolds* number is proposed by *Collins* et al. (1978) and *Bendiksen* (1985):

$$\frac{w(r)}{w_{max}} = 1 - \gamma r^2 - (1-\gamma)r^{2n}, \qquad (1.32)$$

$$\gamma = 7.5/[4.12 + 4.95(\log \text{Re} - 0.743)], \qquad (1.33)$$

$$n = (\gamma - 1)\frac{\log \text{Re} - 0.743}{\log \text{Re} + 0.31}\left(1 - \frac{1}{2}\gamma\right)^{-1} - 1. \qquad (1.34)$$

The resulting friction coefficient is then

$$\frac{1}{\sqrt{\lambda_{fr}}} = 3.5\log \text{Re} - 2.6. \qquad (1.35)$$

1.2.1.3 The von Karman universal velocity profiles

Friction coefficient: Schlichting found that the data of *Nikuradse* for $\text{Re} < 3.4 \times 10^6$ are well reproduced by the velocity profile defined by $w_{max} - w(y) = w*2.5\ln(R/y)$, where $w = w_{max} - 4.07w^*$, resulting in the expression defining the friction factor

$$\frac{1}{\sqrt{\lambda_{fr}}} = 2\log\left(\text{Re}\sqrt{\lambda_{fr}}\right) - 0.8, \qquad (1.36)$$

Schlichting (1982) p. 624.

Velocity profiles: The more accurate mathematical representation of the velocity profiles on Fig. 1.1 is generalized by *von Karman*. The *Prandtl* mixing length theory is used. The effective turbulent cinematic viscosity is assumed to be proportional to the velocity gradient $v' = \ell^2 \left| \dfrac{dw}{dy} \right|$ outside the laminar boundary layer. The proportionality factor, the so-called *mixing length*, is proposed to be proportional to the wall distance,

$$\ell = \kappa y, \tag{1.37}$$

a lucky abstract assumption which turns useful. The constant $\kappa = 0.4$ is called the *von Karman* constant. The share stress in the boundary layer is then

$$\tau = \rho \ell^2 \left| \dfrac{dw}{dy} \right| \dfrac{dw}{dy}. \tag{1.38}$$

This equation can be made dimensionless, $dw^+ = \dfrac{1}{k} d \ln y^+$, with the friction velocity $w^* = ky \dfrac{dw}{dy}$ and then integrated analytically resulting in

$$w^+ = \dfrac{1}{k} \ln y^+ + const. \tag{1.39}$$

Von Karman introduced a buffer zone between the viscous and the fully turbulent layer already introduced by *Prandtl*. In the first zone close to the wall the flow is laminar. In the second and in the third the constants are so computed in order to have smooth profiles. *Schlichting* summarizes the velocity profiles in Table 1.1.

Table 1.1. Velocity profiles in the boundary layer

Sub layer	Defined by	Velocity Profile
Viscous	$y^+ \leq 5$	$w^+ = y^+$
Buffer	$5 \leq y^+ \leq 30$	$w^+ = 5 \ln y^+ - 3.08$
fully turbulent	$30 < y^+$	$w^+ = 2.5 \ln y^+ + 5.5$

Table 1.2 contains some important integrals of the universal profile widely used for several purposes. Γ^+ is the volumetric flow rate per unit width of the wall.

1.2 Steady state flow in pipes with circular cross sections

Table 1.2. Dimensionless volumetric flow per unit width of the surface

Sub layer	$\Gamma^+ = \int_0^{\delta^+} w^+ dy^+$	$\int_0^{\delta^+} \left(\dfrac{dw^+}{dy^+}\right)^2 dy^+$
Viscous	$0.5(\delta^+)^2 \leq 12.5$	δ^+
Buffer	$5\delta^+ \ln \delta^+ - 8.08\delta^+ + 12.664$ ≤ 280.44	$5 + 25\left(\dfrac{1}{5} - \dfrac{1}{\delta^+}\right)$
fully turbulent	$2.5\delta^+ \ln \delta^+ + 3.5\delta^+ - 64.65$	$9.1667 + 6.25\left(\dfrac{1}{30} - \dfrac{1}{\delta^+}\right)$ *for* $\delta^+ \to \infty$, 9.735

Note that a boundary layer Reynolds number defined as

$$\mathrm{Re}_\delta = \bar{w} 4\delta_2 / \nu = \frac{\bar{w}}{w^*} \frac{4\delta_2 w^*}{\nu} = 4\Gamma^+$$

is used for many application of the film flow theory. For film flow analysis it is interesting to find the inversed dependences from Table 1.2, namely $\delta^+ = \delta^+(\Gamma^+)$. Approximations for such dependences using profiles with constant 3.05 instead 3.08 with an error less then 4% was reported by *Traviss* et al. (1973):

$\delta^+ = 0.707 \mathrm{Re}_\delta^{0.5}$ for $0 < \mathrm{Re}_\delta \leq 50$,

$\delta^+ = 0.482 \mathrm{Re}_\delta^{0.585}$ for $50 < \mathrm{Re}_\delta \leq 1125$,

$\delta^+ = 0.095 \mathrm{Re}_\delta^{0.812}$ for $1125 < \mathrm{Re}_\delta$.

After *Prandtl* more complicated expressions for the mixing length are proposed: *Van Driest* (1955) introduced the so-called damping function to give

$$\ell = \kappa y \left[1 - \exp(-y^+/26)\right]. \tag{1.40}$$

It is known that the constant depends actually on the Reynolds number and takes values between 20 and 30. For pipe flow *Nikuradse* (1932) proposed an expression which combined with the *Van Driest* damping factor to give

$$\frac{\ell}{R} = \left[0.14 - 0.08\left(1 - \frac{y}{R}\right)^2 - 0.06\left(1 - \frac{y}{R}\right)^4\right]\left[1 - \exp(-y^+/26)\right]. \tag{1.41}$$

Turbulence in the boundary layer: It was experimentally observed by *Laufer* (1952, 1953) that the fluctuations of the velocity are equilateral in the central part of the pipe but heterogeneous close to the wall (Fig. 1.2). *Laufer*'s data indicate that the fluctuations of the axial velocities in pipe in the wall region are about three times larger than the fluctuations in the other directions–heterogeneous turbulence. The radial fluctuation velocity can be approximated by a *Boltzmann* function

$$u'^+ = \frac{a_1 - a_2}{1 + e^{(y^+ - y_0^+)/dy^+}} + a_2,$$

where $a_1 = -30.33365$, $a_2 = 0.89475$, $y_0^+ = -43.51454$, $dy^+ = 12.72364$.

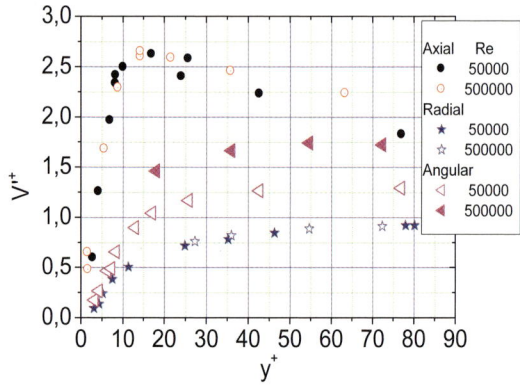

Fig. 1.2. Velocity fluctuations as a function of the distance from the wall measured by *Laufer* (1953)

Observe that the dimensionless radial fluctuation velocity is almost independent on the *Reynolds* number and that for $y^+ > 30$ it is around 0.9 to 1. *Vames* and *Hanratty* (1988) reviewed turbulent measurements in a pipe and reported that close to the wall $r \to R$ the fluctuation velocity is

$$v' \approx 0.9 w^* \tag{1.42}$$

the characteristic time scale of turbulent pulsation $\Delta\tau_e = 0.046 D_h / w^*$ and the eddy diffusivity is $\nu' = 0.037 w^* D$ (note that $\nu' = v'^2 \Delta\tau_e = 0.0414 w^* D$). Knowing the fluctuation of the normal to the wall velocity in a pipe flow is important for analyzing deposition processes in particle loaded flows. This is also essential for post critical heat transfer description in annular flow with droplets in the gas core for pipes and rod bundles. The data obtained for the pipes can be used for bundles due to the systematical experimental observations reported by *Rehme* (1992) p. 572: "…The experimental eddy viscosities normal to the wall are nearly independent on the relative gap width and are comparable to the data of circular tubes by *Reichardt* close to the walls…"

There are attempts to approximate the information presented in Fig. 1.2. *Matida* (1998) proposed the following approximations for the pulsation of the velocity components close to the wall neglecting the dependence on the *Reynolds* number:

$$\frac{u'}{w*} = \frac{0.5241 y^+}{1 + 0.0407 y^{+1.444}}, \tag{1.43}$$

$$\frac{v'}{w*} = \frac{0.00313 y^{+2}}{1 + 0.00101 y^{+2.253}}, \tag{1.44}$$

$$\frac{w'}{w*} = \frac{0.160 y^+}{1 + 0.0208 y^{+1.361}}. \tag{1.45}$$

In this region the fluctuation of the radial velocity measured for large *Reynolds* number by *Laufer* (1953) was approximated by *Lee* and *Durst* (1980) as follows

$$\frac{v'/w*}{\ell/R} = 2.9 \left(\frac{R}{y} \right)^{0.4}. \tag{1.46}$$

Johansen (1991) reported the following approximation for pipe and channels flows with radius or half with R valid not only in the boundary layer but in the entire cross section.

a) The eddy viscosity

$$\frac{v^t}{v} = \left(\frac{y^+}{11.15} \right)^3, \; y^+ < 3, \tag{1.47}$$

$$\frac{v^t}{v} = \left(\frac{y^+}{11.4} \right)^2 - 0.049774, \; 3 \le y^+ < 52.108, \tag{1.48}$$

$$\frac{v^t}{v} = 0.4 y^+, \; 52.108 \le y^+, \tag{1.49}$$

is in agreement with the profiles computed with direct numerical simulation reported by *Kim* et al. (1987).

b) The profile of the time averaged axial velocity in accordance with the above expressions is

$$\frac{w(y)}{w^*} = 11.4\tan^{-1}\left(\frac{y^+}{11.4}\right), \quad y^+ \leq y_0^+ \tag{1.50}$$

$$\frac{w(y)}{w^*} = 15.491 + 2.5\ln\left(\frac{1+0.4y^+}{1+0.4y_0^+}\right), \quad y^+ > y_0^+, \tag{1.51}$$

$$y_0^+ = 52.984. \tag{1.52}$$

c) The fluctuation of the velocity normal to the wall

$$v'^+ = \frac{v'}{u^*} \tag{1.53}$$

approximated with

$$v'^+ = 0.033 y^+ \left[1 - \exp\left(-\frac{y^+}{3.837}\right)\right], \tag{1.54}$$

$$v'^+ = \exp\left[-\left(\frac{y^+}{30}\right)^{7.82} \Big/ 7.82\right], \quad y^+ \leq 30, \tag{1.55}$$

$$v'^+ = v'^+(30) - \left[v'^+(30) - 0.65\right]\frac{y^+ - 30}{R^+ - 30}, \quad 30 \leq y^+ \leq R^+, \tag{1.56}$$

agree with the *Kutateladze* et al. (1979) measurements. The characteristic time scale of the fluctuation is then given by $\Delta\tau_e = v'/v'^2$.

Wall boundary conditions for 3D-modeling: Using $k-\varepsilon$ models in computer codes with large scale discretization is very popular nowadays. In these codes the boundary layer can not be resolved. For computing the bulk characteristics boundary conditions at the wall are required. Usually a point close to the wall e.g. $y_p^+ = 30$, is defined where the profile $w^+ = 2.5\ln y^+ + 5.5$ starts to be valid. At this point the values of the turbulent kinetic energy per unit mass and its dissipation are

$$k_p = 2.5 w^{*3}/y_p^+, \tag{1.57}$$

$$\varepsilon_p = w^{*2}/\sqrt{c_v} \qquad (1.58)$$

with $c_v = 0.09$ coming from the definition equation of the turbulent cinematic viscosity $v' = c_v k^2/\varepsilon$, see for instance *Lee* et al. (1986). *Bradshaw* (1967) found experimentally a useful relationship $k \approx const\ \tau_w/\rho$ that can be used for this purpose as approximation. *Harsha* and *Lee* (1970) provided extensive measurements showing the correctness of the relation $k \approx 3.3 \tau_w/\rho$ for wakes and jets. Computing the *Fanning* factor in $\tau_w = c_w \rho \overline{w}^2/2$ by using appropriate correlation we can approximately estimate the specific turbulent energy at the wall region. Note that *Alshamani* (1978) reported that there is no linearity in the boundary layer $y^+ > 5$ but $k \approx (2.24 w'^+ - 1.13) \tau_w/\rho$, where w'^+ is the dimensionless fluctuation of the axial velocity.

1.2.1.4 The Reichardt solution

In looking for appropriate turbulence description in pipe flow valid not only up to the boundary layer but up to the axis of the pipe *Reichardt* (1951) reproduced the available data for high *Reynolds* numbers in the form of turbulent cinematic viscosity as a function of the distance from the wall:

$$\frac{v'}{w^* R} = \frac{\kappa}{3}\left[\frac{1}{2}+\left(1-\frac{y}{R}\right)^2\right]\left[1-\left(1-\frac{y}{R}\right)^2\right] \qquad (1.59)$$

in

$$\tau = \rho(v+v')\frac{d\overline{w}}{dy},\ \tau/\tau_w = 1 - y/R. \qquad (1.60)$$

Making reasonable approximation *Reichardt* succeeded to obtain a single equation for the velocity profile that covers all the regions from the wall to the axis:

$$\frac{w(y)}{w^*} = 2.5 \ln\left[(1+0.4 y^+)\frac{1.5\left(2-\frac{y}{R}\right)}{1+2\left(1-\frac{y}{R}\right)^2}\right]$$

$$+7.8\left[1-\exp(-y^+/11)-\frac{y^+}{11}\exp(-0.33 y^+)\right]. \qquad (1.61)$$

For $y \ll R$ the expression simplifies to

$$\frac{w(y)}{w^*} = 2.5\ln(1+0.4y^+) + 7.8\left[1-\exp(-y^+/11) - \frac{y^+}{11}\exp(-0.33y^+)\right].$$

The derivative

$$\frac{dw}{dy} = \frac{w^{*2}}{\nu}\left\{\frac{1}{1+0.4y^+} + \frac{7.8}{11}\left[\exp(-y^+/11) + (0.33y^+ - 1)\exp(-0.33y^+)\right]\right\},$$

is very useful for computation of the lift force acting on small particles in the boundary layer at $y = R$. For upward bubbly flow it is directed from the wall into the bulk flow and for droplets from the boundary layer toward the wall.

Note that *Lee* et al. (1986) reported alternative form of the cinematic viscosity as a function of the distance from the wall:

$$\frac{\nu^t}{\nu} = 0.4y^+\left[1 - \frac{11}{6}\left(\frac{y}{R}\right) + \frac{4}{3}\left(\frac{y}{R}\right)^2 - \frac{1}{3}\left(\frac{y}{R}\right)^3\right]\left[1 - \exp\left(-\frac{y^+}{16}\right)\right]^2. \tag{1.62}$$

1.2.2 Transition region

The transition from the hydraulic smooth to complete rough region is defined by

$$5 \leq k^+ \leq 70. \tag{1.63}$$

In this case the velocity profile is $w_{max} - w(y) = w^* 2.5\ln(R/y)$, $\bar{w} = w_{max} - 3.75w^*$. The friction coefficient correlation proposed by *Colebrook* and *White* (1939)

$$\frac{1}{\sqrt{\lambda_{fr}}} = 1.74 - 2\log\left(\frac{k}{R} + \frac{18.7}{\mathrm{Re}\sqrt{\lambda_{fr}}}\right), \tag{1.64}$$

is valid for all roughness regimes. *Avdeev* (1982) proposed to use an explicit approximation of this equation which is more convenient,

$$\frac{1}{\sqrt{\lambda_{fr}}} = 1.74 - 2\log\left(\frac{k}{R} + \frac{49}{\mathrm{Re}^{0.91}}\right). \tag{1.65}$$

1.2.3 Complete rough region

The complete rough region is defined by

$$70 \leq k^+ . \tag{1.66}$$

The velocity profile is defined by $\dfrac{w(y)}{w*} = 5.75 \log\left(\dfrac{y}{k}\right) + 8.5$ resulting in $\bar{w}/w* = \ln(R/k) + 4.75$, $w_{max} - w(y) = 3.75 w*$. The fiction coefficient found by *von Karman* is

$$\frac{1}{\sqrt{\lambda_{fr}}} = 2\log\left(\frac{R}{k}\right) + 1.60 . \tag{1.67}$$

A slight change made by *Schlichting*

$$\frac{1}{\sqrt{\lambda_{fr}}} = 2\log\left(\frac{R}{k}\right) + 1.74 , \tag{1.68}$$

gives the best fit to the *Nikuradse* data.

White (2006) proposed the following approximation of the velocity profile with an offset depending on the type of the roughness

$$\frac{w(y)}{w*} = \frac{1}{\kappa}\ln\left(\frac{yw*}{v}\right) + B - \Delta B ,$$

where $\Delta B = \dfrac{1}{\kappa}\ln\left(1 + c_B \dfrac{kw*}{v}\right)$, $B = 5$, $\kappa = 0.41$, $c_B \approx 0.3$ for sand roughness, $c_B \approx 0.8$ for stationary wavy wall data. The advantage of this approach is that it can also be used for description of the drag coefficient between liquid and gas wavy interface. Comparing with data *Hulburt* et al. (2006) proposed to use $c_{B,2F,base} \approx 0.8$ for gas interaction with the *base* waves and $c_{B,2F,trav} \approx 4.7$ for gas interaction with the *traveling* waves.

1.2.4 Heat transfer to fluid in a pipe

Comparing the momentum and the energy conservation equation for steady developed flow in a pipe

$$\frac{\tau_w}{\rho}\left(1 - \frac{y}{R}\right) = (v + v^t)\frac{dw}{dy} , \tag{1.69}$$

$$\frac{\dot{q}_w''}{\rho c_p}\left(1-\frac{y}{R}\right) = -\left(a+a'\right)\frac{d\overline{T}}{dy}, \tag{1.70}$$

we realize immediately the similarity. Note that in this case

$$\boxed{\frac{\dot{q}_w''/c_p}{\tau_w} = -\frac{a+a'}{v+v'}\frac{d\overline{T}}{d\overline{u}}.} \tag{1.71}$$

For negligible molecular viscosity and conductivity compared to their turbulent counterpart valid for gases and $\Pr' = \frac{v'}{\lambda'/\rho c_p} \approx 1$ *Reynolds* come to the remarkable relation $\dot{q}_w''/\left(\rho \overline{w} c_p \Delta T\right) = \lambda_{fr}/8$, saying the heat transfer between wall and bulk flow is proportional to the friction coefficient. *Reynolds* proposal to consider the turbulent conductivity proportional to the eddy viscosity

$$\Pr' = \frac{\rho c_p v'}{\lambda'} \approx const, \tag{1.72}$$

was very fruitful in obtaining practical correlations for describing the heat transfer in pipes.

So knowing the dependence of the eddy diffusivity on the distance from the wall

$$v'(y) = \frac{\tau_w}{\rho\, d\overline{w}/dy}\left(1-\frac{y}{R}\right) - v,$$

the turbulent thermal conductivity is also known and Eq. (1.70) can be integrated. The result is the temperature distribution as a function of the wall distance. As an example we give below the *Martinelli* solution. The heat flux at the wall is then easily computed because the temperature gradient at the wall is known. This is the method leading many authors to derivation of correlation for the heat transfer at the wall – an important method for the technology achievement of the turbulence theory. The analogy is called *Reynolds* analogy. Expressions for heat transfer coefficient are generated by *Prandtl* and one more accurate by *von Karman* valid up to Pr = 25. Useful correlation set reviewed by *Kirillov* (1985) will be given below.

1.2.4.1 The Martinelli solution for temperature profile

The temperature profiles for heated walls corresponding to the above mentioned three-layers theory are found by *Matrinelli* (1974) who assumed validity of the *Reynolds* analogy. *Martinelli* integrated the energy conservation equation for each of the three regions and obtained the dimensionless temperature

$$T^+(y) = u^* \rho c_p \left[T_w - T(y)\right]/\dot{q}''_w, \tag{1.73}$$

as a function of the dimensionless distance from the wall. The solutions are given below: For the viscous sub layer $y^+ \leq 5$

$$T^+ = \Pr' \Pr y^+. \tag{1.74}$$

For the buffer layer, $5 \leq y^+ \leq 30$, the solution is

$$T^+ = 5\left\{\Pr' \Pr + \ln\left[1 + \Pr' \Pr\left(\frac{y^+}{5} - 1\right)\right]\right\}. \tag{1.75}$$

For the turbulent core $30 < y^+$ the solution is

$$T^+ = 5\left\{\Pr' \Pr + \ln\left(1 + 5\Pr' \Pr\right) + \frac{1}{2}\ln\left(\frac{y^+}{30}\right)\right\} \tag{1.76}$$

These are the equations 24, 25 and 26, respectively, obtained by *Martinelli* (1974). The best comparison with the experimental data was found for turbulent *Prandtl* number set to one, $\Pr' = (\lambda^t/\rho c_p)/v^t \approx 1$. Note that the experimental results obtained by *Ludwig* (1956) indicate $\Pr' \approx 1.5 - r/D_h$. More complex relation is proposed by *Azer* and *Chao* (1960) for $0.6 < Pr < 15$,

$$\Pr' = \frac{1 + 57 \operatorname{Re}^{-0.46} \Pr^{-0.58} \exp\left[-(y/R)^{1.4}\right]}{1 + 135 \operatorname{Re}^{-0.45} \exp\left[-(y/R)^{1.4}\right]}. \tag{1.77}$$

The *Martinelli* solution is important for analyzing nucleate boiling heat transfer at heated walls because it gives the temperature profile in the boundary layer where the babble generation happens. Stable bubble growth is obtained if the temperature at $y = D_{1d}/2$ is larger than the saturation temperature at the local pressure, *Levy* (1967).

Kays (1994) summarized experimental data by several authors leading to $\Pr' \approx 0.7$ to 0.85 for molecular *Prandtl* numbers being between 0.7 and 64 which are well represented by an approximate form of the analytical solution by *Yakhot* et al. (1987) $\Pr' = 0.85 + 0.7(v^t/v)\Pr$. More accurate also for lower *Prandtl* numbers is the correlation $\Pr' = 0.85 + 2(v^t/v)\Pr$, *Kays* (1994) p. 288, with still remaining degree of uncertainty for liquid metals. For air *Kays* and *Crawford* (1993) proposed $\Pr' = 1.07$ for $y^+ < 5$ and

$$\text{Pr}' = \left\{0.5882 + 0.228(v'/v) - 0.0441(v'/v)^2 \left\{1 - \exp\left[-5.165/(v'/v)\right]\right\}\right\}^{-1}$$

for $y^+ > 5$ corresponding to $T^+ = 2.075 y^+ + 3.9$ for $y^+ > 30$. For *water Hollingsworth* et al. (1989) proposed $\text{Pr}' = 1.07$ for $y^+ < 5$ and $\text{Pr}' = 1 + 0.855 - \tanh\left[0.2(y^+ - 7.5)\right]$ for $y^+ > 5$ which is useful also for *Prandtl* numbers up to 64, *Kays* (1994).

1.2.4.2 Practical results from the analogy between momentum and heat transfer

Exhaustive review in this subject is given by *Kirillov* (1985). The data collection by *Kirillov* consists of measurements made by different authors and shows a spread of ± 20%. The main reason for the spread is the surface structures, which may differ due to different physico-chemical influences that are difficult to control. The so called *Reynolds* analogy provided the useful framework for correlating the data. The pioneer analytical solutions in this field are given in Table 1.3. Useful empirical correlations are summarized in Table 1.4. From the 41 discussed correlations *Kirillov* (1985) recommend the simplest expressions valid in a narrow Pr-region.

Table 1.3. Heat transfer to fluid in a pipe described by the *Reynolds* analogy

Result	Valid, method, reference
$\dfrac{Nu}{\text{Re Pr}} = \dfrac{\lambda_{fr}}{8}$	$\text{Pr} = 1$, Reynolds analogy, *Reynolds* (1900)
$\dfrac{Nu}{\text{Re Pr}} = \dfrac{\lambda_{fr}/8}{1 + 1.74\,\text{Re}^{-0.125}(\text{Pr}-1)}$	$\text{Pr} < 10$, two-layer model, *Prandtl* (1910)
$\dfrac{Nu}{\text{Re Pr}} = \dfrac{\lambda_{fr}/8}{1 + 5\sqrt{\dfrac{\lambda_{fr}}{8}}\left\{\text{Pr}-1+\ln\left[1+\dfrac{5}{6}(\text{Pr}-1)\right]\right\}}$	$\text{Pr} < 30$, semi-emp. three-layers model, *Taylor* (1916), *von Karman* (1939)
$\dfrac{Nu}{\text{Re Pr}} = \dfrac{\lambda_{fr}/8}{f_1(\lambda_{fr}) + f_2(\lambda_{fr})\sqrt{\lambda_{fr}/8}\left(\text{Pr}^{2/3}-1\right)}$ $f_1(\lambda_{fr}) = 1 + 3.4\lambda_{fr}$, $f_2(\lambda_{fr}) = 11.7 + 1.8/\text{Pr}^{1/3}$, $\lambda_{fr} = (1.82\log \text{Re} - 1.64)^{-2}$	$0.5 < \text{Pr} < 2000$, $10^4 < \text{Re} < 5\times 10^6$, two-layer model, *Petukhov*, (1970), mean error 1%, except $200 < \text{Pr} < 2000$, $5\times 10^5 < \text{Re} < 5\times 10^6$, 1-2%.
$Nu = \dfrac{0.0396\,\text{Re}^{0.75}\,\text{Pr}}{1 + 1.5\,\text{Re}^{-0.125}\,\text{Pr}^{-1/6}(\text{Pr}-1)}$	*Bühne* (1938), *Hoffman* (1937)

Table 1.4. Heat transfer to fluid in a pipe described by empirical correlations

Correlation	Pr, Re –regions, ref.
$Nu = 0.023 \, Re^{0.8} \, Pr^{0.4}$	$0.25 < Pr < 25$, *Dittus-Boelter* (1930), see Fig. 1.3
$Nu = 0.021 \, Re^{0.8} \, Pr^{0.43}$	$0.6 < Pr < 2500$, $10^4 < Re < 5 \times 10^6$, *Miheev* (1952)
$Nu = 5 + 0.015 \, Re^a \, Pr^b$, $a = 0.88 - 0.24/(4+Pr)$, $b = 0.333 - 0.5 \exp(-0.6 \, Pr)$	*Sleicher* and *Rouse* (1975) $7 < Pr < 1000$
$Nu = \dfrac{0.023 \, Re^{0.8} \, Pr}{1 + 2.14 \, Re^{-0.1} \left(Pr^{2/3} - 1 \right)} \left(\dfrac{Pr}{Pr_w} \right)^{0.25}$	$0.5 < Pr < 200$, *Kutateladze* (1979b)
$\dfrac{Nu}{Re \, Pr} \dfrac{8}{\lambda_{fr}} = \left[1.07 + 12.7 \sqrt{\dfrac{\lambda_{fr}}{8}} \left(Pr^{2/3} - 1 \right) \right]^{-1}$, $\lambda_{fr} = \left[1.82 \log_{10} (Re/8) \right]^{-2}$	*Petukhov* and *Kirillov* (1958)
$\dfrac{Nu}{Re \, Pr} \dfrac{8}{\lambda_{fr}} = \left[1 + y_{lam}^+ \sqrt{\dfrac{\lambda_{fr}}{8}} \left(Pr^{2/3} - 1 \right) + 2.5 \ln Pr \right]^{-1}$, y_{lam}^+, dimensionless laminar sub-layer (≈ 11.5)	*Borstevskij* and *Rudin* (1978) p. 294, $0.5 < Pr < 10000$
$\dfrac{Nu}{Re \, Pr} \dfrac{8}{\lambda_{fr}} = \left[1 + \dfrac{900}{Re} + 12.7 \sqrt{\dfrac{\lambda_{fr}}{8}} \left(Pr^{2/3} - 1 \right) \right]^{-1}$	$0.5 < Pr < 5 \times 10^5$, $4 \times 10^3 < Re < 5 \times 10^6$, *Petukhov et al.* (1974)
$Nu = \dfrac{\sqrt{\dfrac{\lambda_{fr}}{8}} Re \, Pr}{2.12 \ln \left(Re \sqrt{\lambda_{fr}} \right) + 2.12 \ln Pr + 12.5 \, Pr^{2/3} - 10.1}$	$1 < Pr < 10^6$, $5 \times 10^3 < Re < 5 \times 10^6$, *Kader - Yaglom* (1972)
$Nu = 7.83 - \dfrac{5.31}{\log_{10} Re} + 0.01 \, Re^{0.87} \, Pr^n$, $n = 0.34 + \dfrac{0.47}{1 + 2 \, Pr}$	$Pr < 100$, *Bobkov - Gribanov* (1988)

20   1 Some basics of the single-phase boundary layer theory

Equation	Range / Reference
$Nu = \dfrac{\sqrt{\dfrac{\lambda_{fr}}{8}}\operatorname{Re}\operatorname{Pr}}{0.833\left[5\operatorname{Pr}+5\ln(5\operatorname{Pr}+1)+2.5\ln\dfrac{\operatorname{Re}}{60}\sqrt{\dfrac{\lambda_{fr}}{8}}\right]}$	$0.5 < \operatorname{Pr} < 30$, *Kays* (1972)
$Nu = 2.5 + 1.3\log_{10}\left(1+\dfrac{1}{\operatorname{Pr}}\right)+3.9\left(\dfrac{\operatorname{Re}}{1000}\right)^{m}\operatorname{Pr}^{n}$ $m = 0.918 - 0.051\times\log_{10}\left(1+\dfrac{10}{\operatorname{Pr}}\right)$ $n = 0.65 - 0.107\times\log_{10}(1+10\operatorname{Pr})$	$0.01 < \operatorname{Pr} < 100$, $3\times 10^{3} < \operatorname{Re} < 3\times 10^{6}$, *Buleev* (1965)
$\dfrac{Nu}{\operatorname{Re}\operatorname{Pr}} = (k_{1}\ln^{2}Y + k_{2}\ln Y + k_{3})^{-1}$ $k_{1} = 5.75$ $k_{2} = 61.25\operatorname{Pr}^{0.55} - 14.75\ln\operatorname{Pr} - 5.21$ $k_{3} = 42.875\operatorname{Pr}^{0.55} - 10.325\ln\operatorname{Pr} - 32.1$	$0.7 < \operatorname{Pr} < 64$, *Kirillov et al.* (1985)
$Nu = \dfrac{\sqrt{\dfrac{\lambda_{fr}}{8}}\operatorname{Re}\operatorname{Pr}^{n}\left(1+\dfrac{1.75}{9+\operatorname{Pr}}\right)}{1.325\sqrt{\lambda_{fr}}+1}$ $n = 0.33 + 0.266\left(\dfrac{\operatorname{Re}}{10^{4}}\right)^{0.057}\operatorname{Pr}^{-0.383}$ $\quad + 0.047\left(\dfrac{\operatorname{Re}}{10^{4}}\right)^{0.22}\operatorname{Pr}^{-1.384}$	*Migay* (1983)

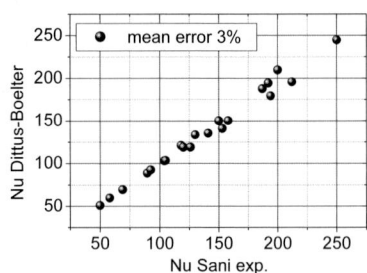

Fig. 1.3. Illustration of the accuracy of the oldest correlation for heat transfer in pipe: data by *Sani* (1960), correlation by *Dittus-Boelter* (1930)

Table 1.5. Water flowing parallel to rods in a bundle

$Nu = f(P/D_{rod}) \text{Re}^{0.8} \text{Pr}^{1/3}$, $f(P/D_{rod}) = 0.026(P/D_{rod}) - 0.006$, triangular – pitch array $f(P/D_{rod}) = 0.042(P/D_{rod}) - 0.024$, square – pitch array	Weisman (1959)

Weisman (1959) found that the heat transfer to water flowing parallel to rod bundles is similar to those in pipes with a linear dependence on the pitch to rod ratio, P/D_{rod}, see Table 1.5.

1.3 Transient flow in pipes with circular cross sections

It is interesting to have a method that answers the question: How acceleration or deceleration of the averaged flow in a pipe influences the wall friction force? *Kawamura* (1975) solved numerically the following system of equations to answer this question.

$$\frac{\partial \bar{w}}{\partial \tau} = -\frac{1}{\rho}\left|\frac{\partial p}{\partial x}\right| + \frac{1}{r}\frac{\partial}{\partial r}\left[(v+v')r\frac{\partial \bar{w}}{\partial r}\right], \qquad (1.78)$$

$$\frac{\partial \bar{T}}{\partial \tau} + \bar{u}\frac{\partial \bar{T}}{\partial x} = \frac{1}{r}\frac{\partial}{\partial r}\left[(a+a')r\frac{\partial \bar{T}}{\partial r}\right], \qquad (1.79)$$

with boundary conditions $r = R$, $\bar{w}(R,\tau) = 0$, $r = 0$, $\frac{\partial \bar{w}}{\partial r} = 0$, $\bar{T}(x,R,\tau) = \bar{T}_w$,

$\dot{q}''_w = -\lambda \left.\frac{d\bar{T}}{dr}\right|_R$; $\frac{d\bar{T}}{dr} = 0$; $T(0,r,\tau) = T_{in}$; $\frac{\partial p}{\partial x} = f(\tau)$ are prescribed. The constitutive relations used are $v' = \ell^2 \left|\frac{dw}{dy}\right|$, $\frac{1}{\ell^2} = \frac{1}{\ell_c^2} + \frac{1}{\ell_w^2}$, $\ell_w = \kappa y\left[1 - \exp(-y^+/26)\right]$,

$\ell_c = 0.045D$, $a' = v'1.5\phi\left[1 - \exp(1/\phi)\right]$, and $\phi = \dfrac{v'/v \Pr}{4.13 + 0.743(v'/v)^{1/2} \Pr^{1/3}}$ (for liquid metals this value has to be multiplied by 2 to obtain heat transfer coefficients experimentally observed). To avoid the defect of the model at the axis the following assumption is made valid for the axis: $\ell\left|\frac{dw}{dy}\right| = 0.01\bar{w}$ in case if

$\ell \left| \dfrac{d\overline{w}}{dy} \right| < 0.01 \overline{w}$. *Kawamura* found the following time scales: The order of magnitude for the momentum exchange between the eddies is $\Delta \tau_e \approx \ell^2 / v'$, the order of magnitude for establishing of the steady state flow is $\Delta \tau_{flow} \approx \dfrac{1}{\lambda_{fr} \operatorname{Re}} D^2 / v$, and therefore $\Delta \tau_e / \Delta \tau_{flow} \approx \lambda_{fr} \operatorname{Re} \left(\dfrac{\ell}{D} \right)^2 \dfrac{v}{v'}$. So for instance at $\operatorname{Re} = 10^5$, $\lambda_{fr} \approx 0.045$, $v/v' = 0.01$ and $\ell/D \approx 0.045$, $\Delta \tau_e / \Delta \tau_{flow} \approx 0.09$. The method of *Kawamura* demonstrates that deceleration has no strong effect on the friction coefficient but fast acceleration increases the friction coefficient.

Kalinin and *Dreitser* (1970) reported very approximate correlation of data with very large scatter by taking into account the influence of the unsteadiness of the flow in the friction coefficient

$$\dfrac{\lambda_{fr,transient}}{\lambda_{fr}} = 1 + const \dfrac{D_h}{\lambda_{fr} \overline{w}^2} \dfrac{d\overline{w}}{d\tau}, \qquad (1.80)$$

which is not recommended. However, the introduction of the dimensionless measure of the acceleration is important. *Marek*, *Mensinger* and *Rehme* (1979) performed careful experiments by accelerating water in pipes. The authors found that only the direct share stress measurement gives useful results. The pressure difference measurement contains acceleration component that is two orders of magnitude larger than the friction component, which makes estimation of the share stress by pressure difference measurements impossible. The authors reported a log-log data plot that is approximated here by

$$\dfrac{\lambda_{fr,transient}}{\lambda_{fr}} = \max \left[1, -1.98671 + 0.01394 \dfrac{D_h}{\lambda_{fr} \overline{w}^2} \dfrac{d\overline{w}}{d\tau} + 9.58394 \times 10^{-8} \left(\dfrac{D_h}{\lambda_{fr} \overline{w}^2} \dfrac{d\overline{w}}{d\tau} \right)^2 \right], \qquad (1.81)$$

valid for $1 < \dfrac{D_h}{\lambda_{fr} \overline{w}^2} \dfrac{d\overline{w}}{d\tau} < 10^5$. Note the considerable increase of the transient friction coefficient $1 \le \lambda_{fr,transient} / \lambda_{fr} < 2500$ with the increase of the acceleration in this region. As far as I know there is no second like this experiment. *Bergant* et al. (1999) reported good comparison with data for pressure wave propagation with small initial velocities 0.1, 0.2, 0.3 m/s using the expression proposed by *Brunone* et al. (1991).

$$\frac{\lambda_{fr,transient}}{\lambda_{fr}} = 1 + f(\text{Re})\frac{D_h}{\lambda_{fr}\overline{w}^2}\left[\frac{\partial \overline{w}}{\partial \tau} + a\,sign(\overline{w})\left|\frac{\partial \overline{w}}{\partial z}\right|\right], \qquad (1.82)$$

in which the function of time averaged *Reynolds* number proposed by *Vardy* and *Brown* (1996)

$$f(\text{Re}) = \left[\frac{7.41}{\text{Re}^{\log(14.3/\text{Re}^{0.05})}}\right]^{1/2}, \qquad (1.83)$$

was used. Note that "a" is the velocity of sound. For *Reynolds* numbers 1875, 3750 and 5600 the $f(\text{Re})$ takes values 0.069, 0.049 and 0.0418, respectively.

Nomenclature

Latin

a	$:= \lambda/(\rho c_p)$, thermal diffusivity, m²/s
a	in transient friction correlations, velocity of sound, m/s
a^t	turbulent thermal diffusivity, m²/s
c_p	specific capacity at constant pressure, J/(kgK)
c_w	$:= \tau_w/(\rho \overline{w}^2/2)$, friction coefficient, *Fanning* factor, dimensionless
$\overline{c_{w,\Delta x}}$	friction coefficient averaged over the length Δx, dimensionless
D	diameter, m
D_h	hydraulic diameter, m
F_w	share force acting on the surface, N
f, f_1, f_2	functions
f_w	share force acting on the surface per unit flow volume, N/m³
h	heat transfer coefficient, W/(m²K)
k	specific turbulent kinetic energy, m²/s²
k_p	specific turbulent kinetic energy at specified distance from the wall at which boundary condition for large scale discretization are prescribed, m²/s²
k	as pipe surface property, roughness, m
k^+	$:= kw*/\nu$, roughness, dimensionless
k_1, k_2, k_3	functions
ℓ	mixing length, m

ℓ_c	mixing length in the central region, m
ℓ_w	mixing length in the wall region, m
Nu	*Nusselt* number, dimensionless
Nu_x	local *Nusselt* number along the x-axis, dimensionless
$Nu_{\Delta x}$	local *Nusselt* number along the Δx, dimensionless
$\overline{Nu_{\Delta x}}$	averaged *Nusselt* number over Δx, dimensionless
$\Pr$	molecular *Prandtl* number, dimensionless
$\Pr'$	turbulence *Prandtl* number, dimensionless
$\Pr_w$	molecular *Prandtl* number at wall temperature, dimensionless
p	pressure, Pa
p_w	static pressure at the wall, Pa
$\dot{q}''_w$	heat flux from the wall into the continuum, W/m²
R	radius, m
R^+	$:= R\, w^*/\nu$, radius, dimensionless
Re	*Reynolds* number, dimensionless
$\text{Re}_{\Delta x}$	*Reynolds* number Δx, dimensionless
r	radius, m
T	temperature, K
T_w	wall temperature, K
$\overline{T}$	averaged temperature, K
T^+	temperature, dimensionless
u	velocity along the x-axis, m/s
u'	fluctuation velocity component along the x-axis, m/s
$u_{\max}$	maximum velocity along the x-axis, m/s
u_∞	velocity far from the surface, m/s
$\overline{u}$	cross section averaged velocity, m/s
u^*	friction velocity, m/s
u^+	velocity, dimensionless
V_{flow}	control volume, m³
v	velocity component in y-direction, m/s
v'	fluctuation of the velocity component in y-direction, m/s
v'^+	fluctuation of the velocity component in y-direction, dimensionless
w	velocity component in z-direction, axial velocity, m/s
w'	fluctuation of the velocity component in z-direction, m/s
w'^+	fluctuation of the velocity component in z-direction, dimensionless
$\overline{w}$	cross section averaged axial velocity, m/s
x	x-coordinate, coordinate parallel to a plate m
Δx	increment of the x-coordinate, m

Y	function
y	y-coordinate, coordinate perpendicular to the wall, m
Δy	increment of the y-coordinate, m
y^+	$:= y w^*/\nu$, y-coordinate, dimensionless
y_0^+	distance from the wall, dimensionless
y_p^+	distance from the wall at which boundary condition for large scale discretization are prescribed, dimensionless

Greek

Γ^+	volumetric flow rate per unit width of the wall, dimensionless
γ	function
δ	boundary layer thickness, m
$\delta_{99\%}$	displacement thickness, m
δ^+	boundary layer thickness, dimensionless
ε	power dissipated irreversibly due to *turbulent pulsations* in the viscous fluid per unit mass of the fluid (dissipation of the specific turbulent kinetic energy), m²/s³
ε_p	continuum dissipation of the specific turbulent kinetic energy at the transition between laminar and turbulent boundary layer, m²/s³
κ	von Karman constant, dimensionless
λ	molecular thermal conductivity, W/(mK)
λ^t	turbulent thermal conductivity, W/(mK)
λ_{fr}	$:= \dfrac{\tau_w}{\rho \bar{w}^2/8} = 4 c_w$, wall friction factor for steady state flow, dimensionless
$\lambda_{fr,transient}$	wall friction factor for transient flow, dimensionless
ν	molecular cinematic viscosity, m²/s
ν^t	turbulent cinematic viscosity, eddy diffusivity, m²/s
ρ	density, kg/m³
τ	time, s
$\Delta \tau_e$	$\approx \ell^2/\nu^t$, time scale for momentum exchange between the eddies, s
$\Delta \tau_{flow}$	$\approx \dfrac{1}{\lambda_{fr} \text{Re}} D^2/\nu$, time scale for establishing the steady state flow in a pipe, s
τ_w	wall share stress, N/m²
ϕ	function

References

Alshamani KMM (1978) Correlation among turbulent share stress, turbulent kinetic energy, and axial turbulence intensity, AIAA Journal, vol 16 no 8 pp 859-861

Avdeev AA (1982) Teploenergetika vol 3 p 23

Azer NZ and Chao BT (1960) A mechanism of turbulent heat transfer in liquid metals, Int. J. Heat Mass Transfer vol 1 pp 121-138

Bendiksen KH (1985) On the motion of long bubbles in vertical tubes, Int. J. Multiphase Flow, vol 11 pp 797-812

Bergant A, Simpson AR and Vitkovsky J (7-9 Sept. 1999) Review of unsteady prediction models in transient pipe flow, 9th IAHR Int. Meeting, Brno, Czech Republic

Bobkov VP and Gribanov YuI (1988) Statisticheskie izmerenija v turbolentnyh potokax, Moskva, Energoatomizdat

Borstevskij JuT and Rudin SN (1978) Upravlenie turbulentnom pograniznom sloe, Visha skola, Kiev

Bradshaw P (1967) The turbulence structure of equilibrium boundary layers, J. Fluid Mechanics, vol 29 pp 625-645

Brunone B, Golia UM and Greco M (1991) Some remarks on the momentum equations for transients, Int. Meeting on Hydraulic Transients with Column Separation, 9th Round Table, IAHR, Valencia, Spain, pp 140-148

Bühne W (1938) Wärme vol 61 p 162

Buleev NI (1965) Trudy 3-j mejdunarodnoj konferenzii po mirnomu ispol'zovaniju atomnoj energii, t. 5, s. 305-313

Colebrook CF (1939) Turbulent flow in pipes with particular reference to the transition region between the smooth and the rough pipe lows, J. Institution Civil Engineers

Collins R, De Moraes FF, Davidson JF and Harrison D (1978) The motion of large gas bubble rising through liquid flowing in a tube, J. Fluid. Mech., vol 89 pp 497-514

Dittus FV and Boelter LMK (1930) Univ. of Calif. Publ. In Engng. Vol 2 no 13 p 443

Hammond GP (1985) Turbulent Prandtl number within a near-wall flow, AIAA Journal vol 23 no 11 pp 1668-1669

Harsha PT and Lee SC (1970) Correlation between turbulent share stress and turbulent kinetic energy, AIAA Journal, vol 8 no 5 pp 1508-1510

Hoffman E (1937) Z. Ges. Kälte-Ind., vol 44 pp 99-107

Hollingsworth DK, Kays WM and Moffat RJ (Sept. 1989) Measurement and prediction of the turbulent thermal boundary layer in water on flat and concave surfaces, Report no. HMT-41, Thermosciences Division, Dep. of Mech. Engr., Stanford Univ., CA

Hurlburt ET, Fore LB and Bauer RC (July 17-20, 2006) A two zone interfacial shear stress and liquid film velocity model for vertical annular two-phase flow, Proceedings of FEDSM2006 2006 ASME Joint U.S. - European Fluids Engineering Summer Meeting July 17-20, Miami, FL, FEDSM2006-98512

Johansen ST (1991) The deposition of particles on vertical walls, Int. J. Multiphase Flow, vol 17 pp 335-376

Kader BA and Yaglom AM (1972) Int. J. Heat Mass Transfer, vol 15 no 12 pp 2329-2351

Kalinin EK and Dreitser GA (1970) Unsteady convective heat transfer and hydrodynamics in channels, Advances in Heat Transfer, vol 6, Eds. GP Hartnet and GS Ervine

Kays WM (1972) Konvektivnyj teplo- i masoobmen, Miskva, Energia, s. 207, translation from English

Kays WM (May 1994) Turbulent Prandtl number – where are we? Transaction of the ASME, vol 116 pp 284-295

Kays WM and Crawford ME (1993) Convective heat and mass transfer, 3d ed., McGraw-Jill, New York
Kawamura H (Juni 1975) Transient hydraulics and heat transfer in a turbulent flow, Kernforschungszentrum Karlsruhe, Report KFK 2166
Kim J, Moin P and Moser R (1987) Turbulent statistics in a fully developed channel flow at low Reynolds number, J. Fluid Mech., vol 177 pp 133-166
Kirillov PL, Markov YuM and Slobodchuk VI (1985) Rasrpredelenie temperatury i zakon teploobmena pri turbulentnom techenii v krugloj trube: Preprint FEI-1703
Kirillov PL (1988) Analiz razchetnyh formul po teploobmenu pri turbulentnom techenii v rtubach, Obsor FEI-0230, Moskva, ZNIIatominform, 80s.
Knudsen JG and Katz DL (1958) Fluid dynamics and heat transfer, McGraw-Hill, New York, p 394, pp 485-486
Kutateladze SS, Khabakhpasheva WM, Orlov VV, Perepelitsa VV and Michailova ES (1979) In Turbulent Share Flows I, pp 91-103, Springer-Verlag, Berlin
Kutateladze SS (1979b) Osnovy teorii teploobmena, 5-e izd., Moskva, Atomizdat
Laufer J (1952) Investigation of turbulent flow in a two-dimensional channel, NACA Report 1053
Laufer J (1953) The structure of turbulence in fully developed pipe flow, NACA Report 1273
Laufer SL (1953) The structure of turbulence in fully developed pipe flow, NASA Technical Note No. 2954
Lee SL and Durst F (1980) On the motion of particles in turbulent flow, US Nuclear Regulatory Commission Report NUREG/CR-1556
Lee SJ, Lahey RT Jr. and Jones OC Jr. (1986) The prediction of two phase turbulence and phase distribution phenomena using k-e model, Int. J. of Multiphase Flow
Levy S (1967) Forced convection subcooled boiling – prediction of the vapor volumetric fraction, Int. J. Heat Mass Transfer, vol 10 pp 951-965
Ludwig HZ (1956) Z. Flugwiss., vol 4 pp 73-81
Martinelli RC (1974) Heat transfer to molten metals, Trans. Am. Soc. Mech. Engrs. vol 69 pp 947-959
Matida EA, Tori A and Nishino K (1998) Proc. Of the 35th Nat. Heat Transfer Symposium of Japan, vol 2 pp 495-496
Marek J, E. Mensinger and Rehme K (März 1979) Experimental friction factors of transient flows through circular tubes (Final report), Primärbericht, Institut für Neutronenphysik und Reaktortechnik, INR-910, Kernforschungszentrum Karlsruhe
Migay VK (1983) Toploobmen v trubah pri turbulentnom techeniy, Tr. ZKTI vyp. 206
Miheev MH (1952) Izv. AN SSSR, no 10 s. 1448-1454
Nikuradse J (1932) Gesetzmäßigkeit der turbulenten Strömung in glaten Rohren, Forsch. Arb. Ing.-Wes. No 1932
Petukhov BS and Kirillov PL (1958) About heat transfer at turbulent fluid flow in tubes, Thermal Engineering, vol 4 pp 63-68, in Russian
Petukhov BS (1970) Heat transfer and friction in turbulent pipe flow with variable physical properties, Advances in Heat Transfer, Academic Press, New York, vol 6 p 503-564
Petukhov BS, Genin LG and Kovalev SA (1974) Teploobmen v jadernych energeticheskih ustanovkax, Moskva, Atomizdat
Prandtl L (1910) Physik. Zeitachrift, vol 11 pp 1072 - 1078 ; (1928) vol 29 , pp 487 – 489
Rehme K (1992) The structure of turbulence in rod bundles and the implications on natural mixing between the subchannels, Int. J. Heat Mass Transfer, vol 35 no 2 pp 567-581
Reichardt H (Juli 1951) Vollständige Darstellung der turbulenten Geschwindigkeiten in glaten Leitungen, Z. angew. Math. Mech., Bd. 31 Nr. 7, S. 208-219
Reynolds O (1974) Proc. Manchester Phil. Soc., reprinted in "Scientific Papers of O. Reynolds" vol. II, 1901 , Cambridge

Sani RleR (4 January 1960) Down flow boiling and non-boiling heat transfer in a uniformly heated tube, University of California, URL-9023, Chemistry-Gen. UC-4, TID-4500 (15th Ed.)

Schlichting H (1982) Grenzschicht-Theorie, Braun, Karlsruhe, 8 Auflage

Sidorov AE (1959) Calculation of resistance and convective heat transfer, Teploenergetika, no 4 pp 79-80

Sleicher CA and Rouse MW (1975) A convenient correlation for heat transfer to constant and variable property fluids un turbulent pipe flow, Int. J. of Heat and Mass Transfer, vol 18 pp 677-683

Taylor GI (1916) Brit. Aeronaut. Comm. Rept. Mem., vol 272 pp 423-429

Traviss DP, Rohsenow WM and Baron AB (1973) Forced convection condensation inside a tubes: A heat transfer equation for condenser design, ASHRAE Trans., no 2272 Rp-63

Vardy AE and Brown JMB (1996) On turbulent, unsteady, smooth-pipe flow, Proc. Int. Conf. on Pressure Surges and Fluid Transients, BHR Group, Harrogate, England, pp 289-311

Vames JS and Hanratty TJ (1988) Turbulent dispersion of droplets for air flow in a pipes, Exports Fluids vol 6 pp 94-104

van Driest ER (1955) On turbulent flow near a wall, Heat Transfer and Fluid Mechanics Institute

von Karman Th (1939) Trans. ASME. vol 61 pp 705-710

Weisman J (1959) Heat transfer to water flowing parallel to tube bundles, Nucl. Sci. Eng., vol 6 p 79

White FM (2006) Viscous fluid flow, 3rd ed, McGraw-Hill, New York

Yakhot Y, Orszag SA and Yakhot A (1987) Heat transfer in turbulent fluids-1, Pipe flow, Int. J. of Heat and Mass Transfer, vol 35 no 1 pp 15-22

2 Introduction to turbulence of multi-phase flows

Single phase turbulence is complex and still considered as a not resolved issue in science. Multiphase flow turbulence is much more complex and of course still far from its final accurate mathematical description. Systematic experimental results for steady states are gained and some theories are developed mainly for low concentration particle- and bubble-flows in simple geometry but not for all flow pattern in transient multi-phase flows in general. However, the need of optimum design of industrial facility operating with multiphase flows dictates intensive activities of many scientists in this field. For such a scientific discipline, that is being in "flow", a summary of the state of the art will help engineers to use what is already achieved and help scientists to understand better where is the lack of physical understanding that has to be filled. This is the reason to write this Section, well knowing the limited range of knowledge accumulated so far. In this work I will review the existing approaches and try to lead them to one that is promising for practical analysis. I will deliberately concentrate my attention to a pragmatic modeling that is far from "decorative mathematics".

2.1 Basic ideas

Let us recall the momentum equations derived in *Kolev* (2007a) Ch. 2 of the dispersed velocity field d surrounded by continuum c

$$\frac{\partial}{\partial \tau}(\alpha_d \rho_d \mathbf{V}_d \gamma_v) + \nabla \cdot (\alpha_d^e \rho_d \mathbf{V}_d \mathbf{V}_d \gamma) + \nabla \cdot \left[\alpha_d^e \left(\rho_d \overline{\mathbf{V}_d' \mathbf{V}_d'}\right)\gamma\right] + \alpha_d^e \gamma \nabla p_d + \alpha_d \gamma_v \rho_d \mathbf{g}$$

$$+ \left(p_d - p_c + \delta_d \sigma_{dc} \kappa_d - \Delta p_c^{d\sigma^*}\right) \nabla \left(\alpha_d^e \gamma\right)$$

$$-\gamma_v \alpha_d \rho_c \left\{ \begin{array}{l} c_{cd}^{vm}\left[\dfrac{\partial}{\partial \tau}\Delta \mathbf{V}_{cd} + (\mathbf{V}_d \cdot \nabla)\Delta \mathbf{V}_{cd}\right] \\[1em] -c_{cl}^L \Delta \mathbf{V}_{cd} \times (\nabla \times \mathbf{V}_c) + \dfrac{1}{D_d}\dfrac{3}{4}c_{cl}^d |\Delta \mathbf{V}_{cd}| \Delta \mathbf{V}_{cd} \end{array} \right\}$$

$$= \gamma_v \sum_{\substack{k=1 \\ k \neq d}}^{3,w} \left(\mu_{kd} \mathbf{V}_k - \mu_{dk} \mathbf{V}_d \right), \tag{2.1}$$

and of the continuous phase velocity field c

$$\frac{\partial}{\partial \tau} \left(\alpha_c \rho_c \mathbf{V}_c \gamma_v \right) + \nabla \cdot \left(\alpha_c^e \rho_c \mathbf{V}_c \mathbf{V}_c \gamma \right) + \nabla \cdot \left[\alpha_c^e \gamma \left(\rho_c \overline{\mathbf{V}_c' \mathbf{V}_c'} - \mathbf{T}_{\eta,c} \right) \right] + \alpha_c^e \gamma \nabla p_c$$

$$+ \alpha_c \gamma_v \rho_c \mathbf{g} + \Delta p_c^{d\sigma*} \nabla \left(\alpha_c^e \gamma \right) - \Delta p_c^{w\sigma*} \nabla \gamma$$

$$+ \gamma_v \alpha_d \rho_c \left\{ \begin{array}{c} c_{cd}^{vm} \left[\dfrac{\partial}{\partial \tau} \Delta \mathbf{V}_{cd} + \left(\mathbf{V}_l \cdot \nabla \right) \Delta \mathbf{V}_{cd} \right] \\ \\ -c_{cd}^L \Delta \mathbf{V}_{cd} \times \left(\nabla \times \mathbf{V}_c \right) + \dfrac{1}{D_d} \dfrac{3}{4} c_{cd}^d \left| \Delta \mathbf{V}_{cd} \right| \Delta \mathbf{V}_{cd} \end{array} \right\}$$

$$+ \gamma_v \rho_c \left[c_w^{vm} \frac{1}{2} \frac{\partial \mathbf{V}_c}{\partial \tau} - c_{cw}^L \mathbf{V}_c \times \left(\nabla \times \mathbf{V}_c \right) + \frac{1}{D_w} \frac{3}{4} c_{cw}^d \left| \mathbf{V}_c \right| \mathbf{V}_c \right]$$

$$= \gamma_v \sum_{\substack{k=1 \\ k \neq c}}^{3,w} \left(\mu_{kc} \mathbf{V}_k - \mu_{ck} \mathbf{V}_c \right). \tag{2.2}$$

The dispersed momentum equation is valid inside the dispersed phase *including the interface*. It includes the interface jump condition. The continuum momentum equation is valid only inside the continuum without the interface.

Remember the notations applied to each field l: α_l is the local volume fraction, α_l^e is the local volume fraction at the interfaces of the control volume (identical to α_l for infinitesimal control volume). The counter part defining the presence of non-flow materials is described by γ_v, the volumetric porosity, the part of the control volume not available for the flow, and γ the volumetric porosity at the interfaces of the control volume called surface permeability (identical to γ_v for infinitesimal control volume).

$\Delta p_c^{d\sigma*}$ is the surface averaged difference between the pressure at the surface $d\sigma$ and the bulk pressure of c. I call this pressure difference *effective stagnation pressure difference*. Analogously, $\Delta p_c^{w\sigma*}$ is defined for the wall. The interfacial interaction coefficients are as follows: c^{vm} is the virtual mass force coefficient, c^d is the drag force coefficient, and c^L is the lift force coefficient. For the estimation of this coefficients for different configurations and flow pattern see *Kolev* (2007b). **g** is the acceleration due to gravity. $\Delta \mathbf{V}_{cd} = \mathbf{V}_c - \mathbf{V}_d$, local velocity difference between the continuous phase c and the disperse phase d.

The intrinsic averaged density is ρ_l. Here p is the thermodynamic pressure. $\mathbf{V}_l$ is the velocity vector of field l and $\mathbf{V}'_l$ is its pulsating component. Note that the molecular fluctuation are much faster than any other processes; so locally external macroscopic forces does not influence the local micro-equilibrium (an hypothesis first stated by *Stokes* in 1845).

How to derive these equations, how to estimate every single term in them, how to integrate them in conjunction with the other conservation laws are the subject of Volume 1 and 2 of this work and will not be repeated here. Our target is to understand how to model the tensor of the turbulent stresses and its influence, of course, on the energy conservation and all other flow processes.

First remember the *Stokes* hypothesis from 1845, *Stokes* Eqs. (8), expressed mathematically

$$\mathbf{T}_{\eta,c} = \eta_c \left[\nabla \mathbf{V}_c + (\nabla \mathbf{V}_c)^T - \frac{2}{3} (\nabla \cdot \mathbf{V}_c) \mathbf{I} \right] = \eta_c \left[2\mathbf{D} - \frac{2}{3} (\nabla \cdot \mathbf{V}_c) \mathbf{I} \right]$$

$$= \eta_c \begin{pmatrix} 2\left(\dfrac{\partial u}{\partial x} - \dfrac{1}{3}\nabla \cdot \mathbf{V}\right) & \dfrac{\partial v}{\partial x} + \dfrac{\partial u}{\partial y} & \dfrac{\partial w}{\partial x} + \dfrac{\partial u}{\partial z} \\ \dfrac{\partial u}{\partial y} + \dfrac{\partial v}{\partial x} & 2\left(\dfrac{\partial v}{\partial y} - \dfrac{1}{3}\nabla \cdot \mathbf{V}\right) & \dfrac{\partial w}{\partial y} + \dfrac{\partial v}{\partial z} \\ \dfrac{\partial u}{\partial z} + \dfrac{\partial w}{\partial x} & \dfrac{\partial v}{\partial z} + \dfrac{\partial w}{\partial y} & 2\left(\dfrac{\partial w}{\partial z} - \dfrac{1}{3}\nabla \cdot \mathbf{V}\right) \end{pmatrix}_c. \quad (2.3)$$

$\nabla \mathbf{V}$ is the dyadic product of the *Nabla* operator and the velocity vector (a second order tensor), T designates the transposed tensor. Note that the *Nabla* operator of the velocity vector,

$$\nabla \mathbf{V} = \mathbf{D} + \mathbf{W} \quad (2.4)$$

consists of a symmetric part

$$\mathbf{D} = \frac{1}{2}\left[\nabla \mathbf{V} + (\nabla \mathbf{V})^T \right], \quad (2.5)$$

called *deformation rate* and a skew part

$$\mathbf{W} = \frac{1}{2}\left[\nabla \mathbf{V} - (\nabla \mathbf{V})^T \right], \quad (2.6)$$

called *spin* or *vortices tensor*. $\nabla \cdot \mathbf{V}$ is the divergence of the velocity vector. The term containing the divergence of the velocity vector

$$\nabla \cdot \mathbf{V} = \frac{\partial u}{\partial x} + \frac{\partial v}{\partial y} + \frac{\partial w}{\partial z}, \qquad (2.7)$$

is called by *Stokes rate of cubic dilatation*. The hypothesis says that the relation between viscous stresses and deformation rate on a control volume is linear and that the proportionality factor is the dynamic viscosity η, that the solid body translations and rotations do not contribute to the viscous forces, that the share stresses are symmetric, and that the relation between volumetric and share viscosity is so that the pressure always equals one-third of the sum of the normal stresses. Each of this point has its ingenious argumentation in the *Stokes* paper. In the multiphase continuous fields, as long as they are fine resolved this stress tensor exists. I recommend to any one having serious intention to understand flows to study the *Stokes* paper.

The term $\rho_l \overline{\mathbf{V}_l' \mathbf{V}_l'}$, called tensor of the turbulent stresses, is obtained for single phase flow after time averaging the momentum equations by *Reynolds* (1894). *Boussinesq* (1877) introduced the idea of turbulent eddy viscosity inside the velocity field so

$$-\rho_l \overline{\mathbf{V}_l' \mathbf{V}_l'} = \eta_l^t \left[2\mathbf{D} - \frac{2}{3}(\nabla \cdot \mathbf{V}_l)\mathbf{I} \right]$$

$$= \eta_l^t \begin{bmatrix} 2\left(\dfrac{\partial u}{\partial x} - \dfrac{1}{3}\nabla \cdot \mathbf{V}\right) & \dfrac{\partial v}{\partial x} + \dfrac{\partial u}{\partial y} & \dfrac{\partial w}{\partial x} + \dfrac{\partial u}{\partial z} \\[6pt] \dfrac{\partial u}{\partial y} + \dfrac{\partial v}{\partial x} & 2\left(\dfrac{\partial v}{\partial y} - \dfrac{1}{3}\nabla \cdot \mathbf{V}\right) & \dfrac{\partial w}{\partial y} + \dfrac{\partial v}{\partial z} \\[6pt] \dfrac{\partial u}{\partial z} + \dfrac{\partial w}{\partial x} & \dfrac{\partial v}{\partial z} + \dfrac{\partial w}{\partial y} & 2\left(\dfrac{\partial w}{\partial z} - \dfrac{1}{3}\nabla \cdot \mathbf{V}\right) \end{bmatrix}_l, \qquad (2.8)$$

that it has the same structure as the *Stokes* hypothesis. The corresponding *Reynolds stresses* are

$$-\rho_l \overline{u_l' u_l'} = \tau_{l,xx}' = \eta_l^t 2\left(\frac{\partial u_l}{\partial x} - \frac{1}{3}\nabla \cdot \mathbf{V}_l\right), \qquad (2.9)$$

$$-\rho_l \overline{v_l' v_l'} = \tau_{l,yy}' = \eta_l^t 2\left(\frac{\partial v_l}{\partial y} - \frac{1}{3}\nabla \cdot \mathbf{V}_l\right), \qquad (2.10)$$

$$-\rho_l w_l' w_l' = \tau_{l,zz}' = \eta_l' 2\left(\frac{\partial w_l}{\partial y} - \frac{1}{3}\nabla \cdot \mathbf{V}_l\right), \tag{2.11}$$

$$-\rho_l u_l' v_l' = \tau_{l,xy}' = \eta_l'\left(\frac{\partial v_l}{\partial x} + \frac{\partial u_l}{\partial y}\right), \tag{2.12}$$

$$-\rho_l u_l' w_l' = \tau_{l,xz}' = \eta_l'\left(\frac{\partial w_l}{\partial x} + \frac{\partial u_l}{\partial z}\right), \tag{2.13}$$

$$-\rho_l v_l' w_l' = \tau_{l,yz}' = \eta_l'\left(\frac{\partial w_l}{\partial y} + \frac{\partial v_l}{\partial z}\right). \tag{2.14}$$

The newly introduced variable, the dynamic turbulent viscosity, η_l', is a flow property. It remains to be modeled. Note that at a given point this is a single value for all directions. Strictly speaking this approach is valid for isotropic turbulence because there is a single eddy viscosity assumed to be valid for all directions.

For isotropic turbulence for which

$$u_l' u_l' = v_l' v_l' = w_l' w_l' = \frac{2}{3}k_l, \tag{2.15}$$

an alternative notation of the term is given here

$$\nabla \cdot \left[\left(\alpha_l^e \rho_l \mathbf{V}_l' \mathbf{V}_l'\right)\gamma\right] = \begin{pmatrix} \frac{\partial \gamma_x \alpha_l^e \rho_l u_l' u_l'}{\partial x} + \frac{\partial \gamma_y \alpha_l^e \rho_l u_l' v_l'}{\partial y} + \frac{\partial \gamma_z \alpha_l^e \rho_l u_l' w_l'}{\partial z} \\ \frac{\partial \gamma_x \alpha_l^e \rho_l v_l' u_l'}{\partial x} + \frac{\partial \gamma_y \alpha_l^e \rho_l v_l' v_l'}{\partial y} + \frac{\partial \gamma_z \alpha_l^e \rho_l v_l' w_l'}{\partial z} \\ \frac{\partial \gamma_x \alpha_l^e \rho_l w_l' u_l'}{\partial x} + \frac{\partial \gamma_y \alpha_l^e \rho_l w_l' v_l'}{\partial y} + \frac{\partial \gamma_z \alpha_l^e \rho_l w_l' w_l'}{\partial z} \end{pmatrix}$$

$$= \begin{pmatrix} \frac{\partial \gamma_y \alpha_l^e \rho_l u_l' v_l'}{\partial y} + \frac{\partial \gamma_z \alpha_l^e \rho_l u_l' w_l'}{\partial z} \\ \frac{\partial \gamma_x \alpha_l^e \rho_l v_l' u_l'}{\partial x} + \frac{\partial \gamma_z \alpha_l^e \rho_l v_l' w_l'}{\partial z} \\ \frac{\partial \gamma_x \alpha_l^e \rho_l w_l' u_l'}{\partial x} + \frac{\partial \gamma_y \alpha_l^e \rho_l w_l' v_l'}{\partial y} \end{pmatrix} + \frac{2}{3}\nabla\left(\gamma \alpha_l^e \rho_l k_l\right)$$

$$= -\hat{\mathbf{S}}_l + \frac{2}{3}\nabla\left(\gamma\alpha_l^e \rho_l k_l\right), \qquad (2.16)$$

where

$$\hat{\mathbf{S}}_l = -\begin{pmatrix} \frac{\partial}{\partial y}\left[\gamma_y\alpha_l^e \rho_l v_l'\left(\frac{\partial v_l}{\partial x} + \frac{\partial u_l}{\partial y}\right)\right] + \frac{\partial}{\partial z}\left[\gamma_z\alpha_l^e \rho_l v_l'\left(\frac{\partial w_l}{\partial x} + \frac{\partial u_l}{\partial z}\right)\right] \\ \frac{\partial}{\partial x}\left[\gamma_x\alpha_l^e \rho_l v_l'\left(\frac{\partial v_l}{\partial x} + \frac{\partial u_l}{\partial y}\right)\right] + \frac{\partial}{\partial z}\left[\gamma_z\alpha_l^e \rho_l v_l'\left(\frac{\partial w_l}{\partial y} + \frac{\partial v_l}{\partial z}\right)\right] \\ \frac{\partial}{\partial x}\left[\gamma_x\alpha_l^e \rho_l v_l'\left(\frac{\partial w_l}{\partial x} + \frac{\partial u_l}{\partial z}\right)\right] + \frac{\partial}{\partial y}\left[\gamma_y\alpha_l^e \rho_l v_l'\left(\frac{\partial w_l}{\partial y} + \frac{\partial v_l}{\partial z}\right)\right] \end{pmatrix}.$$

Here the diagonal symmetric term $\frac{2}{3}\nabla\left(\gamma\alpha_l^e \rho_l k_l\right)$ is considered as "dispersion force" and is directly computed from the turbulent kinetic energy delivered by the turbulence model. Note that in this case the use of Eq. (2.8) together with this term, as done by several authors in the literature, is wrong.

More convenient are notations using the *kinematics viscosity*

$$\nu = \eta/\rho \qquad (2.17)$$

and *eddy diffusivity*

$$\nu' = \eta'/\rho. \qquad (2.18)$$

This type of turbulence is called *Reynolds* turbulence. If you observe the river water after the bridge pillows you will immediately recognize that deformation of the velocity field is one of its reasons. Such turbulence type is available in the continuum velocity field also in multiphase flow, *Brauer* (1980). One of the most popular and up to now most fruitful approaches to describe it is to search for the dependence of the eddy diffusivity on the local parameters of the flow as it will be shown later. Before doing this I will mention at this place that the eddy viscosity can be computed with different degrees of complexity:

- By using the mixing length approaches as consisting of $\nu' = \ell^2 \left|\overline{dw}/dy\right|$ and expressions defining the mixing length as function of the distance from the wall $\ell = \ell(y)$. An example is the combination of *Nikuradse* and *van Driest* expression as discussed in the previous chapter which means that apriority knowledge of the turbulence structure is needed which makes this approach not really predictive. Then, algebraic models

for the particle induced turbulence presented in Chapter 9 have to be used as additives.
- By using *k-eps* transport equation. The contribution of the dispersed phase is then taken into account by introducing addition sources in the differential equations as it will be shown in the next chapter. Note that in this case the effect of the dispersed phase is taken into account and an additive to the eddy viscosity is not necessary.
- By using large eddy simulation algebraic models as shown in Chapter 9.
- By using large eddy simulation *k-eps* models as shown in Chapter 9.

Further understanding of the turbulence can not be reached without attracting the energy conservation in entropy form from Volume 1, *Kolev* (2007a):

$$\frac{\partial}{\partial \tau}(\alpha_l \rho_l s_l \gamma_v) + \nabla \cdot \left\{ \alpha_l^e \rho_l \gamma \left[s_l \mathbf{V}_l - \sum_{i=2}^{i_{max}} (s_{il} - s_{1l}) D_{il}^* \nabla C_{il} \right] \right\}$$

$$-\frac{1}{T_l} \nabla \cdot \left(\alpha_l^e \lambda_l^* \gamma \nabla T_l \right) = \gamma_v \left[\frac{1}{T_l} DT_l^N + \sum_{\substack{m=1 \\ m \neq l}}^{3,w} \sum_{i=1}^{i_{max}} (\mu_{iml} - \mu_{ilm}) s_{il} \right]. \qquad (2.19)$$

Here the new variables are the thermodynamic temperature *T* and the specific field entropy *s*. The diffusion of heat and specie *i* have their turbulent components so that the effective thermal conductivity

$$\lambda_l^* = \delta_l \lambda_l + \rho_l c_{pl} \frac{v_l^t}{\text{Pr}_l^t} = \rho_l c_{pl} \left(\delta_l \frac{v_l}{\text{Pr}_l} + \frac{v_l^t}{\text{Pr}_l^t} \right) \qquad (2.20)$$

and the effective component diffusivity

$$D_{il}^* = \delta_l D_{il} + \frac{v_l^t}{Sc_l^t} = \delta_l \frac{v_l}{Sc_l} + \frac{v_l^t}{Sc_l^t} \qquad (2.21)$$

are considered to be functions of the eddy viscosity by introducing two new flow parameters: the turbulent *Prandtl* number Pr_l^t and the turbulent *Schmidt* number Sc_l^t that have to modeled also.

$$\lambda_l^t = \rho_l c_{pl} v_l / \text{Pr}_l^t \qquad (2.22)$$

is called *turbulent coefficient of thermal conductivity* or *eddy conductivity* and

$$D_{il}^t = v_l^t / Sc_l^t \qquad (2.23)$$

is called *turbulent coefficient of material diffusivity* or *eddy material diffusivity*. The term

$$DT_l^N = \alpha_l \rho_l \left(\delta_l P_{\eta,l} + P_{k,l} + \varepsilon_l \right) + E_l^* + E_l'^* + \dot{q}_l''' + \sum_{i=1}^{i_{max}} \mu_{iwl} \left(h_{iwl} - h_{il} \right)$$

$$+ \sum_{\substack{m=1 \\ m \neq l}}^{l_{max}} \left[\mu_{Mml} \left(h_{Ml}^o - h_{Ml} \right) - \mu_{Mlm} \left(h_{Ml}^o - h_{Ml} \right) + \sum_{n=1}^{n_{max}} \mu_{nml} \left(h_{nm} - h_{nl} \right) \right]$$

$$+ \frac{1}{2} \left[\begin{array}{c} \mu_{wl} \left(\mathbf{V}_{wl} - \mathbf{V}_l \right)^2 - \mu_{lw} \left(\mathbf{V}_{lw} - \mathbf{V}_l \right)^2 + \sum_{m=1}^{3} \mu_{ml} \left(\mathbf{V}_m - \mathbf{V}_l \right)^2 \\ + \mu_{wl} \overline{\left(\mathbf{V}'_{wl} - \mathbf{V}'_l \right)^2} - \mu_{lw} \overline{\left(\mathbf{V}'_{lw} - \mathbf{V}'_l \right)^2} + \sum_{m=1}^{3} \mu_{ml} \overline{\left(\mathbf{V}'_m - \mathbf{V}'_l \right)^2} \end{array} \right], \quad (2.24)$$

requires special attention. $\dot{q}_l'''$ is the thermal energy introduced into the velocity field l per unit volume of the flow including thermal energy released or absorbed during chemical reactions. E_l^* is the irreversible power dissipation caused by the time averaged mass transfer between two regions with different velocities and pressure difference between bulk and interface – usually neglected. $E_l'^*$ is the irreversible power dissipation caused by the time averaged fluctuation of the interface mass transfer between two regions with different velocities and fluctuation of the pressure difference between bulk and interface – usually neglected.

Now we will discuss three important terms reflecting the irreversible part of the dissipation of the mechanical energy. The *irreversibly dissipated power* in the viscous fluid due to *turbulent pulsations* and due to *change of the mean velocity* in space is

$$\gamma_v \alpha_l \rho_l \varepsilon_l = \gamma_v \alpha_l \rho_l \left(\varepsilon_l' + \varepsilon_{\eta,l} \right) = \alpha_l^e \gamma \left[\left(\mathbf{T}'_l : \nabla \cdot \mathbf{V}'_l \right) + \left(\mathbf{T}_{\eta,l} : \nabla \cdot \mathbf{V}_l \right) \right]. \quad (2.25)$$

These components can not be returned back as a mechanical energy of the flow. They express quantitatively the transfer of mechanical energy into thermal energy in the field l. In a notation common for Cartesian and cylindrical coordinates the irreversibly dissipated power in the viscous fluid due to turbulent pulsations is expressed as follows

$$\frac{\varepsilon_l'}{\nu_l} = 2 \left\{ \gamma_r \overline{\left(\frac{\partial u_l'}{\partial r} \right)^2} + \gamma_\theta \overline{\left[\frac{1}{r^\kappa} \left(\frac{\partial v_l'}{\partial \theta} + \kappa u_l' \right) \right]^2} + \gamma_z \overline{\left(\frac{\partial w_l'}{\partial z} \right)^2} \right\}$$

$$+\left[\frac{\partial v'_l}{\partial r}+\frac{1}{r^\kappa}\left(\frac{\partial u'_l}{\partial \theta}-\kappa v'_l\right)\right]\left[\gamma_r\frac{\partial v'_l}{\partial r}+\gamma_\theta\frac{1}{r^\kappa}\left(\frac{\partial u'_l}{\partial \theta}-\kappa v'_l\right)\right]$$

$$+\left[\frac{\partial w'_l}{\partial r}+\frac{\partial u'_l}{\partial z}\right]\left[\gamma_r\frac{\partial w'_l}{\partial r}+\gamma_z\frac{\partial u'_l}{\partial z}\right]+\left[\frac{1}{r^\kappa}\frac{\partial w'_l}{\partial \theta}+\frac{\partial v'_l}{\partial z}\right]\left[\gamma_\theta\frac{1}{r^\kappa}\frac{\partial w'_l}{\partial \theta}+\gamma_z\frac{\partial v'_l}{\partial z}\right]$$

$$-\frac{2}{3}\overline{\left[\frac{\partial u'_l}{\partial r}+\frac{1}{r^\kappa}\left(\frac{\partial v'_l}{\partial \theta}+\kappa u'_l\right)+\frac{\partial w'_l}{\partial z}\right]\left[\gamma_r\frac{\partial u'_l}{\partial r}+\gamma_\theta\frac{1}{r^\kappa}\left(\frac{\partial v'_l}{\partial \theta}+\kappa u'_l\right)+\gamma_z\frac{\partial w'_l}{\partial z}\right]}\geq 0.$$
(2.26)

Similarly, the irreversibly dissipated power in the viscous fluid due to deformation of the mean velocity field in space is expressed as follows

$$\gamma_v\alpha_l\rho_l\varepsilon_{\eta,l}=\alpha_l^e\gamma(\mathbf{T}_{\eta,l}:\nabla\cdot\mathbf{V}_l)$$

$$=\alpha_l^e\rho_l\nu_l\left\{2\left\{\gamma_r\left(\frac{\partial u_l}{\partial r}\right)^2+\gamma_\theta\left[\frac{1}{r^\kappa}\left(\frac{\partial v_l}{\partial \theta}+\kappa u_l\right)\right]^2+\gamma_z\left(\frac{\partial w_l}{\partial z}\right)^2\right\}\\-\frac{2}{3}(\nabla\cdot\mathbf{V}_l)(\gamma\nabla\cdot\mathbf{V}_l)+\tilde{S}_{k,l}^2\right\},\qquad(2.27)$$

where

$$\gamma\nabla\cdot\mathbf{V}_l=\gamma_r\frac{\partial u_l}{\partial r}+\gamma_\theta\frac{1}{r^\kappa}\left(\frac{\partial v_l}{\partial \theta}+\kappa u_l\right)+\gamma_z\frac{\partial w_l}{\partial z},\qquad(2.28)$$

$$\nabla\cdot\mathbf{V}_l=\frac{\partial u_l}{\partial r}+\frac{1}{r^\kappa}\left(\frac{\partial v_l}{\partial \theta}+\kappa u_l\right)+\frac{\partial w_l}{\partial z},\qquad(2.29)$$

$$\tilde{S}_{k,l}^2=\left[\frac{\partial v_l}{\partial r}+\frac{1}{r^\kappa}\left(\frac{\partial u_l}{\partial \theta}-\kappa v_l\right)\right]\left[\gamma_r\frac{\partial v_l}{\partial r}+\gamma_\theta\frac{1}{r^\kappa}\left(\frac{\partial u_l}{\partial \theta}-\kappa v_l\right)\right]$$

$$+\left(\frac{\partial w_l}{\partial r}+\frac{\partial u_l}{\partial z}\right)\left(\gamma_r\frac{\partial w_l}{\partial r}+\gamma_z\frac{\partial u_l}{\partial z}\right)$$

$$+ \left(\frac{1}{r^\kappa} \frac{\partial w_l}{\partial \theta} + \frac{\partial v_l}{\partial z} \right) \left(\gamma_\theta \frac{1}{r^\kappa} \frac{\partial w_l}{\partial \theta} + \gamma_z \frac{\partial v_l}{\partial z} \right). \tag{2.30}$$

Here the *Stokes* hypothesis is used. For single-phase flow, $\alpha_l = 1$, in free three-dimensional space, $\gamma = 1$, the above equation then reducing to the form obtained for the first time by *Rayleigh*. Note that in a turbulent pipe flow in the viscous boundary layer $\varepsilon'_l = 0$ and $\varepsilon_{\eta,l} > 0$. Outside the boundary layer for relatively flat velocity profiles $\varepsilon'_l > 0$ and $\varepsilon_{\eta,l} \to 0$. The specific irreversibly dissipated power per unit viscous fluid mass due to turbulent pulsations $\varepsilon_l = \varepsilon_{\eta,l} + \varepsilon'_l$ is used as important dependent variable characterizing the turbulence in the field. It is a subject of model description. This power is considered to be constantly removed from the specific turbulent kinetic energy per unit mass of the flow field defined as follows

$$k_l = \frac{1}{2}\left(u_l'^2 + v_l'^2 + w_l'^2 \right). \tag{2.31}$$

In fact, Eq. (2.25) is the definition equation for the viscous dissipation rate, ε_l of the turbulent kinetic energy k_l. Here it is evident that ε_l is

(a) a non-negative quadratic form, $\varepsilon_l \geq 0$,
(b) its mathematical description does not depend on the rotation of the coordinate system, and
(c) it contains no derivatives of the viscosity,

compare with *Zierep* (1983) for single phase flow.

This is the second dependent variable for the velocity field which is also a subject of modeling. The term

$$\gamma_v \alpha_l \rho_l P_{k,l} = \alpha_l^e \gamma \cdot (\mathbf{T}'_l : \nabla \cdot \mathbf{V}_l) = \alpha_l^e \left[\nabla \gamma \cdot (\mathbf{T}' \cdot \mathbf{V}_l) - \mathbf{V}_l \cdot (\nabla \gamma \cdot \mathbf{T}') \right]$$

$$= \alpha_l^e \begin{bmatrix} \tau'_{xx}\gamma_x \frac{\partial u}{\partial x} + \tau'_{yy}\gamma_y \frac{\partial v}{\partial y} + \tau'_{zz}\gamma_z \frac{\partial w}{\partial z} \\ + \tau'_{xy}\left(\gamma_y \frac{\partial u}{\partial y} + \gamma_x \frac{\partial v}{\partial x}\right) + \tau'_{zx}\left(\gamma_z \frac{\partial u}{\partial z} + \gamma_x \frac{\partial w}{\partial x}\right) + \tau'_{zy}\left(\gamma_z \frac{\partial v}{\partial z} + \gamma_y \frac{\partial w}{\partial y}\right) \end{bmatrix}. \tag{2.32}$$

is considered to be a generation of turbulent kinetic energy, a turbulence source term. It is removed from the energy conservation and introduced as a source term in the balance equation for the turbulent kinetic energy. Inserting the *Reynolds* stresses by using the *Boussinesq* (1877) hypothesis results in common notation for Cartesian and cylindrical coordinates

$$\gamma_v \overline{P_{k,l}} := \gamma_v \frac{\alpha_l P_{k,l}}{\alpha_l^e v_l^t} = 2\left\{\gamma_r\left(\frac{\partial u_l}{\partial r}\right)^2 + \gamma_\theta\left[\frac{1}{r^\kappa}\left(\frac{\partial v_l}{\partial \theta} + \kappa u_l\right)\right]^2 + \gamma_z\left(\frac{\partial w_l}{\partial z}\right)^2\right\}$$

$$-\frac{2}{3}(\nabla.\mathbf{V}_l)(\gamma\nabla.\mathbf{V}_l) + \tilde{S}_{k,l}^2. \tag{2.33}$$

Compare this expression with Eqs. (2.26) and (2.27) and recognize the difference. An alternative notation of the Eq. (2.33) is given for isotropic turbulence

$$\alpha_l^e \gamma(\mathbf{T}':\nabla\mathbf{V}_l) = -\alpha_l^e \rho_l \frac{2}{3} k_l \gamma\nabla.\mathbf{V}_l + \alpha_l^e \rho_l v_l' \tilde{S}_{k,l}^2. \tag{2.34}$$

Nothing that the pressure pulsation caused the eddies is

$$p' = \rho_l V_l'^2 = \rho_l \frac{2}{3} k_l$$

the term

$$\alpha_l \rho_l \frac{2}{3} k_l \gamma\nabla.\mathbf{V}_l \equiv pd\mathit{Vol}\text{-work}$$

is immediately recognized as the mechanical expansion or compression $pdVol$-work.

If the entropy equation is applied to a single velocity field in a closed system without interaction with external mass, momentum or energy sources, the change in the specific entropy of the system will be non-negative, as the sum of the dissipation terms, $\varepsilon_{\eta,l} + \varepsilon_l'$, is non-negative. This expresses the *second law of thermodynamics*. The second law tells us in what direction a process will develop in nature for closed and isolated systems.

> The process will proceed in a direction such that the entropy of the system always increases, or at best stays the same, $\varepsilon_{\eta,l} + \varepsilon_l' = 0$, – entropy principle.

This information is not contained in the first law of thermodynamics. It results only after combining the three conservation principles (mass, momentum and energy) and introducing a *Legendre* transformation in the form of a *Gibbs* equation. In a way, it is a general expression of these conservation principles.

2.2 Isotropy

If the mean pulsation components in all directions are equal, we have

$$u'^2 = v'^2 = w'^2 = V'^2 = 2k/3, \qquad (2.35)$$

or

$$k_l = (3/2)V_l'. \qquad (2.36)$$

Such type of turbulence is called *isotropic turbulence*. Eq. (2.26) reduces for isotropic turbulence to

$$\varepsilon_l \approx const\ v_l \left(\frac{dV_l'}{dx}\right)^2, \qquad (2.37)$$

Taylor (1935), which is a very important scale which helps to provide the link between the turbulent kinetic energy, its dissipation, and the turbulent cinematic viscosity as given below.

Note that for isotropic turbulence the fluctuating velocity component V' is a random deviate of a *Gaussian* probability distribution

$$f(v') = \frac{1}{\sigma_{V'}\sqrt{2\pi}} e^{-\frac{1}{2}\frac{v'^2}{\sigma_{V'}^2}} \qquad (2.38)$$

with zero mean and variance

$$\sigma_{V'} = V'^2 = 2k/3. \qquad (2.39)$$

For large number of sample of N experimental observations x the square of the mean variance is

$$\sigma_x^2 \cong \frac{\sum_{1}^{N} x^2}{N} - \overline{x}^2 = \frac{\sum_{1}^{N}\left(x^2 - \overline{x}^2\right)}{N} = \overline{x'^2}, \qquad (2.40)$$

a remarkable property of the *Gauss* function. In some applications for description of processes at the heated wall evaporation induces a velocity component $u_{1,blow} = \dot{q}_{wc}''/(\rho_1 \Delta h)$ normal to the wall acting against the pulsation V' and preventing small size velocity fluctuation from reaching the wall, *Scriven* (1969). In this case the statistical average of the difference $v' - u_{1,blow}$

$$\int_{u_{1,blow}}^{\infty} (v' - u_{1,blow}) f(v') dv' = \sigma_{V'} \psi, \qquad (2.41)$$

where

$$\psi = \frac{1}{\sqrt{2\pi}} \exp\left[-\frac{1}{2}\left(\frac{u_{1,blow}}{\sigma_{V'}}\right)^2\right] - \frac{1}{2}\left(\frac{u_{1,blow}}{\sigma_{V'}}\right) erfc\left(\frac{u_{1,blow}}{\sqrt{2}\sigma_{V'}}\right), \qquad (2.42)$$

Pei (1981), is of practical importance for describing boiling critical heat flux in bubbly and in dispersed film flows.

2.3 Scales, eddy viscosity

Some important length and time scales characterizing turbulence are given below. As we will see later these scales are widely used.

2.3.1 Small scale turbulent motion

Assuming that the characteristic velocity pulsation is $V' = \ell_{\mu e} / \Delta \tau_{\mu e}$ and it changes over the distance $\ell_{\mu e}$, a characteristic time scale

$$\Delta \tau_{\mu e, l} \approx \sqrt{12 \frac{v_l}{\varepsilon_l}} \qquad (2.43)$$

can be computed from Eq. (2.37). Here the subscript μe stays for micro-eddy. This time scale is called in the literature *Taylor* time micro-scale of turbulence. For laminar flow the cinematic viscosity is a product of the characteristic velocity of the molecule V' multiplied by the mean free path length ℓ,

$$v = V'\ell, \qquad (2.44)$$

resulting in

$$\text{Re}_{e,l}^l = V_l' \ell_{\mu e, l} / v_l = 1. \qquad (2.45)$$

The length scale resulting from this equation

$$\ell_{\mu e,l} = \nu_l \Big/ \sqrt{\frac{2}{3}k_l} , \qquad (2.46)$$

called *inner scale* or *small scale*, gives the lowest scale for existence of eddies. This length scale is also called in the literature as *Taylor* micro-scale of turbulence. Below this scale eddies dissipate their mechanical energy into heat.

2.3.2 Large scale turbulent motion, *Kolmogorov-Pandtl* expression

Although remote from the reality, the analogy to the laminar flow is frequently transferred to turbulent eddies postulating simply

$$v'_l \approx V'_l \ell_{e,l} , \qquad (2.47)$$

where V'_l is the mean characteristic velocity of pulsation of a large eddy with size $\ell_{e,l}$. If the specific kinetic energy of the turbulent fluctuations is known, a good scale for V'_l is derived for the case in which pulsation in all directions are equal,

$$V'_l \approx \sqrt{\frac{2}{3}k_l} , \qquad (2.48)$$

and therefore

$$v'_l \approx \sqrt{\frac{2}{3}k_l} \ell_{e,l} \approx c'_\eta \sqrt{k_l} \ell_{e,l} , \qquad (2.49)$$

where c'_η is an empirical constant. This formula is called *Kolmogorov–Pandtl* expression, *Kolmogorov* (1942), *Prandtl* (1945). In analogy to the definition expression for the dissipation of the turbulent kinetic energy in isotropic turbulence given by Eq. (2.37), one can write

$$\varepsilon_l \approx \mathrm{const}\, v'_l \left(\frac{V'_l}{\ell_{e,l}}\right)^2 = \mathrm{const}\, v'_l \left(\frac{\sqrt{k_l}}{\ell_{e,l}}\right)^2 , \qquad (2.50)$$

where the constant is in the order of 1. Excluding $\ell_{e,l}$ from (2.49) and (2.50) the link between the turbulent cinematic viscosity and the turbulent characteristics is obtained:

$$v_l^t \approx c_\eta \frac{k_l^2}{\varepsilon_l} . \qquad (2.51)$$

This is a widely used expression. The empirical constant

$$c_\eta = 0.09 \qquad (2.52)$$

is derived from experiments with single phase flow. Inserting Eq. (2.51) in (2.50) results in

$$\varepsilon_l = \sqrt{c_\eta}\, k_l^{3/2}/\ell_{e,l} = 0.3 k_l^{3/2}/\ell_{e,l} = 0.55 V_l'^3/\ell_{e,l} . \qquad (2.53)$$

Kolmogorov (1941, 1949) found for isotropic turbulence from dimensional analysis the same equation

$$\varepsilon_l \cong c_1 V_l'^3/\ell_{e,l} = c_1 \left(\frac{2}{3} k_l\right)^{3/2}/\ell_{e,l} . \qquad (2.54)$$

The order of magnitude of the constant $c_1 \approx 0.55$ is conformed by *Batchelor* (1967), $c_1 \cong 0.35$, and *Hinze* (1955), $c_1 \cong 1$. Using Eqs. (2.51) and (2.53) we obtain interesting expression in terms of the turbulent *Reynolds* number

$$\mathrm{Re}_l' := \frac{V_l' \ell_{e,l}}{v_l'} \approx \sqrt{\frac{2}{3} \frac{c_1}{c_\eta}} \approx 5 . \qquad (2.55)$$

It is considered that eddies smaller than those defined by the above expression

$$\ell_{e,l} = \sqrt{c_\eta}\, k_l^{3/2}/\varepsilon_l \qquad (2.56)$$

start to dissipate. The time scale of the fluctuation of large eddy with size $\ell_{e,l}$ is therefore

$$\Delta \tau_{e,l} = \ell_{e,l}/V' = \sqrt{3 c_\eta/2}\, k_l/\varepsilon_l = 0.37 k_l/\varepsilon_l . \qquad (2.57)$$

The order of magnitude of the constant is experimentally confirmed by *Snyder* and *Lumley* (1971). The authors reported 0.2. Note that *Corrsin* already used in 1963 this value. Close to this result is the result obtained from direct numerical simulation of isotropic flow by *Sawford* (1991). *Sawford* reported 0.19. Some authors used instead: 0.41 *Gosman* et al. (1992); 0.35 *Antal* et al. (1998); 0.27 *Loth* (2001).

Thus, if for isotropic turbulence the turbulent kinetic energy and its dissipation are known in a point, the size of the large eddy and its pulsation period are also known. Next we will discuss the so-called *k-eps* equation describing these quantities as transport properties.

2.4 k-eps framework

Although having several weaknesses the most popular method for describing single phase turbulence is the so-called $k-\varepsilon$ turbulence model. My favorite introduction to this formalism is the book by *Rodi* (1984). The popularity of this model is the main reason for many attempts to extend it to multi-phase flows. In what follows we will mention some of the works. *Akai* (1981) describes separated two phase flow using for each of the both phases a $k-\varepsilon$ model. *Carver* (1983) takes into account approximately the geometry effect on the turbulence by using for each velocity field a $k-\varepsilon$ model, assuming that the velocity field occupies the entire channel. This method remembers the *Martinelli–Nelson* method for modeling of two-phase friction pressure drop in channels. The author pointed out the limitations of their approach.

The $k-\varepsilon$ model is used in a number of papers concerning the modeling of mixtures of gas and solid particles, *Wolkov*, *Zeichik* and *Pershukov* (1994), *Reeks* (1991, 1992), *Simonin* (1991), *Sommerfeld* (1992).

Lahey (1987) successfully extend the single phase $k-\varepsilon$ model to a bubble flow.

The common feature of these works is the concept assuming convection and diffusion of the specific turbulent kinetic energy and its dissipation in the *continuous phase*. For considering the influence of the discrete phase predominantly two approaches are used:

(a) No feedback of the dispersed phase on the continuum turbulence commonly named *one-way coupling*;
(b) The feedback of the dispersed phase on the continuum turbulence is taken into account. This approach is named *two-way coupling*.

The conservation equation for the specific turbulence kinetic energy is derived as follows: Multiply each of the scalar instantaneous momentum equations with the other two *instantaneous* velocity components, respectively. Add the so obtained equations and rearrange the time derivatives and convective terms in order to bring each velocity under the differential sign. Replace the non-averaged velocities with the sum of its averaged values and the pulsation components. Perform time averaging. Thus, the one equation which is the first intermediate result of the derivation is obtained. Next, multiply the *averaged* scalar momentum equations with the *averaged* velocity components in the other two directions, respectively, add the so obtained 6 equations, and rearrange similarly as previously described. Thus, obtain the second equation. Subtract the second equation from the first one, *assume equality of the pulsation components in each direction, isotropy*, multiply by 1/2 and rearrange to obtain the equation for the specific turbulent kinetic energy,

$$\frac{\partial}{\partial \tau}(\alpha_c \rho_c k_c \gamma_v) + \nabla \cdot \left[\alpha_c \rho_c \left(\mathbf{V}_c k_c - v_c^k \nabla k_c\right)\gamma\right]$$

$$= \alpha_c \rho_c \gamma_v \left(\overline{v_c' P_{k,c}} - \varepsilon_c + G_{k,c} + P_{k\mu,c} + P_{kw,c} + P_{kw,c}^\varsigma\right), \tag{2.58}$$

The diffusion coefficient of the turbulent kinetic energy is

$$v_l^k = v_l + v_l' / \mathrm{Pr}_{k,l}^t . \tag{2.59}$$

Pr_k^t is the turbulent *Prandtl* number describing diffusion of the turbulent kinetic energy. The generation of the turbulent kinetic energy $\overline{v_l' P_{k,l}}$ is proportional to the velocity deformation $\overline{P_{k,l}}$. The proportionality factor is the turbulent eddy diffusivity v_l'. For uniform velocity field in space there is no turbulence generation for this particular reason. Note that the part $v_l \overline{P_{k,l}}$ is directly dissipated in heat and is found as a irreversible source term in the continuum energy conservation. With $v_l' \gg v_l$, the generation is much higher than the direct viscous dissipation. $P_{kw,c}$ and $P_{kw,c}^\varsigma$ are the generation of turbulent kinetic energy per unit mass of the continuum due to wall friction and due to local flow obstacles. The dissipation rate ε_l directly reduces the turbulent kinetic energy, see the RHS of Eq. (2.58), $= ... \overline{v_l' P_{k,l}} - \varepsilon_l ...$.

Unlike the derivation of the *k*-equation, the derivation of the ε-equation leads to an equation having large number of terms, see *Besnard* and *Harlow* (1985, 1988), for which it is not known how they all have to be modeled. That is why we write the equation for the *rate of dissipation of the kinetic energy of isotropic turbulence* in analogy with the derivation for single phase flows *intuitive*, without strong proof

$$\frac{\partial}{\partial \tau}(\alpha_c \rho_c \varepsilon_c \gamma_v) + \nabla \cdot \left[\alpha_c \rho_c \left(\mathbf{V}_c \varepsilon_c - v_c^\varepsilon \nabla \varepsilon_c\right)\gamma\right]$$

$$= \alpha_c \rho_c \gamma_v \left\{\frac{\varepsilon_c}{\kappa_c}\left[c_{\varepsilon 1}\left(\overline{v_c' P_{k,c}} + P_{kw,c} + P_{kw,c}^\varsigma\right) - c_{\varepsilon 2}\varepsilon_c + c_{\varepsilon 3} G_{k,c}\right]\right\}. \tag{2.60}$$

Here k/ε_c is characteristic time scale of the dissipation. Note that *Lopez de Bertodano* (1992) proposed especially for the dissipation of the turbulence generated by the particles, $c_{\varepsilon 3} G_{k,c}$, to use different characteristic time scale, namely

$$\frac{2}{3}\frac{c_{vm}}{c_{cd}^d}\frac{D_d}{|\Delta V_{cd}|}.$$

The diffusion coefficient of the dissipation rate of the turbulent kinetic energy is

$$v_l^\varepsilon = v_l + v_l^t / \Pr_{\varepsilon,l}^t . \tag{2.61}$$

$\Pr_\varepsilon^t$ is the turbulent *Prandtl* number describing diffusion of the dissipation of the turbulent kinetic energy. Useful approximation for estimation the source of the dissipation is

$$P_{\varepsilon w,c} \approx \frac{\varepsilon_c}{\kappa_c} c_{\varepsilon 1} \left(P_{kw,c} + P_{kw,c}^\zeta \right).$$

This single form of the *k-eps* model can be applied either

(a) for fine resolution of the bulk flow with spatial treatment of the wall boundary layers ($v_l \overline{P_{k,l}} > 0$, $P_{kw,c} = 0$, $P_{\varepsilon w,c} = 0$) or

(b) for porous body with special treatment of the sources ($v_l \overline{P_{k,l}} = 0$, $P_{kw,c} > 0$, $P_{\varepsilon w,c} > 0$).

For multiphase flow analysis with fine resolution of the bulk flow the velocity deformation term $\overline{P_{k,l}}$ is computed from the mean velocity field and used to constitute the sources. In this case the porous body sources are set to zero, $P_{kw,l} = 0$, $P_{\varepsilon w,l} = 0$.

For multiphase flow analysis in porous body the usually used gross resolution does not allow computing accurately $\overline{P_{k,l}}$. Fortunately for many geometrical arrangements of practical interests we have appropriate empirical information to compute the fictional wall share stress and all other terms required. Therefore the deformation term is set to zero, $\overline{P_{k,l}} = 0$, and the modeling is performed designing appropriately the terms $P_{kw,l} > 0$, $P_{\varepsilon w,l} > 0$, besides the remaining terms.

For single phase flow this equation reduces to the one obtained first by *Hanjalic* and *Launder* (1972). The modeling constants for single phase flow are given in Table 2.1. Special adjustment of the coefficients for pipe flow is given in Table 2.2.

Table 2.1. Coefficients for single phase $k-\varepsilon$ model, see *Rodi* (1984)

$\sigma_t = 0.9$ $c_{\varepsilon 1} = 1.44$ $c_{\varepsilon 2} = 1.92$ $c_{\varepsilon 3} = 1.44$ $c_\eta = 0.09$

$\Pr_k^t = 1.0$

$\Pr_\varepsilon^t = 1.0$

$c_{\varepsilon 3} = 0.432$ for stratified flow, *Maekawa* (1990)

Table 2.2. Coefficients for single phase $k-\varepsilon$ model in pipe flow

Myong-Kasagi (1988a, b)

$v_l^t \approx c_\eta f_\eta \dfrac{k_l^2}{\varepsilon_l}$, $c_\eta = 0.09$, $f_\eta = \left(1+\dfrac{3.45}{\text{Re}^t}\right)\left[1-\exp\left(-\dfrac{y^+}{70}\right)\right]$, $\text{Re}^t = k^2/(v\varepsilon)$

$\sigma_t = 0.9$ $c_{\varepsilon 1} = 1.4$ $c_{\varepsilon 2} = 1.8\left\{1-\dfrac{2}{3}\exp\left[-\left(\dfrac{\text{Re}^t}{6}\right)^2\right]\right\}\left[1-\exp\left(-\dfrac{y^+}{5}\right)\right]^2$

$\Pr_k^t = 1.4$

$\Pr_\varepsilon^t = 1.3$

Zhu and *Songling* (1991), low *Reynolds* number and transition flow:

$v_l^t \approx c_\eta f_\eta \dfrac{k_l^2}{\varepsilon_l}$, $c_\eta = 0.09$, $f_\eta = \left(1+\dfrac{19.5}{\text{Re}^t}\right)\left[1-\exp\left(-0.016\,\text{Re}_y\right)\right]$,

$\text{Re}^t = k^2/(v\varepsilon)$, $\text{Re}_y = k^{1/2}y/v$ for $y<\delta$, $\text{Re}_y = k^{1/2}\delta/v$ for $y>\delta$,

δ boundary layer thickness,

$\sigma_t = 0.9$ $c_{\varepsilon 1} = 1.44\left[1+(0.06/f_\eta)^2\right]$, $c_{\varepsilon 2} = 1.92\left\{1-\exp\left[-(\text{Re}^t)^2\right]\right\}$

$\Pr_k^t = 1$

$\Pr_\varepsilon^t = 1.3$

For two phase bubbly flow in a vertical pipe the modeling coefficients are given in Table 2.3.

Table 2.3. Coefficients for bubbly flow $k-\varepsilon$ model for $30000 < \text{Re}_l < 72000$, see Lahey (1987, 1989)

$\sigma_t = 0.9 \qquad c_{\varepsilon 1} = 1.44 \qquad c_{\varepsilon 2} = 1.92 \qquad c_{\varepsilon 3} = 1.92$

$c_\eta = 0.8 + (0.09 - 0.8)\exp(-100\alpha_l)$

$\text{Pr}_{k,l}^t = 0.037 + 0.21 \times 10^{-5} \text{Re}_l + 0.2 \bigg/ \left\{ 1 + \exp\left[\left(\frac{\text{Re}_l - 65000}{5000}\right)^2\right] \right\}$

$\text{Pr}_\varepsilon^t = 1.3$

For bubbles rising freely in still tank *Lahey* obtains the value for the constant $c_{\eta 3} = 1.92$, which is different to the constant for single-phase flow. He also changed the effective *Prandtl* number for the turbulent kinetic energy diffusion as shown in Table 2.3. *Lee* et al. (1989) show that the coefficient c_η must be considerably higher for a two phase flow in order to predict flatter velocity profile in the central region of a bubble flow in a vertical flow pipe – see Table 2.3.

Troshko and *Hassan* (2001) used for $\text{Pr}_\varepsilon^t = 1.272$.

We will provide more information of the source terms in the next chapters.

Nomenclature

Latin

C_{il} is the mass concentration of the inert component i in the velocity field l, dimensionless

c^{vm} virtual mass force coefficient, *dimensionless*

c^d drag force coefficient, *dimensionless*

c^L lift force coefficient, *dimensionless*

c_p specific heat at constant pressure, $J/(kgK)$

c_η viscosity coefficient, *dimensionless*

$c_{\varepsilon 1}, c_{\varepsilon 2}, c_{\varepsilon 3}$ are the modeling constants for the conservation equation of the energy dissipation, *dimensionless*

c_η model coefficients in k-eps model

D diffusivity, m^2/s

D_l particle size in field l, m

D_{hy} hydraulic diameter (4 times cross-sectional area / perimeter), m

D_{il} $:= v_l/Sc_{il}$, coefficient of molecular diffusion for species i into the field l, m^2/s

D_{il}^t	$:= v_l^t / Sc_{il}^t$ coefficient of turbulent diffusion, m^2/s
D_{il}^*	$:= D_{il} + D_{il}^t$, effective diffusion coefficient, m^2/s
DC_{il}	right-hand side of the non-conservative conservation equation for the inert component, $kg/(sm^3)$
d	total differential
e	specific internal energy, J/kg
$G_{k,l}$	production of turbulent kinetic energy due to bubble relocation in changing pressure field per unit mass of the filed l, W/kg (m²/s³)
$\mathbf{g}$	acceleration due to gravity, m/s^2
h	specific enthalpy, J/kg
Δh	latent heat of evaporation, J/kg
$\mathbf{I}$	unit matrix, *dimensionless*
k	kinetic energy of turbulent pulsation, m^2/s^2
P	irreversibly dissipated power from the viscous forces due to deformation of the local volume and time average velocities in the space, W/kg
p_{li}	$l = 1$: partial pressure inside the velocity field l
	$l = 2,3$: pressure of the velocity field l
p	pressure, Pa
$\Delta p_c^{d\sigma}$	surface averaged difference between the pressure at the surface $d\sigma$ and the bulk pressure of c, effective interfacial stagnation pressure difference in the continuum, Pa
$\Delta p_c^{w\sigma*}$	surface averaged difference between the pressure at the surface $w\sigma$ and the bulk pressure of c, effective wall-continuum stagnation pressure difference in the continuum, Pa
P_k	production of the turbulent kinetic energy per unit mass, W/kg
$\overline{P_{k,l}}$	in $v_{l,ss}^t \overline{P_{k,l}}$ which is the production of the turbulent kinetic energy per unit mass of the velocity field l due to deformation of the velocity field l, W/kg
$P_{kw,l}$	production of turbulent kinetic energy per unit mass of the field l due to friction with the wall, W/kg
$P_{k\mu,l}$	production of turbulent kinetic energy per unit mass of the field l due to friction evaporation or condensation, W/kg
P_ε	production of the dissipation of the turbulent kinetic energy per unit mass, W/kg
$P_{\varepsilon w,l}$	production of the dissipation of the turbulent kinetic energy per unit mass of the field l due to friction with the wall, W/kg
$P_{kw,c}$	generation of turbulent kinetic energy per unit mass of the continuum, W/kg

$P_{\varepsilon w,c}$ "production" of dissipation of the turbulent kinetic energy per unit mass of the continuum, W/kg

$\Pr_l^l$ $:= \rho_l c_{pl} v_l^l / \lambda_l^l$, molecular *Prandtl* number, *dimensionless*

$\Pr_{T,l}^l$ $:= \rho_l c_{pl} v_l^t / \lambda_l^t$, turbulent *Prandtl* number, *dimensionless*

$\Pr_k^t$ turbulent Prandtl number describing diffusion of the turbulent kinetic energy, *dimensionless*

$\Pr_\varepsilon^t$ turbulent Prandtl number describing diffusion of the dissipation of the turbulent kinetic energy, *dimensionless*

$\dot{q}_l'''$ thermal energy introduced into the velocity field l per unit volume of the flow, W/m^3

$\dot{q}_{\sigma l}'''$ $l = 1,2,3$. Thermal power per unit flow volume introduced from the interface into the velocity field l, W/m^3

$\dot{q}_{w\sigma l}'''$ thermal power per unit flow volume introduced from the structure interface into the velocity field l, W/m^3

$\dot{q}_{wc}''$ heat flux from the wall to the continuum, W/m^2

s specific entropy, $J/(kgK)$

Sc^t turbulent *Schmidt* number, *dimensionless*

T temperature, K

T_l temperature of the velocity field l, K

$\mathbf{T}$ shear stress tensor, N/m^2

$u_{1,blow}$ $:= \dot{q}_{wc}'' / \Delta h$, surface averaged evaporation velocity, m/s

$\mathbf{V}$ time and surface averages of the instantaneous fluid velocity with components, u, v, w in $r, \theta,$ and z direction, m/s

$\mathbf{V}'$ pulsating component of $\mathbf{V}$, m/s

$\Delta \mathbf{V}_{lm}$ $\mathbf{V}_l - \mathbf{V}_m$, velocity difference, disperse phase l, continuous phase m carrying l, m/s

v specific volume, m^3/kg

x, y, z coordinates, m

Greek

α_l part of $\gamma_v Vol$ available to the velocity field l, local instantaneous volume fraction of the velocity field l, *dimensionless*

α_{il} the same as α_l in the case of gas mixtures; in the case of mixtures consisting of liquid and macroscopic solid particles, the part of $\gamma_v Vol$ available to the inert component i of the velocity field l, local instantaneous volume fraction of the inert component i of the velocity field l, *dimensionless*

γ_v	the part of $dVol$ available for the flow, volumetric porosity, *dimensionless*
γ	surface permeability, *dimensionless*
Δ	finite difference
δ	small deviation with respect to a given value
δ_l	= 1 for continuous field; = 0 for disperse field, *dimensionless*
∂	partial differential
ε	dissipation rate for kinetic energy from turbulent fluctuation, power irreversibly dissipated by the viscous forces due to turbulent fluctuations, W/kg
η	dynamic viscosity, $kg/(ms)$
η_{vis}	part of the friction energy directly dissipated into heat, *dimensionless*
θ	θ-coordinate in the cylindrical or spherical coordinate systems, *rad*
κ	= 0 for Cartesian coordinates, = 1 for cylindrical coordinates
κ	isentropic exponent
λ	thermal conductivity, $W/(mK)$
λ^t	$:= \rho c_p v^t / \text{Pr}^t$, turbulent thermal conductivity, $W/(mK)$
λ_l^*	$:= \lambda + \lambda^t$, effective thermal conductivity, $W/(mK)$
μ_l	time average of local volume-averaged mass transferred into the velocity field l per unit time and unit mixture flow volume, local volume-averaged instantaneous mass source density of the velocity field l, $kg/(m^3 s)$
μ_{wl}	mass transport from exterior source into the velocity field l, $kg/(m^3 s)$
μ_{il}	time average of local volume-averaged inert mass from species i transferred into the velocity field l per unit time and unit mixture flow volume, local volume-averaged instantaneous mass source density of the inert component i of the velocity field l, $kg/(m^3 s)$
μ_{iml}	time average of local volume-averaged instantaneous mass source density of the inert component i of the velocity field l due to mass transfer from field m, $kg/(m^3 s)$
μ_{ilm}	time average of local volume-averaged instantaneous mass source density of the inert component i of the velocity field l due to mass transfer from field l into velocity field m, $kg/(m^3 s)$
v	cinematic viscosity, m^2/s
v_l^t	coefficient of turbulent cinematic viscosity, m^2/s
v_c^k	$:= v_c + v_c^t / \text{Pr}_{k,c}^t$, diffusion coefficient of the turbulent kinetic energy, m^2/s

ν_c^ε	$:= \nu_c + \nu_c' / \mathrm{Pr}_{\varepsilon,c}'$, diffusion coefficient of the dissipation rate of the turbulent kinetic energy, m^2/s
ρ_l	intrinsic local volume-averaged field density, kg/m^3
ρ_{il}	instantaneous inert component density of the velocity field l, kg/m^3
σ_t	model coefficient in the k-eps models, *dimensionless*
τ	time, s
$\zeta_{fr,co}$	irreversible friction coefficient computed for the total mixture mass flow with the properties of the continuum only, dimensionless

Subscripts

c	continuous
d	disperse
lm	from l to m or l acting on m
w	region "outside of the flow"
e	entrances and exits for control volume Vol
l	velocity field l, intrinsic field average
i	inert components inside the field l, non-condensable gases in the gas field $l = 1$, or microscopic particles in water in field 2 or 3
i	corresponding to the eigenvalue λ_i in Chapter 4
ml	from m into l
iml	from im into il
n	inert component
0	at the beginning of the time step
σ	interface
τ	old time level
$\tau + \Delta\tau$	new time level
0	reference conditions
p,v,s	at constant p,v,s, respectively

Superscripts

'	time fluctuation
d	drag
e	heterogeneous
i	component (either gas or solid particles) of the velocity field
iml	from im into il
i_{max}	maximum for the number of the components inside the velocity field
L	lift
l	intrinsic field average
le	intrinsic surface average
$l\sigma$	averaged over the surface of the sphere
M	non-inert component

ml	from *m* into *l*
n	inert component
m	component
t	turbulent
vm	virtual mass
τ	temporal, instantaneous
	averaging sign

Operators

∇ ·	divergence
∇	gradient

References

Akai M, Inoue A and Aoki S (1981) The prediction of the stratified two-phase flow with a two-equation model of turbulence, Int. J. Multiphase Flow, vol 7 pp 21-29

Amsden AA, Butler TD, O'Rourke PJ and Ramshaw JD (1985) KIVA-A comprehensive model for 2-D and 3-D engine simulations, paper 850554

Antal S, Kurul N, Podowski MZ and Lahey RT Jr (June 8-11, 1998) The development of multidimensional modeling capabilities for annular flows, 3th Int. Conf. On Multiphase Flows, ICMF'98, Lyon, France

Batchelor GK (1967) An introduction to fluid dynamics, Cambridge Univ. Press, Cambs.

Batchelor GK (1988) A new theory of the instability of a uniform fluidized bed, J. Fluid Mechanic v 193 pp 75-110

Besnard DC and Harlow FH (May 1985) Turbulence in two-field incompressible flow, LA-10187-MS, UC-34, Los Alamos National Laboratory

Besnard DC and Harlow FH (1988) Turbulence in multiphase flow, Int. J. Multiphase Flow, vol 6 no 6 pp 679-699

Boussinesq J (1877) Essai sur la théorie des eaux courantes, Mem. Pr's. Acad. Sci., Paris, vol 23 p 46

Brauer H (1980) Turbulence in multiphase flow, Ger. Chem. Eng. 3 pp 149-161

Carver MB (Aug. 1983) Numerical computation of phase distribution in two phase flow using the two-dimensional TOFFEA code, Chalk River Nuclear Laboratories, AECL-8066

Corrsin S (1963) Estimates of the relations between Eulerian and the Lagraian scales in large Reynolds number turbulence, J. Atmos. Sci. vol 20 pp 115-119

Gosman AD et al. (1992) AIChE J. vol 38 pp 1946-1956

Hanjalic K and Launder BE (1972) A Reynolds stress model of turbulence and its application to thin share flows, J. of Fluid Mechanics, vol 52 no 4 pp 609-638

Hinze JO (1955) Fundamentals of hydrodynamics of splitting in dispersion processes, AIChE Journal, vol 1 pp 284-295

Kolev NI (2007a) Multiphase Flow Dynamics, Vol. 1 Fundamentals, 3d extended ed., Springer, Berlin, New York, Tokyo

Kolev NI (2007b) Multiphase Flow Dynamics, Vol. 2 Thermal and mechanical interactions, 3d extended ed., Springer, Berlin, New York, Tokyo

Kolmogoroff AN (1941) The local structure of turbulence in incompressible viscous fluid for very large Reynolds numbers, C. R. Acad. Sci. U.S.S.R., vol 30 pp 825-828

Komogorov AN (1942) Equations of turbulent motion of incompressible fluid, Isv. Akad. Nauk. SSR, Seria fizicheska Vi., no 1-2 pp 56-58
Lahey RT (24-30 May 1987) Turbulence and two phase distribution phenomena in two-phase flow, Proc of Transient Phenomena in Multiphase Flow, Dubrovnik
Lahey RT Jr, Lopez de Bertodano M and Jones OC Jr (1993) Phase distribution in complex geometry conditions, Nuclear Engineering and Design, vol 141 pp 177-201
Lee SL, Lahey RT Jr. and Jones OC Jr. (1989) The prediction of two-phase turbulence and phase distribution phenomena using a k-e model, Japanese J. Multiphase Flow, vol 3 no 4 pp 335-368
Lee MM, Hanratty TJ and Adrian RJ (1989b) An axial viewing photographic technique to study turbulent characteristics of particles, Int. J. Multiphase Flow, vol 15 pp 787-802
Loth E (2001) Int. J. Multiphase Flow vol 27 pp 1051-1063
Maekawa (1990)
Myong K and Kasagi N (1988a) Trans. JSME, Ser.B, 54-507, 3003-3009
Myong K and Kasagi N (1988b) Trans. JSME, Ser.B, 54-508, 3512-3520
Pei BS (1981) Prediction of critical heat flux in flow boiling at low quality, PhD Thesis, University of Cincinnati, Cincinnati, Ohio
Prandtl LH (1945) Über ein neues Formelsystem für die ausgebildete Turbulenz, Nachr. Akad. Wiss., Göttingen, Math.-Phys. Klasse p 6
Reeks MW (1977) On the dispersion of small particles suspended in an isotropic turbulent field, Int. J. Multiphase Flow, vol 3 pp 319
Reeks MW (1991) On a kinetic equation for transport of particles in turbulent flows, Phys. Fluids A3, pp 446-456
Reeks MW (1992) On the continuum equation for dispersed particles in non-uniform. Phys. Fluids A4, pp 1290-1303
Reynolds O (May 1894) On the dynamical theory of incompressible viscous fluids and the determination of the criterion, Cambridge Phil. Trans., pp 123-164
Rodi W (Feb. 1984) Turbulence models and their application in hydraulics – a state of the art review, IId rev ed., University of Karlsruhe
Sawford BL (1991) Phys. Fluids A, vol 3 no 6 p 1577
Scriven LE (1969) Penetration theory modelling, Chem. Engng Educ., vol 3 pp 94-102
Snyder WH and Lumley JL (1971) Some measurements of a particle velocity autocorrelation function in a turbulent flow, J. Fluid Mech. vol 48 pp 41-71
Sommerfeld M (1992) Modeling of particle-wall collisions in confined gas-particle flows. Int. J. Multiphase Flow, vol 26 pp 905-926
Somerfeld M and Zivkovic G (1992) Recent advances in numerical simulation of pneumatic conveying through pipe systems, Computational Methods in Applied Sciences, p 201
Stokes GG, Cambridge Phil. Trans., vol 9 p 57
Stokes GG (1845) On the theories of the internal friction of fluids in motion and of the
Taylor GI (1935) Proc. Roy. Soc. A, vol 151 p 429
Troshko AA and Hassan YA (2001) A two-equations turbulence model of turbulent bubbly flow, Int. J. of Multiphase Flow, vol 27 pp 1965-2000
Wolkov EP, Zeichik LI and Pershukov VA (1994) Modelirovanie gorenia twerdogo topliva, Nauka, Moscow, in Russian.
Zhu H and Songling L (1991) Numerical simulation of transient flow and heat transfer in smooth pipe, Int. J. Heat Mass Transfer, vol 34 no 10 pp 2475-2482
Zierep J (1983) Einige moderne Aspekte der Stroemungsmechanik, Zeitschrift fuer Flugwissenschaften und Weltraumforschung, vol 7 no 6 pp 357-361

3 Sources for fine resolution outside the boundary layer

If the resolution of the computational analyses is fine enough to compute accurately the *deformation* of the velocity field but not fine enough to resolve the boundary layer, the k-ε model is used accomplished with special treatment of the boundary conditions. In such case in general the deformation term is non negative, $\overline{P_{k,l}} \geq 0$ and the porous body source terms are set to zero $P_{kw,l} = 0$, $P_{\varepsilon w,l} = 0$.

3.1 Bulk sources

The bulk source term of turbulent kinetic energy consists of the following.

3.1.1 Deformation of the velocity field

As already mentioned in the previous chapter, inside the velocity field the deformation of a flowing control volume causes not only generation of viscous forces but it also generates turbulence. In the frame of the *k-eps* equation this contribution is manifested in a term proportional to $\overline{P_{k,l}}$ as already defined by Eq. (2.33).

3.1.2 Blowing and suction

Evaporation, condensation, mass injection or removal from a velocity field through the environmental structure influence for sure the generation of turbulence. In each particular cases this contribution have to be investigated – a large field for scientific activities that is not covered up to now. The term reflecting this class of phenomena is generally written as

$$P_{k\mu,l} = \mu_{wl} k_{wl} - \mu_{lw} k_l + \sum_{m=1}^{l_{max}} \left(\mu_{ml} k_{ml}^* - \mu_{lm} k_l \right), \tag{3.1}$$

where k_{wl} is the kinetic energy of turbulent pulsation introduced with the mass source μ_{wl}, k_{ml}^* is the kinetic energy of turbulent pulsation introduced with the mass source μ_{ml}.

3.1.3 Buoyancy driven turbulence generation

For single phase atmospheric flow the buoyancy driven turbulence generation is defined by the term taking into account the *change of the density* of the continuous velocity field

$$G_{k,l} = \frac{v_l^t}{\rho \sigma_t}\left(g_r \frac{\partial \rho_l}{\partial r} + g_\theta \frac{1}{r^\kappa}\frac{\partial \rho_l}{\partial \theta} + g_z \frac{\partial \rho_l}{\partial z}\right). \tag{3.2}$$

For two-phase flow $G_{k,l}$ should be replaced by a term taking into account the much stronger effect of *turbulence generation in the wakes behind the bubbles*. There are two approaches proposed in the literature that describe this phenomenon.

Lahey (1987) and *Lee* et al. (1989) proposed a quantitative relation between the power needed for bubble translation in a liquid with spatially changing local pressure and the part of it generating turbulence. The main idea is summarized below. If single bubble with volume V_b moves along Δz in a liquid across a pressure difference Δp with velocity ΔV_{12}, a technical work $V_b \Delta p$ is performed. Therefore within the time $\Delta \tau = \Delta z / \Delta V_{12}$ the power driving this process is

$$\frac{V_b \Delta p}{\Delta \tau} = \frac{V_b \Delta p}{\Delta z}\frac{\Delta z}{\Delta \tau} = V_b \frac{\Delta p}{\Delta z}\Delta V_{12}. \tag{3.3}$$

For n_1 bubbles per cubic meter of the flow the power density is

$$\alpha_2 G_{k,2} = n_1 V_b \frac{\Delta p}{\Delta z}\Delta V_{12} = \alpha_1 \Delta V_{12} \nabla p. \tag{3.4}$$

This idea is extendible for the three dimensional case

$$\gamma_v \alpha_2 G_{k,2} = -(1-\eta_{vis})\alpha_1\left[(u_1-u_2)\gamma_r\frac{\partial p}{\partial r}+(v_1-v_2)\gamma_\theta\frac{1}{r^\kappa}\frac{\partial p}{\partial \theta}+(w_1-w_2)\gamma_z\frac{\partial p}{\partial z}\right], \tag{3.5}$$

Lee et al. (1989) found that only a $1-\eta_{vis}$ -part of this power, < 17%, is generating large scale turbulence, and that it is a function of the pipe liquid *Reynolds* number

$$1-\eta_{vis} = f(\text{Re}_c) < 0.17, \quad \text{Re}_2 = w_2 D_h / \nu_2, \tag{3.6}$$

$$1-\eta_{vis} = 0.03 + \left(-0.344 \times 10^{-5} \text{Re}_2 + 0.243\right) / \left\{1 + \exp\left[(\text{Re}_2 - 60000)/2000\right]\right\}. \tag{3.7}$$

The last relation is based on data for $30000 < \text{Re}_2 < 72000$, see Fig. 7 in *Lee* et al. (1989). The remaining part generates small scale eddies that dissipate quickly. The problem with this approach for computing $1-\eta_{vis}$ is that is not local but depends on integral variables like Re_2.

3.1.4 Turbulence generated in particle traces

Bataille and *Lance* (1988) assumed again that $1-\eta_{vis}$ part of the power, lost by the continuum to resist the bubble movements with a relative velocity ΔV_{cd}, is transformed into kinetic energy for the generation of wakes behind the bubble (or particle). Part of the turbulent energy is dissipated back into the continuum. If the drag force per unit volume of the mixture is f_{cd}^d, the power lost to resist the bubble is $f_{cd}^d \Delta V_{cd}$. Thus, the $1-\eta_{vis}$ part of this power generates the turbulent wakes behind the bubbles

$$\gamma_v \alpha_c \rho_c G_{k,c} \approx \gamma_v \alpha_d \rho_c (1-\eta_{vis}) \frac{3}{4} \frac{c_{cd}^d}{D_d} |\Delta V_{cd}|^3, \tag{3.8}$$

and the η_{vis} part is dissipated back into the continuum as a heat. *Lopez de Bertodano* (1992) proposed for $(1-\eta_{vis})0.75 c_{cd}^d$ to use 0.25. *Troshko* and *Hassan* (2001) obtained by comparison with the *Wang*'s (1987) data $1-\eta_{vis}$ = 0.45. *Kataoka* and *Serizawa* (1995) derived from his analysis η_{vis} =0.925. Here D_d is the characteristic size of the particles and ΔV_{cd} is the magnitude of the relative velocity. Now let us turn the attention to the computation of $1-\eta_{vis}$. Intuitively for pool flow with negligible effect of the wall on the turbulence it is expected that $1-\eta_{vis}$ is a function of the particle *Reynolds* number

$$\text{Re}_{cd} = \Delta V_{cd} D_d / \nu_c. \tag{3.9}$$

Reichard (1942) derived theoretically limiting *Reynolds* numbers

$$\text{Re}_{1Sph} = 3.73 \left(\frac{\rho_2 \sigma_2^3}{g \eta_2^4} \right)^{0.209} = 3.73 \left(\frac{\Delta \rho_{21}}{\rho_2} Ar^2 \right)^{0.209}, \qquad (3.10)$$

$$\text{Re}_{1St} = 3.1 \left(\frac{\rho_2 \sigma_2^3}{g \eta_2^4} \right)^{0.25} = 3.1 \left(\frac{\Delta \rho_{21}}{\rho_2} Ar^2 \right)^{0.25}, \qquad (3.11)$$

that can be used to construct this function. For laminar liquid and undistorted bubbly flow, $0 \le \text{Re}_1 \le \text{Re}_{1Sph}$,

$$1 - \eta_{vis} \approx 0. \qquad (3.12)$$

For transition regime of periodic deformation of bubbles, $\text{Re}_{1Sph} < \text{Re}_1 \le \text{Re}_{1St}$,

$$1 - \eta_{vis} \approx \left(1 - \eta_{vis,\min}\right) \frac{\text{Re}_1 - \text{Re}_{1SPh}}{\text{Re}_{1St} - \text{Re}_{1SPh}}. \qquad (3.13)$$

Finally, for high *Reynolds* number (stochastic deformation of bubbles, turbulent wake flow), $\text{Re}_1 > \text{Re}_{1St}$.

$$\eta_{vis} \approx \eta_{vis,\min}. \qquad (3.14)$$

Some authors prefer to use instead $(1 - \eta_{vis})(3/4)c_{cd}^d$ a single constant e.g. 0.95 in *Terekhov* and *Pakhomov* (2005). Note the restrain of the constant. It has to be

$$\text{const} \le (1 - \eta_{vis})(3/4)c_{cd}^d. \qquad (3.15)$$

For computation of the drag coefficient c_{cd}^d for different regimes of the bubble flow see Ch. 4 of *Kolev* (2007a and b).

If there is no other source of turbulence like wall effects, velocity gradients etc., the generation equals the dissipation

$$\varepsilon_{c,\infty} = G_{k,c}. \qquad (3.16)$$

Using the *Kolmogorov* relation $\varepsilon_{c,\infty} = 0.3 k_{c,\infty}^{3/2} / \ell_{e,c}$ and the above equation

$$0.3 k_{c,\infty}^{3/2} / \ell_{e,c} \approx \frac{\alpha_d}{\alpha_c} (1 - \eta_{vis}) \frac{3}{4} \frac{c_{cd}^d}{D_d} |\Delta V_{cd}|^3 \qquad (3.17)$$

we obtain the steady state level of turbulence caused by bubble transport only

$$k_{c,\infty} \approx \left[\frac{15}{6}(1-\eta_{vis})c_{cd}^d \frac{\ell_{e,c}}{D_d}\right]^{2/3} \left(\frac{\alpha_d}{\alpha_c}\right)^{2/3} \Delta V_{cd}^2. \qquad (3.18)$$

Slightly different approach to compute the steady developed level of the turbulent kinetic energy generated by bubbles is proposed by *Bataille* and *Lance* (1988). The authors noted that if the length scale associated with the dissipation rate in the bubble wakes

$$(1-\eta_{vis})\frac{\alpha_d}{\pi D_d^3/6} f_{cd}^d \Delta V_{cd} = (1-\eta_{vis})\frac{3\alpha_d}{4D_d} c_{cd}^d \rho_c \Delta V_{cd}^3 \qquad (3.19)$$

for family of bubbles is ℓ_{wake} then the fluctuation velocity obeys

$$V_c'^3/\ell_{wake} \approx (1-\eta_{vis})\frac{3\alpha_d}{4D_d} c_{cd}^d \Delta V_{cd}^3 \qquad (3.20)$$

or

$$V_c'^2 \approx \left[\frac{3}{4}(1-\eta_{vis})c_{cd}^d \frac{\ell_{wake}}{D_d}\right]^{2/3} \alpha_d^{2/3} \Delta V_{cd}^2. \qquad (3.21)$$

Therefore the steady state level of turbulence caused by bubble transport is only

$$k_{c,\infty} = \frac{3}{2}V_c'^2 \approx \frac{3}{2}\left[\frac{3}{4}(1-\eta_{vis})c_{cd}^d \frac{\ell_{wake}}{D_d}\right]^{2/3} \alpha_d^{2/3} \Delta V_{cd}^2, \qquad (3.22)$$

which differs slightly in the constant compared to the previous result. Note that only for small bubble concentrations $\alpha_d/\alpha_c \approx \alpha_d$.

Therefore the turbulent kinetic energy generated by the bubbles is proportional to $\alpha_d^{2/3}$ and ΔV_{cd}^2.

Interesting expression can be derived for the turbulent cinematic viscosity due to the bubbles assuming $v_{cd}^t = V_c' \ell_{wake}$ and using Eqs. (3.18) and (3.22). The result is

$$v_{cd}^t = V_c' \ell_{wake} \approx 1.22 \left[\frac{3}{4}(1-\eta_{vis})c_{cd}^d \frac{\ell_{e,c}}{D_d}\right]^{1/3} \left(\frac{\alpha_d}{\alpha_c}\right)^{1/3} |\Delta V_{cd}| \ell_{wake} \qquad (3.23)$$

and

$$v'_{cd} = V'_c \ell_{wake} \approx \left[\frac{3}{4}(1-\eta_{vis})c^d_{cd}\frac{\ell_{wake}}{D_d}\right]^{1/3} \alpha_d^{1/3} |\Delta V_{cd}| \ell_{wake}, \qquad (3.24)$$

respectively. Comparing with intuitively proposed equations by several authors,

$$v'_{cd} = 0.5\alpha_d D_d |\Delta V_{cd}|, \text{ Sato and Sekoguchi (1975)}, \qquad (3.25)$$

$$v'_{cd} = 0.6\alpha_d D_d(y)|\Delta V_{cd}| f'_{damp}(y^+), \text{ Sekogushi et al. (1979), Sato et al. (1981)}, \qquad (3.26)$$

$$v'_{cd} = 0.4\left(gD_d^4|\Delta V_{cd}|\alpha_d\right)^{1/3}, \text{ Lilienbaum (1983)}, \qquad (3.27)$$

$$v'_{cd} = \sqrt{H(\alpha_d)}D_d|\Delta V_{cd}|, \text{ Batchelor (1988)}, \qquad (3.28)$$

$$v'_{cd} \approx 0.58 D_d^{7/9}|\Delta V_{cd}|\alpha_d^{2/3}, \text{ derived from mixing length hypotheses}, \qquad (3.29)$$

I recommend Eq. (3.23) for any regime except the origination of the bubble which is based on sound physical scaling. Note that the derivation of the *Lilienbaum (1983)* equation is based on the following ideas $\varepsilon_{cd} = \alpha_d |\Delta V_{cd}|\Delta p/\Delta z = \alpha_d |\Delta V_{cd}|g$, $\varepsilon_{cd} = 0.55 V'^3_c/\ell_{e,c} \approx 0.55 V'^3_c/D_d$ and $v'_{cd} = V'_c \ell_{e,c} \approx V'_c D_d$ resulting in $v'_{cd} \approx 1.22\left(gD_d^4|\Delta V_{cd}|\alpha_d\right)^{1/3}$. The constant corresponding to the *Serizawa*'s data was reported to be 1.13. His own data dictated the value of the constant 0.4.

Different approach for computing the bubble induced turbulence source term in the continuum is proposed by *Lopez de Bertodano* et al. (1994). These authors write additional *k*-equation for balancing of the turbulence kinetic energy portion $k_{c,d}$ due to bubble relative motion. The source in this extra *k*-equation is written as relaxation term

$$\gamma_v \alpha_c \rho_c G_{k,c} \approx \gamma_v \alpha_d \rho_c \frac{1}{\Delta \tau_{cd}}\left(k_{c,d\infty} - k_{c,d}\right), \qquad (3.30)$$

where

$$k_{c,d\infty} = \frac{1}{2}c_c^{vm}|\Delta \mathbf{V}_{cd}|^2 \qquad (3.31)$$

is the turbulence kinetic energy association with the fluctuation of the so-called added mass of the continuum. For the relaxation time constant the following expression is used

$$\Delta \tau_{cd} = D_d/|\Delta \mathbf{V}_{cd}|. \qquad (3.32)$$

To my view the decay time constant should be a better relaxation time constant. The argument that steady bubble motion generates turbulence but do not experience virtual mass force speaks against this approach. In any case, the idea that in a transient motion there is more energy dissipated due to virtual and lift forces is important and can be used to revise Eq. (3.23) for transients.

3.2 Turbulence generation due to nucleate boiling

Boiling at hot surfaces can substantially modify the turbulence in the boundary layer depending on the bubble departure diameter. Because the bubble departure diameter D_{1d} is inversely proportional to the square of the velocity, with increasing velocity the diameter decreases. As long as $D_{1d} w^* / v \ll 5$, this influence is negligible. For bubble departure diameters comparable or larger than the viscous sub-layer the influence is important. We identify two mechanisms producing turbulence in this case: a) The expansion work of single bubble in W/m^3 is

$$\frac{\pi}{6} D_{1d}^3 \frac{1}{\rho_2} \left(1 - \frac{\rho''}{\rho_2}\right) \left[p'(T_2) - p\right], \qquad (3.33)$$

Kolev (2007b), being introduced into the surrounding liquid; b) The work for displacement of the surrounding liquid after the bubble departure in W/m^3 is

$$\frac{\pi}{6} D_{1d}^3 \left(V_{21}^d\right)^2 \frac{1}{4}, \qquad (3.34)$$

Kolev (2007b), with a virtual mass coefficient equal to ½. Given the heat perimeter Π of the channel over the section Δz the total amount of turbulence production per unit volume within a boundary layer with thickness δ is

$$\rho_2 P_2^{w,boiling} \left[\frac{W}{kg}\right] = \frac{f_{1w} n_{1w}''}{\delta} \rho_2 \frac{\pi}{6} D_{1d}^3 \left\{\left(1 - \frac{\rho''}{\rho_2}\right)\frac{1}{\rho_2}\left[p'(T_2) - p\right] + \frac{1}{4}\left(V_{21}^d\right)^2\right\}.$$

$$(3.35)$$

Here f_{1w} is the bubble departure frequency and n_{1w}'' is the number of the active size per unit surface.

Avdeev (1982, 1983) proposed to consider bubbly boundary layer as a surface with equivalent roughness being part of the local bubble size $k = 0.257 D_1$ for use in the modified Colebrook and White (1939) relation

$$\frac{1}{\sqrt{\lambda_{fr}}} = 1.74 - 2\log\left(\frac{k}{R_h} + \frac{49}{\text{Re}^{0.91}}\right) = 1.74 - 2\log\left(0.514\frac{D_1}{D_h} + \frac{49}{\text{Re}^{0.91}}\right). \qquad (3.36)$$

3.3 Treatment of the boundary layer for non boiling flows

It is a common practice to treat the boundary layer as a plate or pipe flow boundary layer based on zero pressure gradient and constant properties of the fluid near the wall. Using the existing knowledge for the distribution of the parameters inside it, one can compute the parameters at the prescribed distance y_p from the wall. For computer codes it is advisable to write a pre-processor that computes: (a) $y_p = \min\left(y_{p,x}, y_{p,y}, y_{p,z}\right)$ with components being the smallest distances between the point of interest and the neighboring walls in the three co-ordinate directions; (b) the vector $\mathbf{n}_{yp} := \left(n_{p,x}, n_{p,y}, n_{p,z}\right)$ storing the orientation of the wall; and (c) $w_{c,p}$ being the velocity parallel to the wall

$$w_{c,p} = \mathbf{V}_c - \left(\mathbf{n}_{yp} \cdot \mathbf{V}_c\right) \mathbf{n}_{yp}.$$

The velocity as a function of the wall distance within

$$30 < \left(y_p^+ := w_{c,p}^* y_p / v_c\right) < 100$$

obeys the *von Karman* logarithmic low

$$w_{c,p}^+ := w_{c,p} / w_c^* = \frac{1}{\kappa} \ln y_p^+ + 5.5 = \frac{1}{\kappa} \ln\left(e^{\kappa 5.5} y_p^+\right) = \frac{1}{\kappa} \ln\left(E y_p^+\right). \qquad (3.37)$$

Here $w_c^* = \sqrt{\tau_c^w / \rho_c}$ is the friction velocity, $\kappa = 0.41$ is the *von Karman* constant and $E \approx 9.5$. We have here the inversed task: we know the velocity $w_{c,p}$ far from the wall at the distance y_p. From Eq. (3.37) the friction velocity w_c^* can be computed and consequently the wall share stress, see *Launder* and *Spalding* (1974). Of course one can compute the friction coefficient and the wall share stress for any specific geometry and wall roughness using appropriate specific correlation. The dissipation of the turbulent kinetic energy at y_p can be computed using the definition equation

$$\varepsilon_{c,p} = \frac{\tau_c^w}{\rho_c} \frac{dw_c}{dy} = w_c^{*2} \frac{dw_c}{dy} \qquad (3.38)$$

and the derivative

$$\frac{dw_c}{dy} = \frac{w_c^*}{\kappa y_p} \qquad (3.39)$$

3.3 Treatment of the boundary layer for non boiling flows

from the boundary layer momentum equation with the *Prandtl* mixing length hypothesis. The result is

$$\varepsilon_{c,p} = \frac{w_c^{*3}}{\kappa y_p}. \qquad (3.40)$$

The corresponding value for the turbulent kinetic energy can then be approximated using $\varepsilon_c = \sqrt{c_\eta}\, k_c^{3/2}/\ell_{e,c}$ with *Prandtl* mixing length $\ell_{e,c} = \kappa y_p$. The result is

$$k_{c,p} = w_c^{*2}/c_\eta^{1/3}. \qquad (3.41)$$

This approach is used by many authors e.g. *Rody* (1984) p. 45, *Lahey* (1987). Troshko and Hassan (2001) proposed to use similar formalism for two-phase bubbly flow replacing κ with the effective κ for two phase flow and $y_p = 30$, see the model of these authors in Ch. 9.

Eliminating the friction velocity from the both equations results in the relation

$$k_c c_\eta^{1/3} = \left(\kappa y_p \varepsilon_{c,p}\right)^{2/3}, \qquad (3.42)$$

which is also used in the literature. *Launder* and *Spalding* (1974) proposed to use instead the volume averaged dissipation of the turbulent kinetic energy inside the layer with thickness y_p

$$\frac{\Pi_h}{Vol}\int_0^{y_p}\varepsilon_c dy = \frac{\Pi_h}{Vol} c_\eta \frac{k_p^{3/2}}{\kappa} \ln\left(Ey_p \frac{c_\eta^{1/4} k_{c,p}^{1/2}}{v_c}\right). \qquad (3.43)$$

Here Π_h is the wetted perimeter and *Vol* is the control volume. For two phase bubble flow *Lee* et al. (1989) used Eq. (3.42) and modified Eq. (3.40) by introducing additional dissipation equal to the generation due to the relative motion of bubbles

$$\varepsilon_{c,p} = \left(w_{c,p}^*\right)^3 / \left(\kappa y_p^+\right) + \frac{G_{k2,p}}{1-\alpha_{1,p}}, \qquad (3.44)$$

where

$$\alpha_{1,p} = \alpha_1 \left[5.3\left(\frac{\text{Re}_2}{100000}\right)^2 - 5.9\left(\frac{\text{Re}_2}{100000}\right) + 0.99\right] \qquad (3.45)$$

For computing the generation term due to the relative motion of bubbles in the liquid, G_{k2}, see Eq. (3.5) through (3.7). Remember that this approach is not local because it depends on the global quantity Re_2.

For adiabatic flow *Borodulja* et al. (1980) proposed to approximate the profile between $y = 0$ and $y = y_p$ as follows

$$\alpha_1 = \alpha_{1p}\left[3\left(\frac{y}{y_p}\right)^2 - 2\left(\frac{y}{y_p}\right)^3\right] \text{ for } 0 < y \le y_p/2,$$

$$\alpha_1 = \alpha_{1p}\left[1 - 3\left(1-\frac{y}{y_p}\right)^2 - 2\left(1-\frac{y}{y_p}\right)^3\right] \text{ for } y_p/2 < y.$$

In case of single phase flow at a heated wall *Launder* and *Spalding* (1974) computed the heat flux $\dot{q}_c^{\prime\prime w}$ using again appropriate for the specific geometry empirical correlation. Knowing the heat flux the temperature at y_p is then defined by

$$\frac{\rho_c c_{p,c}(T_{c,p} - T_w)}{\dot{q}_c^{\prime\prime w}} c_\eta^{1/4} k_{c,p}^{1/2}$$

$$= \frac{\Pr_c'}{\kappa} \ln\left(E y_p \frac{c_\eta^{1/4} k_{c,p}^{1/2}}{\nu_c}\right) + \Pr_c' \frac{\pi/4}{\sin \pi/4}\left(\frac{c_{\text{van Driest}}}{k}\right)^{1/2}\left(\frac{\Pr_c}{\Pr_c'} - 1\right)\left(\frac{\Pr_c'}{\Pr_c}\right)^{1/4}. \quad (3.46)$$

Here $c_{\text{van Driest}}$ is the *van Driest*'s constant, equal to 26 for smooth wall. $\Pr_c$ and $\Pr_c'$ are the *molecular* and the *turbulent Prandtl* numbers, respectively. The last term is the so-called *resistance of the molecular sub layer*. There are not enough experimental data for the estimation of the turbulent *Prandtl* number in complex situations in general. If one has no better choice for the flow near the wall, $\Pr_t' \approx 0.9$ is appropriate. For jets and vortices the approximation $\Pr_t' \approx 0.5$ is appropriate.

That such approach can be used in much more complicated geometries is a consequence of the small boundary layer thickness compared to the other geometrical dimensions. In the thin boundary layer the pressure gradient effect, the change of the properties of the fluid and the mass transfer in the field have secondary influence. For the same reason the influence of the wall curvature can be neglected in the immediate neighborhood of the wall. Thus, the wall function for a plane wall can be used for walls with arbitrary geometries without changes.

For more detailed analyses allowing for heterogeneous turbulence at the wall the *Reynolds* stress boundary condition given by *Launder* et al. (1975) is useful

$$-\overline{V_c' V_c'} = \begin{vmatrix} 5.1 & 0 & 1 \\ 0 & 2.3 & 0 \\ 1 & 0 & 1 \end{vmatrix} w^{*2}. \quad (3.47)$$

The kinetic energy of turbulence is the trace of this expression,

$$k_c = 4.2 w^{*2}, \qquad (3.48)$$

and its dissipation is given by Eq. (3.40).

3.4 Initial conditions

Usually engineers cut part of the system for their analyses and replace the remaining part by boundary conditions. The boundary conditions contain some degree of arbitrariness because exact knowledge of the remaining part is seldom available. Some sound engineering intuitions is needed here. Then, the initial conditions in the flow region can be computed so as to satisfy the steady state distribution. This can be done analytically for $k_{1,\infty}$ and $\varepsilon_{1,\infty}$. If it is done numerically, meaningful initial conditions are necessary. Appropriate assumptions are used in the literature of which some are listed below.

Kinetic energy: The specific turbulent kinetic energy can take 0.1 to 10% of the averaged kinetic energy of the field in given points of the integration domain.

Dissipation: The estimation of the initial values of the dissipation of the specific turbulent kinetic energy is associated with much bigger uncertainty. If for the particular case it is possible to estimate the mixing length $\ell_{e,l}$, the corresponding values of ε_l can be estimated by using the *Kolmogorov* relation

$$\varepsilon_l = 0.3 k_l^{3/2} / \ell_{e,l} = 0.21 V_l'^3 / \ell_{e,l}. \qquad (3.49)$$

For instance postulating 5% turbulence:

$$V_1' = V_2' = 0.05 V, \qquad (3.50)$$

$$k_2 = k_1 = \frac{3}{2} V'^2 = 3.75 \times 10^{-3} V^2. \qquad (3.51)$$

Then

$$\varepsilon_2 = \varepsilon_1 = 2.625 \times 10^{-5} \ V^3 / D_h. \qquad (3.52)$$

Alternatively ε_l can be computed by assuming cinematic turbulent viscosity v_l' satisfying

$$\mathrm{Re}^t := \ell_{e,l} k_l^{1/2} / v_l' = 500, \qquad (3.53)$$

and then using the *Prandtl–Kolmogorov* hypothesis $v_l' = c_{nl} k_l^2 / \varepsilon_l$.

Nomenclature

Latin

c_1	constant in the Kolmogorov equation, dimensionless
c_{cd}^{vm}	coefficient for the virtual mass force or added mass force acting on a dispersed particle, dimensionless
c_{cd}^{d}	coefficient for the drag force or added mass force acting on a dispersed particle, dimensionless
$c_{\varepsilon 1}, c_{\varepsilon 2}, c_{\varepsilon 3}$	empirical coefficients in the source term of the ε-equation
c_η	empirical constant or function connecting the eddy cinematic diffusivity with the specific turbulent kinetic energy and its dissipation
D	diameter, m
D_1	bubble diameter, m
D_{1d}	bubble departure diameter, m
D_d	diameter of dispersed particle, m
D_h	hydraulic diameter, m
$D_{h,l}$	hydraulic diameter of the "tunnel" of field *l* only, m
f_{1w}	bubble departure frequency, 1/s
f_{cd}^d	drag force experienced by the dispersed phase from the surrounding continuum, N/m³
$G_{k2,p}$	production of turbulent kinetic energy due to bubble relocation in changing pressure field per unit mass of the liquid at the transition to the viscous boundary layer, W/kg
$G_{k,l}$	production of turbulent kinetic energy due to bubble relocation in changing pressure field per unit mass of the filed *l*, W/kg (m²/s³)
g	gravitational acceleration, m/s²
k	specific turbulent kinetic energy, m²/s²
k_0	initial specific turbulent kinetic energy, m²/s²
$k_{c,d}$	specific turbulence kinetic energy due to bubble-liquid relative motion only, m²/s²
$k_{c,d\infty}$	specific turbulence kinetic energy of the continuum due the fluctuation of the dispersed particles associated with the so-called added mass of the continuum, m²/s²
$k_{c,p}$	specific continuum turbulent kinetic energy at boundary to the viscous boundary layer used as boundary condition for large scale simulations, m²/s²

Nomenclature 67

k_l	$:= \frac{1}{2}\left(u_l'^2 + v_l'^2 + w_l'^2\right)$, specific turbulent kinetic energy of field l, m²/s²
$k_{l,\infty}$	specific steady state developed turbulent kinetic energy of field l, m²/s²
k_{ml}^*	kinetic energy of turbulent pulsation introduced with the mass source μ_{ml}, m²/s²
k_p	specific turbulent kinetic energy at the transition between laminar and turbulent boundary layer used as a boundary condition for large scale simulations outside the boundary layer, m²/s²
k_{wl}	kinetic energy of turbulent pulsation introduced with the mass source μ_{wl}, m²/s²
k_∞	specific steady state developed turbulent kinetic energy, m²/s²
ℓ_{mix}	mixing length, m
ℓ_e	size of the large eddy, m
$\ell_{\mu e,l}$	lowest spatial scale for existence of eddies in field l called *inner scale* or *small scale* or *Taylor* micro-scale (μ) of turbulence, m
n_1	number of bubbles in unit mixture volume, 1/m³
n_{1w}''	number of the activated seeds at a heated wall producing bubbles, 1/m²
$P_2^{w,boiling}$	production of turbulent kinetic energy per unit mass of the flow due to bubble generation and departure from the wall, W/kg
P_k	production of the turbulent kinetic energy per unit mass, W/kg
$\overline{P_{k,l}}$	production of the turbulent kinetic energy per unit mass of the velocity field l due to deformation of the velocity field l, W/kg
P_{kw}	irreversibly dissipated power per unit flow mass outside the viscous fluid due to turbulent pulsations equal to production of turbulent kinetic energy per unit mass of the flow, W/kg (m²/s³)
$P_{kw,1}^\zeta$	production of turbulent kinetic energy per unit mass of the gas due to irreversible singularity, W/kg
P_{kw}^ζ	production of turbulent kinetic energy per unit mass due to irreversible singularity, W/kg
$P_{kw,l}$	production of turbulent kinetic energy per unit mass of the field l due to friction with the wall, W/kg
$P_{k\mu,l}$	production of turbulent kinetic energy per unit mass of the field l due to friction evaporation or condensation, W/kg
P_ε	production of the dissipation of the turbulent kinetic energy per unit mass, W/kg
$P_{\varepsilon w}$	production of the dissipation of the turbulent kinetic energy per unit mass due to friction with the wall, W/kg

$P_{\varepsilon w,l}$	production of the dissipation of the turbulent kinetic energy per unit mass of the field l due to friction with the wall, W/kg
Pr^t	turbulence Prandtl number, dimensionless
Pr^t_k	turbulent Prandtl number describing diffusion of the turbulent kinetic energy, dimensionless
Pr^t_ε	turbulent Prandtl number describing diffusion of the dissipation of the turbulent kinetic energy, dimensionless
p	pressure, Pa
p_a	atmospheric pressure, Pa
p_c	pressure inside the continuum, Pa
p_d	pressure inside the dispersed phase, Pa
Δp	pressure difference, Pa
R	radius, m
R_d	radius of the dispersed particle, m
Re	Reynolds number, dimensionless
Re_{cd}	Reynolds number based on relative velocity, continuum properties and size of the dispersed phase, dimensionless
Re_{1Sph},	Reynolds number defining transition regime of periodic deformation of bubbles, dimensionless
Re_{1St}	
Re^t_c	turbulence continuum Reynolds number, -
Re_{co}	$:= D_h (\rho \bar{w})/\eta_c$ Reynolds number computed so that all the two phase mass flow possesses the properties of the continuum, dimensionless
r	radius, m
$r*$	radius, dimensionless
T	temperature, K
$T_{c,p}$	continuum temperature at the boundary of the viscous layer used as a boundary condition for large scale simulations outside the boundary layer, K
T_w	wall temperature, K
$\overline{T}$	averaged temperature, K
u	radial velocity component, m/s
u'	fluctuation of the radial velocity, m/s
u^+	radial velocity, dimensionless
$u*$	radial friction velocity, m/s
$\bar{u}$	cross section averaged radial velocity, m/s
u_1	bubble radial velocity, m/s
u'_1	fluctuation of the bubble radial velocity, m/s
u_2	liquid radial velocity, m/s

u_l	radial velocity of field l, m/s
u'_l	fluctuation of the radial velocity of field l, m/s
V'	fluctuation of the velocity, m/s
$\mathbf{V}$	velocity vector, m/s
V_{21}^d	difference between liquid and gas velocity, m/s
V_b	bubble departure volume, m³
ΔV_{12}	difference between gas and liquid velocity, m/s
$\Delta \mathbf{V}_{ml}$	difference between m- and l-velocity vectors, m/s
Vol	control volume, m³
v	velocity component in angular direction, m/s
v'	fluctuation of the velocity component in angular direction, m/s
$\overline{v'}$	time average of the angular velocity fluctuation, m/s
w	axial velocity, m/s
w^*	friction velocity, dimensionless
$\overline{w}$	cross section averaged friction velocity, m/s
w^+	axial velocity, dimensionless
w_1	bubble axial velocity, m/s
w_2	liquid axial velocity, m/s
$w_{2,far}$	liquid velocity far from the wall, m/s
w_2^+	liquid axial velocity, dimensionless
w'_1	fluctuation of the axial bubble velocity, m/s
w'_2	fluctuation of the axial liquid velocity not taking into account the influence of the bubble, m/s
w''_2	fluctuation of the axial liquid velocity caused only by the presence of bubble, m/s
Δw_{12}	local axial velocity difference between bubbles and liquid, m/s
$\overline{\Delta w_{12}}$	cross section averaged axial velocity difference between bubbles and liquid, m/s
$\Delta w_{12\infty}$	steady state axial bubble rise velocity in liquid, m/s
w_c	continuum axial velocity, m/s
$\overline{w}_c$	averaged axial continuum velocity, m/s
w_c^*	continuum axial friction velocity, m/s
$w_{c,p}$	continuum axial velocity at the boundary of the viscous layer used as a boundary condition for large scale simulations outside the boundary layer, m/s

$w_{c,p}^*$	continuum axial friction velocity at the boundary of the viscous layer used as a boundary condition for large scale simulations outside the boundary layer, dimensionless
w_l	axial velocity of field l, m/s
w_l'	fluctuation of the axial velocity of field l, m/s
$\overline{w}_l$	cross section axial velocity of field l, m/s
Δw_{cd}	axial velocity difference between dispersed and continuous phase, m/s
x	x-coordinate, m
y	y-coordinate, distance from the wall, m
y_0	distance between the bubble and the wall, m
$y_{\lim}$	virtual distance from the wall in which almost all the viscous dissipation is lumped, m
y_{lt}^+	viscous boundary layer limit, dimensionless
y_p	distance from the wall marking the end of the boundary layer, m
y_p^+	distance from the wall marking the end of the boundary layer, dimensionless
$y_{p,x}$	distance from the closest wall in x-direction marking the end of the boundary layer (for use in porous body concepts), m
$y_{p,y}$	distance from the closest wall in y-direction marking the end of the boundary layer (for use in porous body concepts), m
$y_{p,z}$	distance from the closest wall in z-direction marking the end of the boundary layer (for use in porous body concepts), m
$y_{sym.\,lines}$	distance from the wall to the symmetry line in the bundles, m
y^+	distance from the wall, dimensionless
z	axial coordinate, m
Δz	finite of the axial distance, m

Greek

α	volumetric fraction, dimensionless
β	in Lance and Bataille equation: part of the power lost by the continuum to resist the bubble generating kinetic energy in the wakes behind the bubble, dimensionless
γ	surface permeability defined as flow cross section divided by the cross section of the control volume (usually the three main directional components are used), dimensionless
γ_v	volumetric porosity defined as the flow volume divided by the considered control volume, dimensionless
δ	boundary layer with thickness, m

δ_l	= 1 in case of continuous field l; = 0 in case of disperse field l
ε	power dissipated irreversibly due to *turbulent pulsations* in the viscous fluid per unit mass of the fluid (dissipation of the specific turbulent kinetic energy), m²/s³
ε_0	initial value of the dissipation of the specific turbulent kinetic energy, m²/s³
$\varepsilon_{c,p}$	continuum dissipation of the specific turbulent kinetic energy at the transition between laminar and turbulent boundary layer, m²/s³
$\varepsilon_{l,\infty}$	specific steady state developed dissipation of the turbulent kinetic energy of field l, m²/s²
ε_p	in sense of two group theory: dissipation of the large scale motion group, m²/s²
ε_T	in sense of two group theory: irreversible friction dissipation of the transition eddies, m²/s²
η	dynamic viscosity, kg/(ms)
η^t	turbulent or eddy dynamic viscosity, kg/(ms)
η^l	molecular dynamic viscosity, kg/(ms)
η_{vis}	part of the mechanical energy directly dissipated into heat after a local singularity and not effectively generating turbulence, dimensionless
θ	angular coordinate, rad
κ	= 0, Cartesian coordinates; = 1, cylindrical coordinates, - or von Karman constant, -
λ	thermal conductivity, W/(mK)
λ_{fr}	friction coefficient, dimensionless
$\lambda_{fr,12}$	friction coefficient for the liquid–gas interface, dimensionless
$\lambda_{fr,co}$	friction coefficient computed for the total mixture mass flow with the properties of the continuum only, dimensionless
$\lambda_{fr,eff}$	effective friction coefficient, dimensionless
μ_{lm}	mass transferred from field l into field m per unit time and unit mixture volume, kg/(sm³)
μ_{ml}	mass transferred from field m into field l per unit time and unit mixture volume, kg/(sm³)
μ_{lw}	mass exhausted from field l through the wall w per unit time and unit mixture volume, kg/(sm³)
μ_{wl}	mass injected into the field l through the wall w per unit time and unit mixture volume, kg/(sm³)
ν_l	cinematic viscosity, m²/s
ν_l^t	turbulent or eddy cinematic viscosity of field l, m²/s

$\nu_{l,eff}$	effective cinematic viscosity of field l, m²/s
ν_{12}^t	turbulent or eddy cinematic viscosity of the liquid caused by the bubbles only, m²/s
$\nu_{c,y}^t$	turbulent or eddy cinematic viscosity of field l in direction y, m²/s
$\nu_{c,z}^t$	turbulent or eddy cinematic viscosity of field l in direction z, m²/s
ν_l^k	total cinematic diffusivity of the turbulent kinetic energy, m²/s
ν_l^ε	total cinematic diffusivity of the dissipation of the turbulent kinetic energy, m²/s
ν_l^*	$:= \nu_l^t + \delta_l \nu_l$, effective cinematic viscosity, m²/s
ξ	function in the Wang expression for the lift coefficient, dimensionless
ρ	density, kg/m³
$\Delta\rho_{21}$	liquid–gas density difference, kg/m³
σ	surface tension, N/m
τ	time, s
τ_l	share stress in field l, N/m²
τ_w	wall share stress, N/m²
$\Delta\tau$	time interval, s
$\Delta\tau_k$	time constant for the decay of the turbulent kinetic energy, s
$\Delta\tau_e$	time scale of fluctuation of large eddy, s
$\Delta\tau_{\mu e,l}$	time scale corresponding to the spatially lowest scale for existence of eddies in field l called *inner time scale* or *small time scale* or *Taylor* micro-scale of turbulence, s

Subscripts

1	gas
2	liquid
3	droplet
c	continuum
d	disperse
l	field l
m	field m
e	eddy
μ	associated to mass transfer or micro-scale
r	radial direction
θ	angular direction
z	axial direction
w	wall
k	axial discretization index

References

Avdeev AA (1982) Teploenergetika vol 3 p 23

Avdeev AA (1983) Gidrodynamika turbulentnyih techeniy puzyrkovoj dwuchfasnoj smesi, Teplofisika visokih temperature, vol 21 no 4 pp 707-715

Bataille J and Lance M (1988) Turbulence in multiphase flows, Proc. Of the first world congress on Experimental Heat Transfer, Fluid Mechanics, and Thermodynamics, held Sept. 4-9, 1988 in Dubrovnik, Yugoslavia, Elsevier, Shah RK, Ganic´EN and Yang KT eds.

Batchelor GK (1988) A new theory of the instability of a uniform fluidized bed, J. Fluid Mechanic v 193 pp 75-110

Borodulja WA, Kosmowski I, Lilienbaum W, Chodan IW and Pyljow SA (1980) Mechanische Austauschvorgänge bei einem Flüssigkeits-Gasgemisch in einem geneigten Strömungskanal, Wissenschaftliche Zeitschrift der TH Oto von Guerike Magdeburg, vol 24 no 4 pp 95-97

Colebrook CF (1939) Turbulent flow in pipes with particular reference to the transition region between the smooth and the rough pipe lows, J. Institution Civil Engineers

Kataoka I and Serizawa A (1995) Modeling and prediction of bubbly two phase flow, Proc. 2nd Int. Conf. Multiphase Flow, Kyoto, pp MO2 11-16

Kolev NI (2007a) Multiphase Flow Dynamics, Vol. 1 Fundamentals, 3d extended ed., Springer, Berlin, New York, Tokyo

Kolev NI (2007b) Multiphase Flow Dynamics, Vol. 2 Thermal and mechanical interactions, 3d extended ed., Springer, Berlin, New York, Tokyo

Lahey RT (24-30 May 1987) Turbulence and two phase distribution phenomena in two-phase flow, Proc of Transient Phenomena in Multiphase Flow, Dubrovnik

Launder BE and Spalding DB (1974) The numerical computation of turbulent flows, Computer methods in applied mechanics and engineering, vol 3 pp 269-289

Launder BE, Reece GJ and Rodi W (1975) Progress in development of a Reynolds stress turbulence closure, J. Fluid Mech. Vol 68 pp 537-566

Lee SL, Lahey RT Jr. and Jones OC Jr. (1989) The prediction of two-phase turbulence and phase distribution phenomena using a k-e model, Japanese J. Multiphase Flow, vol 3 no 4 pp 335-368

Lilienbaum W (1983) Turbulente Blasenströmung im geneigten Kanal, Technische Mechanik vol 6 Heft 1 S. 68-77

Lopez de Bertodano M (1992) Turbulent bubbly two-phase flow in triangular duct, Ph.D. thesis (Nuclear Engineering) Rensselaer Polytechnic Institute

Lopez de Bertodano M, Lahey RT Jr. and Jones OC (1994) Phase distribution of bubbly two-phase flow in vertical ducts, Int. J. Multiphase Flow, vol 20 no 5 pp 805-818

Reichardt H (Mai/Juni 1942) Gesetzmäßigkeiten der freien Turbulenz, VDI-Forschungsh. Nr. 414, Beilage zu "Forschung auf dem Gebiet des Ingenieurwesens", Ausgabe B, Band 13

Rodi W (Feb. 1984) Turbulence models and their application in hydraulics – a state of the art review, IId rev ed., University of Karlsruhe

Sato Y and Sekoguchi K (1975) Liquid velocity distribution in two phase bubbly flow, Int. J. Multiphase Flow, vol 2 pp 79-95

Sato Y, Sadatomi M and Sekoguchi K (1981) Momentum and heat transfer in two-phase bubble-flow-I, Theory, Int. J. Multiphase Flow, vol 7 pp 167-177

Sekogushi K, Fukui H and Sato Y (1979) Flow characteristics and heat transfer in vertical bubble flow, Two-Phase Flow Dynamics, Japan-U.S. Seminar, eds. Bergles AE and Ishigai S, Hemisphere Publishing Corporation, Washington

Serizawa A, Kataoka I, Mishiyoshi I (1975) Turbulence structure at air-water bubbly flow, Part 1, 2, 3, Int. J. Multiphase Flow, vol 2 no 3 pp 221-259

Terekhov VI and Pakhomov MA (2005) Numerical study of downward bubbly flow in a vertical tube, 4th Int. Conf. On Computational Heat and Mass Transfer, Paris

Troshko AA and Hassan YA (2001) A two-equations turbulence model of turbulent bubbly flow, Int. J. of Multiphase Flow, vol 27 pp 1965-2000

Wang SK, Lee SJ, Jones OG and Lahey RT (1987) 3D-turbulence structure in bubbly two-phase flows, Int. J. Multiphase Flow, vol 13 no 3 pp 327-343

4 Source terms for *k-eps* models in porous structures

Internal structures of heat exchangers, nuclear reactors, chemical reactors filled with dispersed materials etc. can be considered as a porous structure. Resolving each detail of such structures is technically not feasible. Normally a detailed analysis of the boundary layers is performed and the results are used to compute integral turbulence sources for a large scale analysis. Note the necessary condition that such approach works: The computational volume contains a structure with all its representative parameters but not part of it.

4.1 Single phase flow

Laufer (1953) experimentally investigated the structure of turbulence in fully developed pipe flow. From his Fig. 17 it was clearly visible that about 70% of the viscous dissipation happens in the boundary sub-layer defined by $y^+ < 11.5$. The production of the turbulence starts at about $y^+ > 5$ and has its maximum again at $y^+ \approx 11.5$. At this point the production equals the dissipation.

4.1.1 Steady developed generation due to wall friction

Let us consider single phase flow with a predominant direction. The friction force per unit flow volume can be computed in this case using the microscopic correlations for the friction coefficient

$$f_w = \frac{\lambda_{fr}}{D_h} \frac{1}{2} \rho \overline{w}^2 .$$
(4.1)

The power per unit flow volume needed to overcome this force is $f_w \overline{w}$. Part of this power is dissipated in the laminar boundary sub-layer, $y^+ \leq 5$, and by the turbulent eddies outside of this layer. In the laminar sub-layer the velocity is linear function of the wall distance

$$w = y \frac{\overline{w}^2}{\nu} \frac{\lambda_{fr}}{8} \quad (w^+ = y^+), \tag{4.2}$$

and consequently

$$\frac{dw}{dy} = \frac{\overline{w}^2}{\nu} \frac{\lambda_{fr}}{8}. \tag{4.3}$$

The *irreversible dissipated power* per unit flow *volume* caused by the viscous forces *due to deformation of the mean values of the velocities* in the space is then approximately

$$\frac{\Pi \Delta z}{F_{flow} \Delta z} \int_0^{y_{\lim}} \rho \nu \left(\frac{d\overline{w}}{dy}\right)^2 dy = \frac{4}{D_h} \rho w^{*3} \int_0^{y^+_{\lim}} \left(\frac{du^+}{dy^+}\right)^2 dy^+ = \frac{4 y^+_{\lim}}{D_h} \rho \frac{\overline{w}^4}{\nu} \left(\frac{\lambda_{fr}}{8}\right)^2$$

$$= \frac{4 y^+_{\lim}}{D_h} \rho (w^*)^3. \tag{4.4}$$

The difference

$$\rho \gamma_v P_{kw} = \gamma_v \left[f_w \overline{w} - \frac{\Pi \Delta z}{F_{flow} \Delta z} \int_0^{y_{\lim}} \rho \nu \left(\frac{d\overline{w}}{dy}\right)^2 dy \right] = \gamma_v \frac{\lambda_{fr}}{D_h} \frac{1}{2} \rho \overline{w}^3 \left(1 - y^+_{\lim} \sqrt{\frac{\lambda_{fr}}{8}} \right), \tag{4.5}$$

is the *irreversibly dissipated power* per unit flow volume *outside* the viscous fluid due to *turbulent pulsations*. Therefore

$$P_{kw} = \frac{\lambda_{fr}}{D_h} \frac{1}{2} \overline{w}^3 \left(1 - y^+_{\lim} \sqrt{\frac{\lambda_{fr}}{8}} \right) = \frac{\lambda_{fr}}{\ell_{e,l}} \frac{1}{2} \overline{w}^3. \tag{4.6}$$

Here $\eta_{vis} = y^+_{\lim} \sqrt{\frac{\lambda_{fr}}{8}}$ is the part of the mechanical energy required to overcome the friction that is dispersed in heat. *Chandesris* et al. (2005) proposed to consider P_{kw} as a net generation of turbulent kinetic energy per unit time and unit mass. Note that the dimensions of the production is in *W/kg* i.e. m^2/s^3. *Chandesris* et al. considered

$$\ell_{e,l} = D_h \bigg/ \left(1 - y_{\lim}^+ \sqrt{\frac{\lambda_{fr}}{8}}\right) \tag{4.7}$$

as a characteristic macroscopic length scale of turbulence production in channels. The "production" source of the turbulent energy dissipation is modeled by *Chandesris* et al. (2005) as

$$P_{\varepsilon w, l} = c_{\varepsilon 2} P_{kw}^2 / k_{l,\infty} = c_{\varepsilon 2} \left(\varepsilon_{l,\infty} / k_{l,\infty}\right) P_{kw}. \tag{4.8}$$

For the case without convection and diffusion the dissipation equation results in

$$d\varepsilon / d\tau = c_{\varepsilon 2} \left(P_{kw}^2 / k_\infty - \varepsilon^2 / \kappa\right) \tag{4.9}$$

which is a relaxation form leading with the time to

$$\varepsilon_\infty = P_{kw}. \tag{4.10}$$

For the computation of the equilibrium kinetic energy of turbulence *Chandesris* et al. used the *Kolmogorov*'s Eq. (2.53) in the form

$$\varepsilon_{l,\infty} = const \frac{k_{l,\infty}^{3/2}}{\ell_{e,l}}, \tag{4.11}$$

valid for isotropic turbulence. The constant was considered as a geometry dependent modeling constant. Assuming $\varepsilon_{l,\infty} = P_{kw}$ in the above equation results in

$$const\, k_{l,\infty}^{3/2} = \lambda_{fr} \frac{1}{2} \overline{w}^3 \tag{4.12}$$

or

$$k_{l,\infty} = \left(\frac{\lambda_{fr}}{2 const}\right)^{2/3} \overline{w}^2 = \left(\frac{\lambda_{fr}}{0.3164}\right)^{2/3} c_k \overline{w}^2 \tag{4.13}$$

or using the *Blasius* law

$$\lambda_{fr} = 0.3164 / \mathrm{Re}^{1/4}, \tag{4.14}$$

$$k_{l,\infty} = \left(\frac{0.1582}{const}\right)^{2/3} \overline{w}^2 \, \text{Re}^{-1/6} = c_k \overline{w}^2 \, \text{Re}^{-1/6}, \qquad (4.15)$$

so that

$$const = 0.1582/c_k^{1.5}. \qquad (4.16)$$

The resulting turbulence decay time constant is then

$$\Delta \tau_{l,\infty} = k_{l,\infty}/\varepsilon_{l,\infty} = \frac{2c_k}{\lambda_{fr} \, \text{Re}^{1/6}} \frac{\ell_{e,l}}{\overline{w}}. \qquad (4.17)$$

Table 4.1. Modeling constants for k-eps source terms in porous structures *Chandesris* et al. (2005)

Geometry	y_{lim}^+	c_k	$const = 0.1582/c_k^{1.5}$	$1/(2const)^{2/3}$
channels	8	0.0306	29.55	0.066
pipes	7	0.0367	22.50	0.079
rod bundles	16	0.0368	22.41	0.079

The constants are selected after comparison with direct numerical simulation results and given in Table 4.1. Note that if we take the von *Karman*'s universal three layers representation of the velocity profile an exact expression is obtained for the generation of the turbulence

$$P_{kw} = \frac{\lambda_{fr}}{D_h} \frac{1}{2} \overline{w}^3 \left(1 - 9.735 \sqrt{\frac{\lambda_{fr}}{8}}\right). \qquad (4.18)$$

Therefore the order of magnitude for y_{lim}^+ estimated by *Chandesris* et al. (2005) is correct.

We learn from this analysis that if a technical facility has to be designed to promote turbulence then it has to have as less friction surface as possible. This reduces the irreversible viscous dissipation in the viscous boundary layer.

The cross section averaged time scale of the fluctuation of *large* eddies can then be computed:

$$\Delta \tau_{e,l,\infty} = 0.37 k_{l,\infty}/\varepsilon_{l,\infty} = 0.37 \frac{2c_k}{\lambda_{fr} \, \text{Re}^{1/6}} \frac{\ell_{e,l}}{\overline{w}}. \qquad (4.19)$$

4.1.2 Heat transfer at the wall for steady developed flow

The cross section averaged time scale of the fluctuation of *small* eddies in developed pipe flow is therefore

$$\Delta \tau_{\mu e,l} \approx b \sqrt{v_l / \varepsilon_{l,\infty}} \,. \tag{4.20}$$

Before jumping apart from the wall, the turbulent eddies stay at the wall during the time $\Delta \tau_{\mu e,l}$, and receive heat from the wall by heat conduction. Therefore, the average heat flux at the wall follows the analytical solution of the *Fourier* equation averaged over the period $\Delta \tau_{\mu e,l}$

$$\dot{q}''_{wl,\infty} = \frac{1}{\Delta \tau_{\mu e,l,\infty}} \int_0^{\Delta \tau_{\mu e,l,\infty}} \dot{q}''_{wl}(\tau) d\tau = 2 \left(\frac{\lambda_l \rho_l c_{pl}}{\pi \Delta \tau_{\mu e,l,\infty}} \right)^{1/2} (T_w - T_l), \tag{4.21}$$

or

$$\frac{\dot{q}''_{wl,\infty} D_h}{(T_w - T_l) \lambda_l} = \frac{2^{3/4}}{\pi^{1/2} b^{1/2}} \left[\lambda_{fr} \left(1 - y^+_{\lim} \sqrt{\frac{\lambda_{fr}}{8}} \right) \right]^{1/4} \text{Re}^{3/4} \text{Pr}_l^{1/2}. \tag{4.22}$$

This is the general form in which

$$\text{Re} = D_h \bar{w} / v_l \,. \tag{4.23}$$

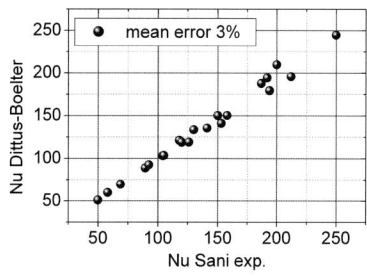 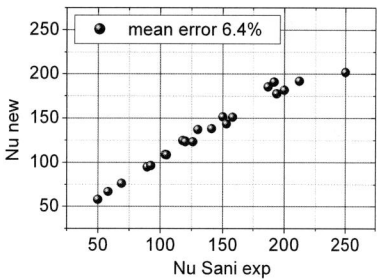

Fig. 4.1. Predicted *Nusselt* number as a function of the computed: a) *Dittus-Boelter* (1930) correlation; b) Small eddy wall renewal hypothesis using the *Blasius* equation for the friction pressure loss. Data by *Sani* (1960)

Note the similarity of the *Dittus-Boelter* correlation

$$\frac{\dot{q}''_{wl,\infty} D_h}{(T_w - T_l)\lambda_l} = 0.024 \operatorname{Re}^{0.8} \operatorname{Pr}_l^{0.4}. \tag{4.24}$$

Comparison of the *Dittus-Boelter* correlation with the *Sani*'s (1960) data gives a mean error of 3% – see Fig. 4.1a. A special form for flow in pipe with smooth wall is obtained by using the *Blasius* correlation for friction. Comparison with the same data with $b = 98.32$ results in mean error of 6.4% and increasing divergence for lower *Prandtl* numbers. Therefore it seems more accurate to compute $\Delta\tau_{\mu e,l,\infty}$ from

$$\frac{\dot{q}''_{wl,\infty} D_h}{(T_w - T)\lambda_l} = 2\left(\frac{\lambda \rho c_p}{\pi \Delta\tau_{\mu e,l,\infty}}\right)^{1/2} \frac{D_h}{\lambda_l} = 0.024 \operatorname{Re}^{0.8} \operatorname{Pr}_l^{0.4}, \tag{4.25}$$

resulting in

$$\Delta\tau_{\mu e,l,\infty} = D_h^2 \bigg/ \left[\pi\left(0.012 \operatorname{Re}^{0.8} \operatorname{Pr}_l^{0.4}\right)^2 a_l\right]. \tag{4.26}$$

4.1.3 Heat transfer at the wall for non developed or transient flow

Now let us answer the question: How does transient turbulence influence the heat transfer at the wall? Knowing that the instant heat transfer obeys

$$\frac{\dot{q}''_{wl} D_h}{(T_w - T)\lambda_l} = 2\left(\frac{\lambda \rho c_p}{\pi \Delta\tau_{\mu e,l}}\right)^{1/2} \frac{D_h}{\lambda_l}, \tag{4.25}$$

we obtain the ratio

$$\frac{\dot{q}''_{wl}}{\dot{q}''_{wl,\infty}} = \left(\frac{\Delta\tau_{\mu e,l,\infty}}{\Delta\tau_{\mu e,l}}\right)^{1/2}. \tag{4.26}$$

> Therefore, increasing the frequency of the turbulence with respect to the steady developed flow increases the heat transfer by following a square root function.

4.1.4 Singularities

In case of local structures with complex geometry that can not be resolved by the selected discretization, the irreversible energy dissipation has to be added as additional source of turbulence kinetic energy,

$$\rho \gamma_v P_{kw}^\zeta = f_w^\zeta \overline{w} = \zeta_{fr} \frac{1}{2} \rho \overline{w}^3 \left(1 - \eta_{vis}\right), \qquad (4.27)$$

in the particular control volume in which the singularity is located. Here η_{vis} is the part of the energy directly dissipated into heat and not effectively generating turbulence. The corresponding dissipation source is then

$$c_{\varepsilon 1} P_{kw}^\zeta / \left(\varepsilon_l / k_l\right). \qquad (4.28)$$

This approach is already used by *Windecker* and *Anglart* (1999) with $\eta_{vis} = 0$. *Serre* and *Bestion* (2005) proposed to use for the dissipation

$$c_{\varepsilon 1} P_{kw}^\zeta / \Delta \tau_e \qquad (4.29)$$

instead of the turbulence decay time constant ε_l / k_l, the time scale of the fluctuation of large eddy $\Delta \tau_e = 0.37 k_l / \varepsilon_l$, and Eq. (2.57).

4.2 Multi-phase flow

4.2.1 Steady developed generation due to wall friction

In analogous way to those used for the single phase flow we will derive an expression for generation of turbulent kinetic energy due to friction. The pressure loss due to friction is usually expressed in terms of the pressure loss of fictive flow consisting of the same mass flow rate

$$\rho \overline{w} = \sum_{l=1}^{l_{max}} \alpha_l \rho_l \overline{w}_l \qquad (4.30)$$

but having the properties of the continuum. The so computed pressure loss is then modified by the so-called two-phase friction multiplier Φ_{co}^2 or *Martinelli-Nelson* multiplier,

$$\left(\frac{dp}{dz}\right)_{fr} = f_w = \frac{\lambda_{fr,co}\Phi_{co}^2}{D_h}\frac{1}{2}\frac{(\rho\overline{w})^2}{\rho_c} = \frac{4\tau_w}{D_h}. \tag{4.31}$$

Here

$$\tau_w = \frac{\lambda_{fr,co}\Phi_{co}^2}{8}\frac{(\rho\overline{w})^2}{\rho_c} \tag{4.32}$$

is the wall share stress. *co* stays for "continuum only." The effective friction coefficient is

$$\lambda_{fr,eff} = \lambda_{fr,co}\Phi_{co}^2. \tag{4.33}$$

The multi-phase friction velocity is then

$$w^* = \sqrt{\frac{\tau_w}{\rho_c}} = \frac{(\rho\overline{w})}{\rho_c}\sqrt{\frac{\lambda_{fr,co}\Phi_{co}^2}{8}}. \tag{4.34}$$

The effective boundary layer dimensionless velocity and wall distance are

$$w^+ = \frac{(\rho w)}{\rho_c w^*} \tag{4.35}$$

and

$$y^+ = y\frac{w^*}{\nu}, \tag{4.36}$$

respectively. The presence of vapor in the liquid for instance increases considerably the friction pressure loss. Recovering the profiles from averaged parameters is difficult. In any case the boundary layer thickness in which the irreversible viscous dissipation happens should be smaller. Therefore we introduce intuitively

$$y_{\lim}^+ = \alpha_c y_{\lim,co}^+ \tag{4.37}$$

in

$$P_{kwc} = \frac{\lambda_{fr,co}\Phi_{co}^2}{\alpha_c D_{hyd}}\frac{1}{2}\frac{(\rho\overline{w})^3}{\rho_c^3}\left(1 - \alpha_c y_{\lim,co}^+\sqrt{\frac{\lambda_{fr,co}\Phi_{co}^2}{8}}\right). \tag{4.38}$$

In the limiting case of continuum only we have $y^+_{\lim} = y^+_{\lim,co}$. The validity of this formalism remains to be checked with data for the averaged level of turbulence which are still not available. Here the division with the continuous volume fraction reflects the assumption that all of the dissipation is put into the continuum.

4.2.2 Heat transfer at the wall for forced convection without boiling

Now let us try to compute the heat transfer coefficient at the wall for two phase flow by using the idea that the renewal period for the eddies at the wall is dictated by turbulence

$$\Delta \tau_{\mu e,c} \approx b \sqrt{v_c / \varepsilon_{c,\infty}} , \qquad (4.39)$$

where all the increased turbulence generation is imposed into the continuum

$$\varepsilon_{c,\infty} = P_{kwc} = \frac{\lambda_{fr,co} \Phi^2_{co}}{\alpha_c D_{hyd}} \frac{1}{2} \frac{(\rho \overline{w})^3}{\rho^3_c} \left(1 - \alpha_c y^+_{\lim,co} \sqrt{\frac{\lambda_{fr,co} \Phi^2_{co}}{8}} \right). \qquad (4.40)$$

Inserting into the averaged solution of the *Fourier* equation

$$\dot{q}''_{wc} = \frac{1}{\Delta \tau_{\mu e,c,\infty}} \int_0^{\Delta \tau_{\mu e,c,\infty}} \dot{q}''_{wc}(\tau) d\tau = 2 \left(\frac{\lambda_c \rho_c c_{pc}}{\pi \Delta \tau_{\mu e,c,\infty}} \right)^{1/2} (T_w - T_c) \qquad (4.41)$$

results in

$$\frac{\dot{q}''_{wc} D_h}{(T_w - T_c) \lambda_c} = \frac{2^{3/4}}{\sqrt{\pi b}} \left[\frac{\lambda_{fr,co} \Phi^2_{co}}{\alpha_c} \left(1 - \alpha_c y^+_{\lim,co} \sqrt{\frac{\lambda_{fr,co} \Phi^2_{co}}{8}} \right) \right]^{1/4} \operatorname{Re}_{co}^{3/4} \operatorname{Pr}_c^{1/2} , \qquad (4.42)$$

where the effective *Reynolds* number is computed so that all the two phase mass flow possesses the properties of the continuum

$$\operatorname{Re}_{co} = D_h (\rho \overline{w}) / \eta_c . \qquad (4.43)$$

The friction coefficient

$$\lambda_{fr,co} = \lambda_{fr,co} (\operatorname{Re}_{co}, etc.) \qquad (4.44)$$

is also a function of this *Reynolds* number. Now let us build the ratio of the two-phase *Nusselt* number to the *Nusselt* number computed so that all the two phase mass flow consists of continuum only. The result is

$$Nu = \left(\frac{\Phi_{co}^2}{\alpha_c} \frac{1 - \alpha_c y_{\lim,co}^+ \sqrt{\frac{\lambda_{fr,co} \Phi_{co}^2}{8}}}{1 - y_{\lim}^+ \sqrt{\frac{\lambda_{fr,co}}{8}}} \right)^{1/4} Nu_{co} \approx \left(\frac{\Phi_{co}^2}{\alpha_c} \right)^{1/4} Nu_{co}. \qquad (4.46)$$

> We immediately recognize that the increase of the turbulence leading to increase of the friction pressure drop is responsible for the increased heat transfer from or to the wall in the two phase flow region.

Note the difference to the *Chen* result obtained in 1963,

$$Nu \approx \Phi_c^{0.89} Nu_c. \qquad (4.47)$$

The *Nusselt* number Nu_c was computed assuming that the continuum part of the flow occupies the total cross section i.e.

$$\text{Re}_c = X_c D_h (\rho \bar{w}) / \eta_c, \qquad (4.48)$$

$$\text{Pr}_c = v_c / a_c. \qquad (4.49)$$

Observe that instead of the *Martinelli-Nelson* multiplier Φ_{co}^2 used here, the *Lockhart* and *Martinelly* multiplier Φ_c^2 were used by *Chen*.

Note also the difference to the *Collier* (1972) result

$$Nu = \frac{1}{\alpha_c^{0.8}} Nu_{co}.$$

4.2.3 Continuum–continuum interaction

Consider film flow in a pipe. The mechanical gas–film interaction creates interfacial force. For turbulent gas bulk the maximum on turbulent energy generation is

$$P_{kw} = \frac{\lambda_{fr,12}}{D_{h,1}} \frac{1}{2} |\Delta w_{12}|^3 (1 - \eta_{vis}), \qquad (4.50)$$

where

$$D_{h,1} = D_h \sqrt{1-\alpha_1} \qquad (4.51)$$

is the hydraulic diameter of the gas flow. The part of this power directly dissipated in the gas, η_{vis}, has to be derived from experiments delivering the turbulence structure of film flow. There is a good reason to consider as a good approximation $\eta_{vis} \approx 0$ because the ripples and the roll waves at the film surface do not allow laminar boundary layer and therefore viscous dissipation in a classical sense. It is still not clear how much of this power goes to generate turbulence in the denser fluid.

4.2.4 Singularities

The two phase friction multiplier can be used as a first approximation for computation of the local pressure losses due to flow obstacles different than wall friction.

$$\left(\frac{dp}{dz}\right)_{fr,\varsigma} = f_w^\varsigma = \varsigma_{fr,co} \Phi_{co}^2 \frac{1}{2} \frac{(\rho \overline{w})^2}{\rho_c}. \qquad (4.52)$$

The effective friction coefficient is then

$$\varsigma_{fr,eff} = \varsigma_{fr,co} \Phi_{co}^2. \qquad (4.53)$$

This approach is appropriate. *Neykov* et al. (2005) defined a benchmark consisting of boiling flow in a rod bundle with 7-spacer grids as shown in Fig. 4.2a and b typical for the so-called nuclear boiling water reactors. Using the measured friction coefficient for single phase flow (fitted with 1% mean error), and Eq. (4.52) with the *Friedel* (1979) correlation for computing of the two phase multiplier, results in an agreement for the pressure drop as shown in Fig. 4.2c.

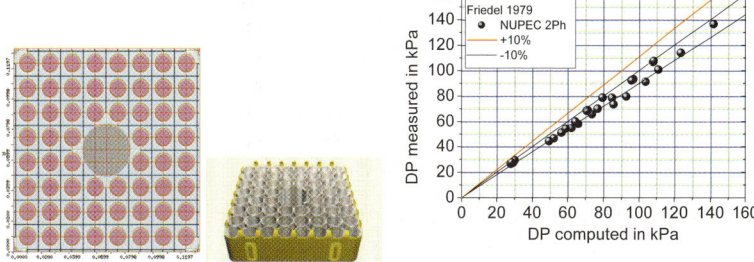

Fig. 4.2. a) Bundle design; b) Ferrule spacer grid; c) Measured versus computed pressure drop for boiling channels. Pressure level about 70 bars. Two-phase friction multiplier by *Friedel* (1979), mean error 7.38%

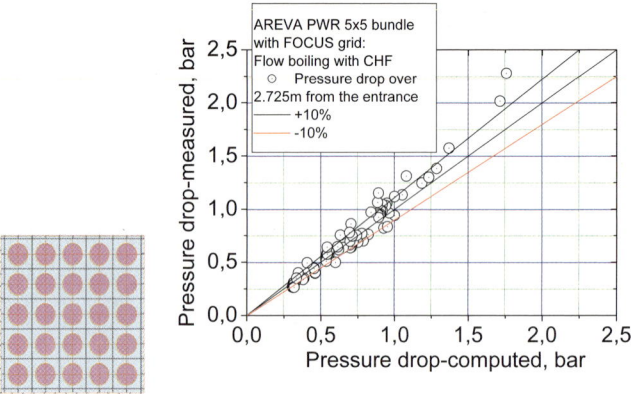

Fig. 4.3. a) AREVA PWR bundle with FOCUS grid; b) Measured, *Vogel* et al. (1991), versus computed total pressure drop in a 5x5 bundle with FOCUS grid. Pressure level about 160 bar

Similarly, applying the same method to a bundle typical of the so-called nuclear pressurized water reactors results in the agreement given in Fig. 4.3.

The source generating turbulent kinetic energy in the mixture is then

$$\rho P_{kw}^{\zeta} = \zeta_{fr,co} \Phi_{co}^2 \frac{1}{2} \frac{(\rho \overline{w})^3}{\rho_c^2} (1-\eta_{vis}). \tag{4.54}$$

Note that this source is computed per unit *mixture* volume. In case of bubbly or churn turbulent flow the term goes for producing turbulence in the continuum only

$$\alpha_c \rho_c P_{kw}^{\zeta} = \alpha_c \rho_c \left[\zeta_{fr,co} \Phi_{co}^2 \frac{1}{2} \frac{\rho_c}{\rho} \frac{(\rho \overline{w})^3}{\rho_c^3} (1-\eta_{vis}) \right] = \alpha_c \rho_c P_{kw,c}^{\zeta}. \tag{4.55}$$

In case of two continua we have for sure some redistribution between the two continua. How local singularities redistribute this generation into parts going into the liquid and into the gas is a very complex three-dimensional problem. In the framework of the large scale averaging we may use a volume fraction weighting

$$\frac{\alpha_2}{\alpha_1+\alpha_2} \rho P_{kw}^{\zeta} = \alpha_2 \rho_2 \left[\frac{1}{\alpha_1+\alpha_2} \frac{\rho_c}{\rho_2} \zeta_{fr,co} \Phi_{co}^2 \frac{1}{2} \frac{(\rho \overline{w})^3}{\rho_c^3} (1-\eta_{vis}) \right] = \alpha_2 \rho_2 P_{kw,2}^{\zeta}, \tag{4.56}$$

$$\frac{\alpha_1}{\alpha_1+\alpha_2}\rho P_{kw}^\zeta = \alpha_1\rho_1\left[\frac{1}{\alpha_1+\alpha_2}\frac{\rho_c}{\rho_1}\zeta_{fr,co}\Phi_{co}^2\frac{1}{2}\frac{(\rho\bar{w})^3}{\rho_c^3}(1-\eta_{vis})\right] = \alpha_1\rho_1 P_{kw,1}^\zeta, \quad (4.57)$$

as an ad hoc approach that has to be improved in the future.

4.2.5 Droplets deposition at walls for steady developed flow

In vertical pipe flow the deposition of droplets due to turbulent fluctuations is a very important process. A general expression defining the mass flow rate of the deposed droplets is given in *Kolev* (2007b)

$$(\rho w)_{32} = \frac{1-\chi}{1+\chi}\sqrt{\frac{2}{\pi}}\left(1-e^{-\Delta\tau/\Delta\tau_{13}}\right)u_1'\rho_{3c}. \quad (4.58)$$

Here χ is the *reflection coefficient*, $\Delta\tau_{13}$ is the particle relaxation time. The time interval $\Delta\tau$ is the minimum of the life time of the large eddy inside the gas flow

$$\Delta\tau_{1e} = 0.37 k_1/\varepsilon_1 \quad (4.59)$$

and the time between two particle collisions

$$\Delta\tau_{33,col} \approx (D_3/\Delta V_{33})/\alpha_3. \quad (4.60)$$

u_1' is the gas pulsation velocity and ρ_{3c} is the droplet density.

Note that interesting 2D-experiments for increased fine droplet deposition ($D_3 = 25\,\mu m$, $u = 14.5\,m/s$, $\alpha_3\rho_3 = 0.2\times10^{-3}\,kg/m^3$) due to a obstacle at a surface is reported by *Ryjkov* and *Hmara* (1976).

4.2.6 Droplets deposition at walls for transient flow

It is the obvious that

$$\frac{(\rho w)_{32}}{(\rho w)_{32,\infty}} \approx \frac{u_1'}{u_{1,\infty}'} \approx \sqrt{\frac{k_1}{k_{1,\infty}}}. \quad (4.61)$$

> Therefore, increasing the turbulent kinetic energy of the continuum with respect to the steady developed flow increases the droplet deposition by following a square root function.

Nomenclature

Latin

C_{dc}^t	response coefficient, dimensionless
C_{dc}^{t*}	response coefficient for single particle, dimensionless
D_h	hydraulic diameter, m
f_w	$:= \dfrac{\lambda_{fr}}{D_h}\dfrac{1}{2}\rho \bar{w}^2$, friction force per unit flow volume, N/m³
f_w^ζ	resistance force caused by local flow obstacles per unit volume of the flow mixture, N/m³
$\ell_{e,l}$	characteristic macroscopic length scale of turbulence production in channels, m
P_{kw}	irreversibly dissipated power per unit flow mass outside the viscous fluid due to turbulent pulsations equal to production of turbulent kinetic energy per unit mass of the flow, W/kg (m²/s³)
$P_{kw,1}^\zeta$	production of turbulent kinetic energy per unit mass of the gas due to irreversible singularity, W/kg
P_{kw}^ζ	production of turbulent kinetic energy per unit mass due to irreversible singularity, W/kg
$P_{kw,l}$	production of turbulent kinetic energy per unit mass of the field l due to friction with the wall, W/kg
Pr	Prandtl number, dimensionless
P_ε	production of the dissipation of the turbulent kinetic energy per unit mass, W/kg
$P_{\varepsilon w}$	production of the dissipation of the turbulent kinetic energy per unit mass due to friction with the wall, W/kg
$P_{\varepsilon w,l}$	production of the dissipation of the turbulent kinetic energy per unit mass of the field l due to friction with the wall, W/kg
$\dot{q}''$	heat flux, W/m²
$\dot{q}_c''^w$	heat flux from the wall into the continuum, W/m²
$\dot{q}_{w2}''$	heat flux from the wall into the liquid, W/m²
Re	$:= D_h \bar{w}/\nu$, Reynolds number, dimensionless
Re_{co}	$:= D_h(\rho\bar{w})/\eta_c$ Reynolds number computed so that all the two phase mass flow possesses the properties of the continuum, dimensionless
T	temperature, K
$\overline{u'}$	rms of the radial velocity fluctuation, m/s
V'	fluctuation of the velocity, m/s

V	velocity vector, m/s
w	axial velocity, m/s
w^*	friction velocity, dimensionless
$\overline{w}$	cross section averaged friction velocity, m/s
w^+	axial velocity, dimensionless
w_c	continuum axial velocity, m/s
$\overline{w}_c$	averaged axial continuum velocity, m/s
w_c^*	continuum axial friction velocity, m/s
X_1	gas mass concentration, m/s
X_{tt}	Lockhart–Martinelli parameter, dimensionless
x	x-coordinate, m
y	y-coordinate, distance from the wall, m
$y_{\lim}$	virtual distance from the wall in which almost all the viscous dissipation is lumped, m
$y_{\lim}^+$	virtual distance from the wall in which almost all the viscous dissipation is lumped, dimensionless
$y_{\lim,co}^+$	virtual distance from the wall in which almost all the viscous dissipation is lumped for the total mass flow considered as consisting of *Continuum Only*, dimensionless
y^+	distance from the wall, dimensionless
z	axial coordinate, m
Δz	finite of the axial distance, m

Greek

α	volumetric fraction, dimensionless
γ	surface permeability defined as flow cross section divided by the cross section of the control volume (usually the three main directional components are used), dimensionless
γ_v	volumetric porosity defined as the flow volume divided by the considered control volume, dimensionless
ε	power dissipated irreversibly due to *turbulent pulsations* in the viscous fluid per unit mass of the fluid (dissipation of the specific turbulent kinetic energy), m²/s³
ε_0	initial value of the dissipation of the specific turbulent kinetic energy, m²/s³
ζ_{fr}	irreversible friction coefficient, dimensionless
$\zeta_{fr,co}$	irreversible friction coefficient computed for the total mixture mass flow with the properties of the *Continuum Only*, dimensionless

$\zeta_{fr,eff}$	$:= \zeta_{fr,co} \Phi_{co}^2$, effective irreversible friction coefficient, dimensionless
η	dynamic viscosity, kg/(ms)
η_{vis}	part of the energy directly dissipated into heat and not effectively generating turbulence, dimensionless
χ	reflection coefficient, dimensionless
λ	thermal conductivity, W/(mK)
λ_{fr}	friction coefficient, dimensionless
$\lambda_{fr,co}$	friction coefficient computed for the total mixture mass flow with the properties of the continuum only, dimensionless
$\lambda_{fr,eff}$	effective friction coefficient, dimensionless
v	cinematic viscosity, m²/s
Π_h	wetted perimeter, m
ρ	density, kg/m³
$(\rho w)_{32}$	mass flow rate of the deposed droplets, kg/(m²s)
$\rho \overline{w}$	$:= \sum_{l=1}^{l_{max}} \alpha_l \rho_l \overline{w}_l$, mixture mass flow rate, kg/(m²s)
$\Delta \rho_{21}$	liquid–gas density difference, kg/m³
σ	surface tension, N/m
τ	time, s
τ_w	wall share stress, N/m²
$\Delta \tau$	time interval, s
$\Delta \tau$	time interval, minimum of the life time of the large eddy inside the gas flow and the time between two particle collisions, s
$\Delta \tau_{le}$	$:= 0.37 k_1 / \varepsilon_1$ life time of the large eddy inside the gas flow, s
$\Delta \tau_{33,col}$	$\approx (D_3 / \Delta V_{33}) / \alpha_3$, time between two particle collisions, s
$\Delta \tau_{12}$	droplet relaxation time, s
$\Delta \tau_{13}$	particle relaxation time, s
$\Delta \tau_{cd}$	$:= D_d / |\Delta \mathbf{V}_{cd}|$ particle relaxation time, s
$\Delta \tau_{\mu e,l}$	time scale corresponding to the spatially lowest scale for existence of eddies in field l called inner time scale or small time scale or Taylor microscale of turbulence, s

Subscripts

∞	steady, developed flow
1	gas
2	liquid
3	droplet

c	continuum
d	disperse
l	field l
m	field m
e	eddy
μ	associated to mass transfer or microscale
r	radial direction
θ	angular direction
z	axial direction
w	wall
k	axial discretization index

References

Chandesris M, Serre G and Sagaut (2005) A macroscopic turbulence model for flow in porous media suited for channel, pipe and rod bundle flows, 4th Int. Conf. On Computational Heat and Mass Transfer, Paris

Collier JG (1972) Convective boiling and condensation, McGraw-Hill, New York

Dittus FV and Boelter LMK (1930) Univ. of Calif. Publ. In Engng. Vol 2 no 13 p 443

Friedel L (1979) New friction pressure drop correlations for upward, horizontal, and downward two - phase pipe flow. Presented at the HTFS Symposium, Oxford, September 1979 (Hoechst AG Reference No. 372217/24 698)

Kolev NI (2007a) Multiphase Flow Dynamics, Vol. 1 Fundamentals, 3d extended ed., Springer, Berlin, New York, Tokyo

Kolev NI (2007b) Multiphase Flow Dynamics, Vol. 2 Thermal and mechanical interactions, 3d extended ed., Springer, Berlin, New York, Tokyo

Laufer J (1953) The structure of turbulence in fully developed pipe flow, NACA Report 1273

Neykov B, Aydogan F, Hochreiter L, Ivanov K, Utsuno H, Fumio K, Sartori E, Martin M (November 2005) NUPEC BWR full-size fine-mesh bundle test (BFBT) benchmark, vol 1: Specifications, US NRC, OECD Nuclear Energy Agency, NEA/NSC/DOC (2005) 5

Ryjkov SB and Hmara OM (1976) Intensificazija osajdenija kapel v dvuchfaznom pogranicnom sloe karotkoj plastiny, Teploenergetika, no 10 pp 78-80 (in Russian)

Sani RleR (4 January 1960) Down flow boiling and non-boiling heat transfer in a uniformly heated tube, University of California, URL-9023, Chemistry-Gen. UC-4, TID-4500 (15th Ed.)

Serre G and Bestion D (Oct. 2-6, 2005) Progress in improving two-fluid model in system code using turbulence and interfacial area concentrations, The 1th Int. Top. Meeting on Nuclear Reactor Thermal-Hydraulics (NURETH-11), Avignon, France

Windecker G and Anglart H (Oct. 3-8, 1999) Phase distribution in BWR fuel assembly and evaluation of multidimensional multi-fluid model, The 9th Int. Top. Meeting on Nuclear Reactor Thermal-Hydraulics (NURETH-9), San Francisco, California

Vogel, Bruch, Wang (19. Aug. 1991) SIEMENS Test Section 52 (DTS52) Description of experiments, KWZ BT23 1991 e 244, Erlangen, proprietary

5 Influence of the interfacial forces on the turbulence structure

The interfacial forces are in interaction with all other forces in the momentum equations. Experiments for investigation of the turbulence are concentrated always on some collective action of the forces. Therefore they have to be described correctly. Of course, performing simple analytical experiments by isolating only one force is what is always needed but very difficult. We discuss in this chapter the role of some of the forces which are still under investigation world wide.

5.1 Drag forces

In *Kolev* (2007b) constitutive relation for computation of the interfacial drag forces for variety of flow pattern and regimes are available. In the previous chapter we saw clearly that the drag forces contribute to the generation of turbulence in the trace of the bubbles. So using these two references the effect of the drag forces can be calculated.

5.2 The role of the lift force in turbulent flows

Note on particle rotation: A rotating sphere obeys the law

$$I_d \frac{d\omega_d}{d\tau} = -C_d^\omega \left(\frac{D_d}{2}\right)^5 \frac{1}{2} \rho_d |\omega_d| \omega_d .$$

Here the particle rotation velocity is ω_d and I_d is the particle's moment of inertia.

$$C_d^\omega = \frac{c_1}{\left(\text{Re}_{cd}^\omega\right)^{1/2}} + \frac{c_2}{\text{Re}_{cd}^\omega} + c_3 \, \text{Re}_{cd}^\omega$$

is a coefficient depending on the rotation *Reynolds* number

$$\text{Re}_{cd}^{\omega} = \left(\frac{D_d}{2}\right)|\omega_d|/\nu_c .$$

The c-coefficients are given by *Yamamotto* et al. (2001) in the following table:

Re_{cd}^{ω}	0 to 1	1 to 10	10 to 20	20 to 50	>50
c_1	0	0	5.32	6.44	6.45
c_2	50.27	50.27	37.2	32.2	32.1
c_3	0	0.0418	5.32	6.44	6.45

We learn from this dependence that both small and light particles can be easier to rotation compared to heavy and large particles. The following three main idealizations gives an idea for origination of the so-called lift force:

a) Rotating symmetric particle in symmetric flow of continuum experiences a lift force called *Magnus* force (a Berlin physicist *Gustav Magnus* 1802-1870). The curiosity of Lord *Rayleigh* to explain the trajectory of the tennis ball lead him in 1877 to the corresponding explanation. The force was analytically estimated by *Jukowski* and independently by *Kutta*, see in *Albring* (1970) p. 75.
b) Non-rotating symmetric particle in non-symmetric continuum flow experiences lift force, *Jukowski*
c) Non rotating asymmetric particle in symmetric continuum flow experiences lift force, *Jukowski*.

A combination of the radial liquid and gas momentum equations results to

$$p(r) = p(R) - (1-\alpha_1)\rho_2 \overline{u'^2} - \int_R^r \frac{(1-\alpha_1)\rho_2 \left(\overline{u'^2} - \overline{v'^2}\right)}{r*} dr* . \qquad (5.1)$$

The above equation reduces to the Eq. 4a in *Laufer* (1953) for zero-void. Later it was found that this is valid for bubbles with small sizes. If fluctuation velocities in the radial and in the azimuthally directions, respectively, are obtained from experiment and the void profile, the pressure variation along the pipe radius can be computed. If it fits to the measured, there is no need for other forces to explain the physics. But if there are differences, they may come from the so-called lift, lubrication and dispersion forces. The steady state momentum equations for bubbles and liquid are

$$\alpha_1 \frac{dp}{dr} - \frac{1}{r}\frac{d}{dr}(\alpha_1 r \tau_{1,rr}) + \frac{1}{r}\alpha_1 \tau_{1,\theta\theta} - f_{r,21}^L = 0 , \qquad (5.2)$$

$$(1-\alpha_1)\frac{dp}{dr} - \frac{1}{r}\frac{d}{dr}\left[r(1-\alpha_1)\tau_{2,rr}\right] + \frac{1}{r}(1-\alpha_1)\tau_{2,\theta\theta} + f_{r,21}^L = 0. \quad (5.3)$$

The lift force can be computed from the above equation if the pressure profile along the radius is known and the share terms are estimated as follows $\tau_{1,rr} \approx 0$, $\tau_{1,\theta\theta} \approx 0$, $\tau_{2,rr} = -\rho_2 \overline{u_2'^2}$, $\tau_{2,\theta\theta} = -\rho_2 \overline{v_2'^2}$. Using this approach *Wang* et al. (1987) observed that close to the wall a) the velocity gradient has a maximum, b) the velocity fluctuations have maximum and c) the static pressure has a minimum. Therefore small bubbles tend to occupy regions with higher turbulence unless other forces drive them away.

As already mentioned, objects with *negligible rotation* in share flow exert a lift force. The lift force acting on a bubble if it does not rotate is defined as follows

$$\mathbf{f}_{21}^L = c_{21}^L \alpha_1 \rho_2 (\mathbf{V}_2 - \mathbf{V}_1) \times \nabla \times \mathbf{V}_2 = c_{21}^L \alpha_1 \rho_2 (w_2 - w_1)\frac{dw_2}{dr}. \quad (5.4)$$

Some authors use for c_{21}^L a constant: *Troshko* and *Hassan* (2001) 0.06, *Lopez de Bertodano* (1992) 0.1, *Morel* (1997) ½.

Staffman (1965, 1968) derived for negligible particle rotation, negligible particle *Reynolds* number and small gradients of the continuum velocity the analytical expression for the shear lift force

$$c_{21}^L = 3.084 v_2^{1/2} \bigg/ \left(D_1 \left|\frac{dw_2}{dr}\right|^{1/2}\right). \quad (5.5)$$

Inside the boundary layer of bubbly flow having $w_1 > w_2$ and $dw_2/dr < 0$ the lift force is the ascertaining force for the bubbles toward the wall. Note that the spatial resolution in discrete analyses has to be fine enough in order to accurately compute the rotation of the continuous velocity field. Bad resolution like those used in the so-called sub-channel analyses produces only useless noise that makes the use of this force meaningless.

Mei (1992) proposed an expression that can be used for larger particle *Reynolds* numbers

$$c_{21}^L = Mei\ 3.084 v_2^{1/2} \bigg/ \left(D_1 \left|\frac{dw_2}{dr}\right|^{1/2}\right), \quad (5.6)$$

where

$$Mei = \left(1 - 0.3314\omega^{1/2}\right)\exp(-0.1\,\text{Re}_{12}) + 0.3314\omega^{1/2}, \quad \text{Re}_{12} \leq 40, \quad (5.7)$$

$$Mei = 0.0524(\omega \text{Re}_{12})^{1/2}, \quad \text{Re}_{12} > 40, \quad (5.8)$$

and

$$\mathrm{Re}_{12} = \Delta w_{12} D_1 / \nu_2,\qquad(5.9)$$

$$\omega = \frac{D_1/2}{w_2 - w_1}\left|\frac{dw_2}{dr}\right|.\qquad(5.10)$$

In a later work *Klausner* et al. (1993) found that the lift force on a bubble attached to a wall can be computed using

$$c_{21}^L = \frac{16}{3} 3.877 \omega^{3/2} \left(0.014\omega^2 + \mathrm{Re}_{12}^{-2}\right)^{1/4},\qquad(5.11)$$

which is valid for larger *Reynolds* numbers than previous relation. In a later work *Mei* and *Klausner* (1995) proposed to use interpolation between the *Stafman*'s results for small *Reynolds* numbers and *Auton*'s results, (1987), for large *Reynolds* numbers:

$$c_{21}^L = \frac{3}{8\omega^{1/2}}\left\{\frac{16}{9}\omega + \left[\frac{1.72 J \sqrt{2\omega/\mathrm{Re}_{12}}}{\mathrm{Re}_{12}^{1/2}}\right]^{-2}\right\}^{1/2},\qquad(5.12)$$

$$J \approx 0.6765\left\{1 + \tanh\left[2.5\left(\log_{10}\sqrt{2\omega/\mathrm{Re}_{12}} + 0.191\right)\right]\right\}$$

$$\times \left\{0.667 + \tanh\left[6\left(\log_{10}\sqrt{2\omega/\mathrm{Re}_{12}} - 0.32\right)\right]\right\}.\qquad(5.13)$$

Moraga et al. (1999) proposed

$$c_{21}^L = \left[0.12 - 0.2\exp\left(\frac{\mathrm{Re}_{12}\,\mathrm{Re}_{2,rot}}{36\times 10^4}\right)\right]\exp\left(\frac{\mathrm{Re}_{12}\,\mathrm{Re}_{2,rot}}{3\times 10^7}\right)$$

where $\mathrm{Re}_{2,rot} = rot\mathbf{V}_c D_1^2/\nu_c$, and $rot\mathbf{V}_c$ is the local vorticity e.g. dw_2/dr in axially symmetric flow. This equation possesses a sign inversion at large *Reynolds* numbers. For bubbly flow at atmospheric conditions the order of magnitude of c_{21}^L is around 0.1.

There are other expressions for the lift force on a single bubble. *Tomiyama* et al. (2002) measured trajectories of single bubbles in simple share flows of glycerol–water solution. They obtained the following empirical correlation

$$c_{21}^L = \min\left[0.288\tanh\left(0.121\mathrm{Re}_{12}\right), f\left(E\ddot{o}_{1m}\right)\right]\quad\text{for }E\ddot{o}_{1m} < 4,\qquad(5.14)$$

$$c_{21}^L = f(E\ddot{o}_{1m}) = 0.00105 E\ddot{o}_{1m}^3 - 0.0159 E\ddot{o}_{1m}^2 - 0.0204 E\ddot{o}_{1m} + 0.474$$

for $\quad 4 \leq E\ddot{o}_{1m} \leq 10.7$, (5.15)

$$c_{21}^L = -0.29 \text{ for } 10.7 < E\ddot{o}_{1m},$$ (5.16)

based on experiments within the region of parameters defined by $1.39 \leq E\ddot{o}_{1m} \leq 5.74$, $-5.5 \leq \log_{10} Mo_{12} \leq -2.8$, and $0 < |\nabla \times \mathbf{V}_2| \leq 8.3 s^{-1}$. The lift coefficient varied in this region between about 0.3 and -0.3. Here modified *Eötvös* and *Morton* numbers are used built with the horizontal bubble size

$$E\ddot{o}_{1m} = g(\rho_2 - \rho_1) D_{1,\max}^2 / \sigma_{12},$$ (5.17)

$$Mo_{12} = g(\rho_2 - \rho_1) \eta_2^4 / (\rho_2^2 \sigma_{12}).$$ (5.18)

The aspect ratio of the bubble is computed by using the *Wellek* et al. (1966) correlation

$$D_{1,\max} / D_{1,\min} = 1 + 0.163 E\ddot{o}_{1m}^{0.757}.$$ (5.19)

The lift coefficient for a bubble with diameter 3mm in an air–water system in accordance with the *Tomiyama* et al. correlation equals to 0.288. *Zun* (1980) performed measurements and estimated a value for small bubbles of about 0.3 very similar to *Sato* et al. (1977). *Naciri* et al. (1992) experimentally measured the lift coefficient of a bubble in a vortex and reported the value 0.25.

It should be emphasized that the above reviewed considerations are for single object in the share flow. The presence of multiple objects in the share flow is found to influence this force too.

The importance of the findings by *Tomiyama* et al. (2002) is in the observation that for large bubbles the lift force changes the sign. *Krepper, Lucas* and *Prasser* (2005) observed experimentally that in vertical bubbly flow the void profile depends on the bubble size spectrum. For spectrum with predominant small size bubbles a wall void peaking is observed. For spectrums having predominant large bubbles the central void packing is observed. This effect was reproduced by *Krepper* et al. (2005) by using lift force applied on multiple groups with $c_{21}^L = 0.05$ for $D_1 < 0.006m$ and $c_{21}^L = -0.05$ for $D_1 \geq 0.006m$. The improvement was considerable going from 1 to 2 groups. No substantial change was reported if more than 8 size groups are used.

Wang et al. (1987) introduced the influence of the local volume fraction into the lift coefficient

$$c_{21}^L(\xi) = 0.01 + \frac{0.49}{\pi} \cot^{-1} \frac{\log \xi + 9.3168}{0.1963},$$ (5.20)

as a function of

$$\xi = \exp(-\alpha_1) 2\omega \left(\frac{D_1}{D_h} \frac{1}{\text{Re}_{12}}\right)^2 \left(\frac{w_1}{\Delta w_{12\infty}}\right)^2, \quad (5.21)$$

where

$$\Delta w_{12\infty} = 1.18(g\sigma/\rho_2)^{1/4}. \quad (5.22)$$

This coefficient varies between 0.01 and 0.1 in accordance with the *Wang*'s et al. data. The disadvantage of this approach is that due to the dependence $\xi = \xi(D_h)$ the correlation depends on global geometry characteristics and can not be applied locally.

Conclusions: (a) The spatial resolution in finite volume analyses has to be fine enough in order to accurately compute the rotation of the continuous velocity field. Bad resolution like those used in the so-called sub-channel analyses produces only useless noise that makes the use of this force meaningless. (b) There is no method known to me that is based on local conditions and that allows taking into account the effect of multiple objects on the lift force. (c) The other problem is that small bubbles will probably rotate and the application of lift force derived for non-rotating objects in share flows is questionable. (d) Heavy solid particles carried by gas are rather subject to lift force because they hardly will "see" the rotation of the surrounding continuum.

There is at least one other force acting toward the equalizing of the void profiles – the so-called dispersion force which will be discussed in a moment.

5.3 Lubrication force in the wall boundary layer

For adiabatic flows no bubbles are observed at the wall. This led *Antal* et al. (1991) to the conclusion that there is a special force at the wall similar to the *lubrication* force pushing the bubbles away from the surface,

$$\mathbf{f}_{cd}^{Lw} = \frac{\alpha_d \rho_c |\hat{\mathbf{V}}|^2}{R_d} \left(-0.104 - 0.06|\Delta V_{cd}| + 0.147 \frac{R_d}{y_0}\right) \mathbf{n}_w \geq 0, \quad (5.23)$$

where y_0 is the distance between the bubble and the wall and $\mathbf{n}_w$ is the unit outward normal vector on the surface of the wall and

$$\hat{\mathbf{V}} = \mathbf{V}_d - \mathbf{V}_c - \left[\mathbf{n}_w \cdot (\mathbf{V}_d - \mathbf{V}_c)\right] \mathbf{n}_w \quad (5.24)$$

is the velocity difference component parallel to the wall. *Lopez de Bertodano* (1992) add to the above expressions $-0.0075\alpha_d \rho_c |\mathbf{V}_c| \mathbf{V}_d / R_d$ in case of $y \le R_d$. Instead of the two constants in the above expression *Lopez de Bertodano* (1992) used 0.2 and 0.12, and *Troshko* and *Hassan* (2001) 0.02 to 0.03 and 0.04 to 0.06, respectively. *Krepper*, *Lucas* and *Prasser* (2005) used

$$\mathbf{f}_{cd}^{Lw} = \alpha_d \rho_c |\hat{\mathbf{V}}|^2 \max\left(-\frac{0.0064}{D_d} + \frac{0.016}{y_0}, 0\right)\mathbf{n}_w \tag{5.25}$$

instead.

Tomiyama (1998) reported the following empirical correlation for the wall force

$$\mathbf{f}_{cd}^{Lw} = c_{cd}^{Lw} \alpha_d \rho_c |\hat{\mathbf{V}}|^2 R_d \left(\frac{1}{y_0^2} - \frac{1}{(D_h - y_0)^2}\right)\mathbf{n}_w, \tag{5.26}$$

where

$$c_{cd}^{Lw} = \exp(-0.933 E\ddot{o} + 179) \quad \text{for} \quad 1 \le E\ddot{o} \le 5, \tag{5.27}$$

$$c_{cd}^{Lw} = 0.007 E\ddot{o} + 0.04 \quad \text{for} \quad 5 \le E\ddot{o} \le 33. \tag{5.28}$$

5.4 The role of the dispersion force in turbulent flows

5.4.1 Dispersed phase in laminar continuum

It is known that even low-velocity potential flow over a family of spheres is associated with natural fluctuations of the continuum. The produced oscillations of the laminar continuum are called *pseudo-turbulence* by some authors. The averaged pressure over the dispersed particles surface is smaller than the volume averaged pressure. Therefore in flows with spatially changing concentration of the disperse phase an additional force acts toward the concentrations gradients. For bubbly flow *Nigmatulin* (1979) obtained the analytical expression

$$-\alpha_c^e \mathbf{T}_{pseudo\ turbulence, c} = \alpha_d^e \rho_c |\Delta \mathbf{V}_{cd}|^2 \begin{vmatrix} \frac{4}{20} & 0 & 0 \\ 0 & \frac{3}{20} & 0 \\ 0 & 0 & \frac{3}{20} \end{vmatrix}, \tag{5.29}$$

see also *Van Wijngaarden* (1982).

5.4.2 Dispersed phase in turbulent continuum

For dispersed phase in turbulent continuum it is observed that the turbulence tends to smooth the volumetric concentrations of the dispersed phase. In other words pulsations of the continuum produce force that drives particles to move from places with higher concentration to places with smaller one. To illuminate this force let us recall once again the terms in the local volume and time averaged momentum equations (2.1) and (2.2) for the disperse and the continuous field. We will use as a framework for the discussion the *k-eps* model. Remember that for isotropic turbulence we have

$$\nabla \cdot \left[\left(\alpha_l^e \rho_l \mathbf{V}_l' \mathbf{V}_l' \right) \gamma \right] = -\tilde{\mathbf{S}}_l + \frac{2}{3} \nabla \left(\gamma \alpha_l^e \rho_l k_l \right). \qquad (5.30)$$

Introducing the above equation into the momentum equations after differentiating the last term, interpreting the normal *Reynolds* stresses as turbulent pressure fluctuation

$$p_l' = \frac{2}{3} \rho_l k_l , \qquad (5.31)$$

and rearranging we obtain

$$\frac{\partial}{\partial \tau} \left(\alpha_d \rho_d \mathbf{V}_d \gamma_v \right) + \nabla \cdot \left(\alpha_d^e \rho_d \mathbf{V}_d \mathbf{V}_d \gamma \right) - \tilde{\mathbf{S}}_d + \alpha_d^e \gamma \nabla \left(p_d + p_d' \right) + \alpha_d \gamma_v \rho_d \mathbf{g}$$

$$+ \left(p_d - p_c + \delta_d \sigma_{dc} \kappa_d - \Delta p_c^{d\sigma*} + p_d' \right) \nabla \left(\alpha_d^e \gamma \right) \ldots = \ldots, \qquad (5.32)$$

$$\frac{\partial}{\partial \tau} \left(\alpha_c \rho_c \mathbf{V}_c \gamma_v \right) + \nabla \cdot \left(\alpha_c^e \rho_c \mathbf{V}_c \mathbf{V}_c \gamma \right) - \tilde{\mathbf{S}}_c - \nabla \cdot \left(\alpha_c^e \gamma \mathbf{T}_{\eta,c} \right) + \alpha_c^e \gamma \nabla \left(p_c + p_c' \right)$$

$$+ \alpha_c \gamma_v \rho_c \mathbf{g} + \left(\Delta p_c^{d\sigma*} + p_c' \right) \nabla \left(\alpha_c^e \gamma \right) - \Delta p_c^{w\sigma*} \nabla \gamma \ldots = \ldots \qquad (5.33)$$

respectively. Remember that the dispersed momentum equation is valid inside the dispersed phase *including the interface*. It includes the interface jump condition. The continuum momentum equation is valid only inside the continuum that is outside the interface.

5.4.2.1 Bulk-interface pressure difference

In accordance with *Stuhmiller* (1977) the interface averaged pressure at the site of the continuum is smaller with

$$\Delta p_c^{l\sigma *} = -0.37 c_{cd}^d \rho_c |\Delta \mathbf{V}_{cd}|^2, \qquad (5.34)$$

than the continuum bulk averaged pressure.

> Therefore the force
>
> $$... = ... -0.37 c_{cd}^d \rho_c |\Delta \mathbf{V}_{cd}|^2 \nabla (\alpha_d^e \gamma) \qquad (5.35)$$
>
> in the momentum equation of the dispersed phase leads to positive acceleration of the bubbles toward the negative void gradient.

This facilitates dispersion of the particles. Note that the numerical constant takes values of ½ for potential flow around sphere, *Lamb* (1932). Lance and Bataille (1991) reported values between 0.5 and 0.7 for 5mm oblate spheroid bubbles in water.

Hwang and Schen [22] (1992) reported a method for computing the pressure distribution around a sphere from measured data for *Reynolds* numbers larger than 3000. Using this distribution the authors computed the surface averaged pressure over the sphere in the three directions

$$\Delta p_m^{l\sigma *} = -f(c_{cd}^d) \frac{1}{4} \rho_m |\Delta \mathbf{V}_{ml}|^2 \qquad (5.36)$$

and found that it is non isotropic, i.e. $f(c_{cd}^d)$ takes different values in the lateral directions compared to the direction of the relative velocity. For high *Reynolds* numbers the authors reported

$$\Delta p_m^{l\sigma *} = \rho_m |\Delta \mathbf{V}_{ml}|^2 \begin{vmatrix} \frac{1}{20} & 0 & 0 \\ 0 & -\frac{2}{5} & 0 \\ 0 & 0 & -\frac{2}{5} \end{vmatrix}. \qquad (5.37)$$

In the limiting case for vanishing *Reynolds* number the authors reported

$$\Delta p_m^{l\sigma*} = \rho_m |\Delta \mathbf{V}_{ml}|^2 \begin{vmatrix} \dfrac{81}{160} & 0 & 0 \\ 0 & \dfrac{27}{160} & 0 \\ 0 & 0 & \dfrac{27}{160} \end{vmatrix}. \tag{5.38}$$

5.4.2.2 Turbulence dispersion force

Other reason causing redistribution of the bubbles is the turbulence in the continuum. The higher the local turbulence is the lower the local pressure. Therefore small bubbles tend to occupy regions with higher turbulence unless other forces drive them away. Let us see how turbulence in the continuum influence dispersion. Considering two velocity fields

$$\alpha_c = 1 - \alpha_d, \tag{5.39}$$

and bubbly flow for which

$$k_d \approx k_c, \tag{5.40}$$

will result in

$$p'_c = \frac{2}{3}\rho_c k_c, \tag{5.41}$$

$$p'_d = \frac{2}{3}\frac{\rho_d}{\rho_c}\rho_c k_c, \tag{5.42}$$

and therefore in *asymmetric dispersion force components* into the both momentum equations for constant permeability

$$\ldots + \frac{\rho_d}{\rho_c}\frac{2}{3}\rho_c k_c \nabla(\alpha_d^e \gamma)\ldots = \ldots, \tag{5.43}$$

$$\ldots - \frac{2}{3}\rho_c k_c \nabla(\alpha_d^e \gamma)\ldots = \ldots. \tag{5.44}$$

Whether it is allowed to lump the pulsation pressure component into the overall bulk pressure as proposed by some authors is not proven. Therefore I prefer to stay with this notation. Now let us see what is available in the literature in this field. *Lopez de Bertodano* (1992) proposed for bubbly flow different form of this force naming it dispersion force as a *symmetric* force in the both momentum equations

$$\text{gas: } \ldots + c_{cd}^t \rho_c k_c \nabla \alpha_d \ldots = \ldots, \tag{5.45}$$

$$\text{liquid: } \ldots - c_{cd}^t \rho_c k_c \nabla \alpha_d \ldots = \ldots, \tag{5.46}$$

with $c_{cd}^t = 0.1$ proposed by *Lahey* et al. (1993). The same approach was used later other authors e.g. *Windecker* and *Anglart* (1999), *Krepper* and *Egorov* (2005). *Krepper*, *Lucas* and *Prasser* (2005) used $c_{cd}^t = 0.5$, *Morel* (1997) 0.01 to 0.1, *Lopez de Bertodano* (1992) 0.1, *Troshko* and *Hasan* (2001) 0.01 to 0.03. *Antal* et al. (1998) used instead

$$c_{cd}^t = \frac{\Delta \tau_c}{\Delta \tau_{cd}} \frac{\Delta \tau_c}{\Delta \tau_c + \Delta \tau_{cd}} \tag{5.47}$$

where

$$\frac{1}{\Delta \tau_c} = \frac{1}{\Delta \tau_{ce}} + \frac{1}{\Delta \tau_{cr}}, \tag{5.48}$$

and

$$\Delta \tau_{ce} = 0.35 k_c / \varepsilon_c, \tag{5.49}$$

$$\Delta \tau_{cd} = \frac{4}{3} \frac{c_{cd}^{vm}}{c_{cd}^d} \frac{D_d}{|\Delta V_{cd}|}, \tag{5.50}$$

$$\Delta \tau_{cr} = \frac{1}{2} 0.35 k_c^{3/2} \Big/ \left(\varepsilon_c |\Delta V_{cd}| \right). \tag{5.51}$$

Shi et al. (2005) performed time averaging of the already volume averaged drag force term

$$\frac{3}{4} \rho_c \frac{c_{cd}^d}{(1-\alpha_d)} |\Delta \mathbf{V}_{cd}| \overline{\alpha_d' V_c'}. \tag{5.52}$$

Postulating

$$\overline{\alpha'_d V'_{cd}} \approx -\frac{v'_c}{\Pr'_{\alpha,c}} \nabla \alpha_d, \qquad (5.53)$$

the authors obtained additional force component acting as a dispersion force:

$$\text{gas: } \ldots + \frac{3}{4}\frac{c^d_{cd}}{(1-\alpha_d)} |\Delta V_{cd}| \frac{\rho_c v'_c}{\Pr'_{\alpha,c}} \nabla \alpha_d \ldots = \ldots, \qquad (5.54)$$

$$\text{liquid: } \ldots - \frac{3}{4}\frac{c^d_{cd}}{(1-\alpha_d)} |\Delta V_{cd}| \frac{\rho_c v'_c}{\Pr'_{\alpha,c}} \nabla \alpha_d \ldots = \ldots . \qquad (5.55)$$

Conclusions: Small bubbles tend to occupy regions with higher turbulence in which the averaged static pressure is lower unless "dispersion forces" drive them away. The name "dispersion force" is used in the literature with no unique meaning. Three phenomena are identified as source for this force: (a) The difference between the surface and volume averaged pressure combined with concentration gradient is a real force driving bubbles toward low volume concentrations; (b) The spatial variation of the turbulence energy in the dispersed field

$$\frac{2}{3}\nabla\left(\gamma \alpha_l^e \rho_l k_l\right) \qquad (5.56)$$

acts as a dispersion force. This form of the force is a consequence of the isotropy assumption; c) The fluctuation of the superficial velocity,

$$\overline{\alpha'_d V'_c} \qquad (5.57)$$

is considered to give additional drug compared to the steady state drug force.

Nomenclature

Latin

c^L_{21} lift force coefficient, dimensionless
$D_{1,\max}$ maximum bubble diameter, m
D_h hydraulic diameter, m
$Eö_{1m}$ $:= g(\rho_2 - \rho_1) D^2_{1,\max}/\sigma_{12}$, Eötvös number build with the horizontal bubble size, dimensionless

Nomenclature

$\mathbf{f}_{21}^L$	lift force vector caused by the liquid 2 acting on the bubble, field 1, per unit flow volume, N/m³
$\mathbf{f}_{cd}^{Lw}$	lubrication force vector pushing the bubbles away from the surface,
$f_{r,21}^L$	lift force caused by the liquid 2 acting on the bubble, field 1, into r-direction per unit flow volume, N/m³
g	gravitational acceleration m/s²
k	specific turbulent kinetic energy, m²/s²
Mey	corrector for the lift force coefficient, dimensionless
Mo_{12}	$:= g(\rho_2 - \rho_1)\eta_2^4 / (\rho_2^2 \sigma_{12})$, Morton number, dimensionless
$\mathbf{n}_w$	unit outward normal vector on the surface of the wall, dimensionless
I	unity matrix, -
p	pressure, Pa
$\Delta p_c^{l\sigma*}$	continuum bulk averaged pressure minus interface averaged pressure at the site of the continuum, Pa
Pr^t	turbulence Prandtl number, dimensionless
R	pipe radius, m
r	radius, m
r*	radius, m
Re_{12}	$:= |\Delta w_{12}| D_1 / \nu_2$, Reynolds number based on the bubble liquid relative velocity, dimensionless
$\overline{u'^2}$, $\overline{v'^2}$	normal turbulent stress components in r and θ direction
V	velocity vector, m/s
ΔV_{cd}	relative velocity of the dispersed phase with respect to the continuum, m/s
$\hat{V}$	$:= \mathbf{V}_d - \mathbf{V}_c - [\mathbf{n}_w \cdot (\mathbf{V}_d - \mathbf{V}_c)]\mathbf{n}_w$, velocity difference component parallel to the wall, m/s
V'	fluctuations of the velocity vector, m/s
V	velocity vector, m/s
w	axial velocity, m/s
$\Delta w_{12\infty}$	$:= 1.18(g\sigma/\rho_2)^{1/4}$, bubble free rising velocity, m/s
y_0	distance between the bubble and the wall, m

Greek

γ	surface permeability, dimensionless
γ_v	volumetric porosity, dimensionless
$\Delta \tau_{cd}$	particle relaxation time, s
$\Delta \tau_{ce}$	time scale of fluctuation of large eddy of the continuum, s

θ	angular coordinate, rad
ε	power dissipated irreversibly due to turbulent pulsations in the viscous fluid per unit mass of the fluid (dissipation of the specific turbulent kinetic energy), m²/s³
α	volumetric fraction, dimensionless
ν	cinematic viscosity, m²/s
ρ	density, kg/m³
$\tau_{l,xy}$	force per unit surface normal to x directed into y-direction, N/m²
ω	$:= \dfrac{D_1/2}{w_2 - w_1} \left|\dfrac{dw_2}{dr}\right|$, liquid field rotation, dimensionless
σ_{12}	surface tension between gas and liquid, N/m
η	dynamic viscosity, kg/(ms)
σ	surface tension, N/m
κ	curvature

Superscript

′   fluctuation component

Subscripts

1   gas
2   liquid

References

Albring W (1970) Angewandte Strömungslehre, Theodor Steinkopff, Dresen
Antal SP, Lahey RT Jr and Flaherty JE (1991) Int. J. Multiphase Flow, vol 17 no 5 pp 635-652
Auton RT (1987) The lift force on a spherical body in rotating flow, J. Fluid Mechanics, vol 183 pp 199-218
Boussinesq J (1877) Essai sur la théorie des eaux courantes, Mem. Pr´s. Acad. Sci., Paris, vol 23 p 46
Hwang GJ, Schen HH (Sept. 21-24, 1992) Tensorial solid phase pressure from hydrodynamic interaction in fluid-solid flows. Proc. of the Fifth International Topical Meeting On Reactor Thermal Hydraulics, NURETH-5, Salt Lake City, UT, USA, IV pp 966-971
Klausner JF, Mei R, Bernhard D and Zeng LZ (1993) Vapor bubble departure in forced convection boiling, Int. J. Heat Mass Transfer, vol 36 pp 651-662
Kolev NI (2007a) Multiphase Flow Dynamics, Vol. 1 Fundamentals, 3d extended ed., Springer, Berlin, New York, Tokyo
Kolev NI (2007b) Multiphase Flow Dynamics, Vol. 2 Thermal and mechanical interactions, 3d extended ed., Springer, Berlin, New York, Tokyo
Krepper E, Lucas D and Prasser H-M (2005) On the modeling of bubbly flow in vertical pipes, Nuclear Engineering and Design, vol 235 pp 597-611

Krepper E and Egorov Y (May 16-20, 2005) CFD-Modeling of subcooled boiling and application to simulate a hot channel of fuel assembly, 13th Int. Conference on Nuclear Engineering, Beijing, China

Lahey RT Jr, Lopez de Bertodano M and Jones OC Jr (1993) Phase distribution in complex geometry conditions, Nuclear Engineering and Design, vol 141 pp 177-201

Lahey R Jr and Drew DA (2001) The analysis of two-phase flow and heat transfer using a multi-dimensional four-field, two fluid model, Nuclear Engineering and Design, vol 2004 pp 29-44

Lance M and Bataille J (1991) Turbulence in the liquid phase of a uniform bubbly air-water flow, J. of Fluid Mechanics, vol 22 pp 95-118

Lamb H (1932) Hydrodynamics, Dover, New York

Laufer J (1953) The structure of turbulence in fully developed pipe flow, NACA Report 1273

Lopez de Bertodano M (1992) Turbulent bubbly two-phase flow in triangular duct, PhD Thesis, Renssaelaer Polytechnic Institute, Troy, NY

Mei R (1992) An approximate expression for the shear lift force on spherical particle at finite Reynolds number, Int. J. Multiphase Flow, vol 18 no 1 pp 145-147

Mei R and Klausner JF (1995) Shear lift force on spherical bubbles, Int. J. Heat FluidFlow, vol 15 pp 62-65

Moraga FJ, Bonetto FJ, Lahey RT Jr (1999) Lateral forces on sphere in turbulent uniform shear flow, Int. J. Multiphase Flow, vol 25 pp 1321-1372

Morel C (1997) Turbulence modeling and first numerical simulation in turbulence two-phase flow, Report SMTH/LDMS/97-023, October, CEA/Grenoble, France

Naciri A (1992) Contribution à l'étude des forces exercées par un liquide sur une bulle de gaz: portance, masse ajoutée et interactions hydrodynamiques, Doctoral Dissertation, École Central de Lyon, France

Nigmatulin RI (1979) Spatial averaging in the mechanics of heterogeneous and dispersed systems, Int, J. Multiphase Flow, vol 4 pp 353-385

Sato Y, Honda T, Saruwatari S and Sekoguchi K (1977) Two-phase bubbly flow, Part 2 (in Japanese) Trans. Jap. Soc. Mech. Engrs. Vol 43 no 370 pp 2288-2296

Shi J-M, Burns AD and Prasser H-M (May 16-20, 2005) Turbulent dispersion in poly-dispersed gas-liquif flows in a vertical pipe, 13th Int. Conf. on Nuclear Engineering, Beijing, China, ICONE 13

Staffman PG (1965) The lift on a small sphere in a slow shear flow, J. Fluid Mech. vol 22 pp 385-400

Staffman PG (1968) Corrigendum to "The lift on a small sphere in a slow shear flow", J. Fluid Mech. vol 31 pp 624

Stuhmiller JH (1977) The influence of the interfacial pressure forces on the character of the two-phase flow model, Proc. of the 1977 ASME Symp. on Computational Techniques for Non-Equilibrium Two-Phase Phenomena, pp 118-124

Tomiyama A (1998) Struggle with computational bubble dynamics, Proc. Of the 3rd Int. Conf. on Multiphase Flow ICMF-98, France

Tomiyama A et al. (2002) Transverse migration of single bubbles in simple shear flows, Chemical Engineering Science, vol 57 pp 1849-1858

Troshko AA and Hassan YA (2001) A two-equations turbulence model of turbulent bubbly flow, Int. J. of Multiphase Flow, vol 27 pp 1965-2000

van Wijngaarden L (1998) On pseudo turbulence, Theoretical and Computational Fluid Dynamics, vol 10 pp 449–458

Wang SK, Lee SJ, Jones OC and Lahey RT Jr (1987) 3-D turbulence structure and phase distribution measurements in bubbly two-phase flows, Int. J. Multiphase Flow, vol 13 no 3 pp 327-343

Wellek RM, Agrawal AK and Skelland AHP (1966) Shapes of liquid drops moving in liquid media, AIChE J. vol 12 p 854

Yamamotto Y, Potthoff M, Tanaka T, Kajishima and Tsui Y (2001) Large-eddy simulation of turbulent gas-particcle flow in a vertical channel : effect of considering inter-particle collisions, J. Fluid Mechanics, vol 442 pp 303-334

Zun I (1980) The transferees migration of bubbles influenced by walls in vertical bubbly flow, Int. J. Multiphase Flow, vol 6 pp 583-588

6 Particle–eddy interactions

$k-\varepsilon$ models for analysis of transport of solid particles in gas flows are reported in a number of papers. *Reeks* (1991, 1992), *Simonin* (1991), *Sommerfeld* (1992), *Wolkov, Zeichik* and *Pershukov* (1994), and the references given there are good sources to start with this issue. The common feature of these works is the concept assuming convection and diffusion of the specific turbulent kinetic energy and its dissipation in the continuous phase. For considering the influence of the discrete phase predominantly two approaches are used:

(a) No feedback of the dispersed phase on the continuum turbulence commonly named *one-way coupling*;
(b) The feedback of the dispersed phase on the continuum turbulence is taken into account. This approach is named *two-way coupling*.

6.1 Three popular modeling techniques

Several mathematical techniques are developed to describe dispersed particles and continua. In one of them the continuum is described in *Euler* coordinates and each particle is traced in *Lagrangian* manner. This class of methods is called *Euler–Lagrange* method. Its best advantage is that collisions are computed in a natural way. The description of turbulence in such methods can again be done in different way. One of them is called *Random Dispersion Model* (RDM). In this approach the *k-eps* equation for the continuum are solved. Usually assumption of isotropic turbulence is used. Then the fluctuating velocity component V' is assumed to be random deviates of a *Gaussian* probability distribution with zero mean and variance. So, using the random generator in each cell a magnitude and direction of the fluctuation component are generated and overlaid to the mean velocity field. The tracing of the particle trajectory is then done in the next time step using this continuum velocity. The method is successfully used in many small scale applications. The need to solve as many particle conservation equations as particles are present in the integration domain is the main limitation of this method.

> Therefore *Euler-Euler* description of continuum and disperse phase is much more effective. This is the reason why I recommend such type of methods for solving daily engineering problems.

The time scales for interaction of the particles with eddies are discussed below. This information is used in many model elements and variety of applications.

6.2 Particle–eddy interaction without collisions

It is interesting to consider the behavior of a particle entrapped by large eddy with size $\ell_{e,c}$. If the particle has a large mass, it can not follow the pulsation of the continuum. Only small particles can follow the eddy oscillations. The ratio

$$C_{dc}^{\prime*} := V_d' / V_c' \qquad (6.1)$$

reached after the time $\Delta\tau$ counted from the beginning of the interaction is called *response coefficient* for *single* particle. It characterizes the ability of the particles to response on a pulsation of the continuum. Therefore for isotropic turbulence we have

$$k_d \approx k_c C_{dc}^{\prime*2}. \qquad (6.2)$$

6.2.1 Response coefficient for single particle

Integrating the momentum equation written for the interaction between the continuum eddy and the particle over the time $\Delta\tau$ results in

$$C_{dc}^{\prime*} := V_d' / V_c' = 1 - e^{-\Delta\tau/\Delta\tau_{cd}}. \qquad (6.3)$$

For $\Delta\tau \ll \Delta\tau_{cd}$,

$$C_{dc}^{\prime*} = \left(1 + \Delta\tau_{cd}/\Delta\tau\right)^{-1}. \qquad (6.4)$$

which well compares with the *Reeks* (1977) expression

$$C_{dc}^{\prime*} = \left(1 + 0.7\,\Delta\tau_{cd}/\Delta\tau\right)^{-1} \qquad (6.5)$$

obtained for droplets carried by gas in vertical pipes. Equation (6.5) was experimentally confirmed by *Lee* et al. (1989b).

The characteristic time for the response of a single particle used above is

$$\Delta\tau_{cd} := \frac{4}{3} \frac{\rho_d + \rho_c c_{cd}^{vm}}{\rho_c} \frac{D_d}{c_{cd}^d \left|\Delta\mathbf{V}_{cd}\right|} \qquad (6.6)$$

For particles in gas we have

$$\Delta \tau_{cd} = \frac{\left(\rho_d + \rho_c c_{cd}^{vm}\right) D_d^2}{18 \eta_c \psi(\text{Re}_{cd})}, \qquad (6.7)$$

where for the *Stokes* regime

$$\psi(\text{Re}_{cd}) = 1. \qquad (6.8)$$

For larger velocity differences

$$\psi(\text{Re}_{cd}) = 1 + 0.15 \text{Re}_{cd}^{0.687} \text{ for } \text{Re}_{cd} \leq 10^3, \qquad (6.9)$$

$$\psi(\text{Re}_{cd}) = 0.11 \text{Re}_{cd}/6 \text{ for } \text{Re}_{cd} > 10^3, \qquad (6.10)$$

Zaichik et al. (1998). *Zaichik* and *Alipchenkov* (1999) obtained the following general formalism for estimation of the response coefficient for dispersed flow in pipes:

$$C_{dc}^{t*2} = \frac{1 + A\Delta\tau_{cd}/\Delta\tau_{L,cd}}{1 + \Delta\tau_{cd}/\Delta\tau_{L,cd}}, \quad A = \frac{\left(1 + c_{cd}^{vm}\right)\rho_c/\rho_d}{1 + c_{cd}^{vm}\rho_c/\rho_d}. \qquad (6.11)$$

The *Lagrangian* time scale of turbulence averaged over the cross section is

$$\Delta\tau_{L,c} = 0.04 D_h/w_c^*. \qquad (6.12)$$

The *Lagrangian* eddy–droplet interaction time

$$\Delta\tau_{L,cd} = \Delta\tau_{L,cd}(St = 0) + \left[\Delta\tau_{L,cd}(St \to \infty) - \Delta\tau_{L,cd}(St = 0)\right] f(St) \quad (6.13)$$

is expressed as a function of the interaction time for very small

$$\Delta\tau_{L,cd}(St = 0) = \Delta\tau_{L,c} \frac{4(3a + 3a^2/2 + 1/2)}{5a(1+a)^2}, \qquad (6.14)$$

and very large particles

$$\Delta\tau_{L,cd}(St \to \infty) = \Delta\tau_{L,c} \frac{6(2+b)}{5(1+b)^2}. \qquad (6.15)$$

Here $St = \Delta\tau_{cd}/\Delta\tau_{L,c}$ is the *Stokes* number, $b = |w_c - w_d|/w_c^*$ is the dimensionless drift parameter,

$$a = \left(1 + b^2 + \frac{2}{\sqrt{3}}b\right)^{1/2}, \qquad (6.16)$$

and the interpolation function is

$$f(St) = \frac{St}{1+St} - \frac{5St^2}{4(1+St)^2(2+St)}. \qquad (6.17)$$

6.2.2 Responds coefficient for clouds of particles

The response coefficient for particles with large concentrations C_{dc}^t may differ from the response coefficient for single particle. It is therefore a function of the local volume fraction of the dispersed phase. Proposal for the estimation of C_{dc}^t is given by *Rusche* (2000)

$$\frac{C_{dc}^t - 1}{C_{dc}^{t*} - 1} = \exp\left[-\left(180\alpha_d - 4710\alpha_d^2 + 4.26 \times 10^4 \alpha_d^3\right)\right], \qquad (6.18)$$

where according to *Wang* (1994) the response coefficient for single particle is

$$C_{dc}^{t*} = \frac{3}{2} \frac{c_{cd}^d |\Delta V_{cd}| \ell_{e,c}}{v_c D_d \operatorname{Re}_c^t}, \qquad (6.19)$$

with $\ell_{e,c} = \sqrt{c_\eta}\, k_c^{3/2}/\varepsilon_c$, $V_c' \approx \sqrt{\frac{2}{3}k_c}$ and $\operatorname{Re}_c^t := \frac{V_c' \ell_{ec}}{v_c} \approx \sqrt{\frac{2}{3}}c_\eta \frac{k_c^2}{\varepsilon_c v_c}$. This method is used by *Lo* (2005).

6.2.3 Particle–eddy interaction time without collisions

For the case of no particle–particle collisions some authors considered the existence time of the eddy $\Delta\tau_{e,c}$ as the duration of the interaction with the eddy

$$\Delta\tau = \Delta\tau_{e,c}. \qquad (6.20)$$

This is valid for flows with small concentrations of the dispersed phase for which the particle–particle collision time is larger than the existence time of the eddy of the continuum.

Other authors set the duration of the interaction with the eddy equal to the time required to cross an eddy. If the size of the eddy is $\ell_{e,c}$ and the macroscopic particle velocity with respect to the continuum velocity is $|\Delta \mathbf{V}_{cd}|$, the interaction time is set to

$$\Delta \tau_{ed,\text{int}} = -\Delta \tau_{cd} \ln\left(1 - \frac{\ell_{e,c}}{\Delta \tau_{cd} |\Delta \mathbf{V}_{cd}|}\right). \tag{6.21}$$

For $\frac{\ell_{e,c}}{\Delta \tau_{cd} |\Delta \mathbf{V}_{cd}|} > 1$ the particle is captured by the eddy and the assumption $\Delta \tau_{ed,\text{int}} = \Delta \tau_{cd}$ can be used, *Gosman* and *Ioannides* (1981).

6.3 Particle–eddy interaction with collisions

For larger particle volume concentrations collisions break the particle dragging process. After each particular collision the dragging process starts again. Therefore the time elapsed between two collisions should be used as the eddy–particle interaction time. We own this conclusion to *Hanratty* and *Dykhno* (1977). In what follows we discuss how to compute the time elapsed between two successive collisions.

Smoluchowski obtained in 1918 for the number of collisions per unit volume

$$\dot{n}_{d,col} = n_{d1} n_{d2} \frac{\pi}{4} \left(\frac{D_{d1} + D_{d2}}{2}\right)^2 \Delta V_{d1d2}, \tag{6.22}$$

for collision of two groups of particles with sizes D_{d1} and D_{d2}. Here n_{d1} and n_{d2} are particle number densities (number of particles per cubic meter) and ΔV_{d1d2} is the relative velocity between two groups in the same control volume. For only one group of particles

$$\dot{n}_{d,col} = n_d n_d \frac{\pi}{4} D_d^2 \Delta V_{dd} = f_{d,col} n_d \tag{6.23}$$

with collision frequency of a *single* particle

$$f_{d,col} = n_d \frac{\pi}{4} D_d^2 \Delta V_{dd} = \frac{3}{2} \alpha_d \frac{\Delta V_{dd}}{D_d}. \tag{6.24}$$

The time elapsed between two collisions is approximately

$$\Delta \tau_{d,col} = f_{d,col}^{-1} = \frac{2}{3}(D_d/\Delta V_{dd})/\alpha_d \,. \tag{6.25}$$

For *Maxwellian* distribution of the fluctuation velocities the *averaged* fluctuation velocity of the continuum is

$$\overline{V_c'} = \left(\frac{16}{3\pi}\right)^{1/2} \sqrt{k_c} \,, \tag{6.26}$$

Zeichik (1998). Using the relaxation coefficient, the fluctuation component of the particles can be computed and used as ΔV_{dd}. For very small particles that completely follow the pulsation of the carrier phase *Staffman* and *Turner* (1956) found

$$\Delta V_{dd} = \frac{8}{3}\sqrt{\frac{2}{\pi}} D_d \sqrt{\frac{\varepsilon_c}{v_c}} \,. \tag{6.27}$$

For particles that can not follow the turbulent pulsation of the carrier phase *Somerfeld* and *Zivkovic* (1992) reported

$$\Delta V_{dd} = \frac{16}{\sqrt{\pi}}\sqrt{\frac{2}{3}k_d} \,. \tag{6.28}$$

Zeichik proposed in 1998 general solution to this problem which is valid for all sizes of the particles.

Nomenclature

Latin

b	$:= \|w_c - w_d\|/w_c^*$ drift parameter, dimensionless
C_{dc}^{t*}	$:= V_d'/V_c'$, response coefficient for single particle, dimensionless
c_{cd}^{vm}	coefficient for the virtual mass force or added mass force acting on a dispersed particle, dimensionless
D_d	diameter of dispersed particle, m
D_{d1}	size of the dispersed particles belonging to group 1, m
D_{d2}	size of the dispersed particles belonging to group 2, m

$f_{d,col}$	collision frequency of single particle, 1/s
k	specific turbulent kinetic energy, m²/s²
ℓ_e	size of the large eddy, m
$\ell_{\mu e,l}$	lowest spatial scale for existence of eddies in field l called inner scale or small scale or Taylor micro-scale (μ) of turbulence, m
$\Delta \ell$	characteristic size of the geometry, m
n_d	number of dispersed particles per unit mixture volume, 1/m³
n_{d1}	number of dispersed particles per unit mixture volume belonging to group 1 with common averaged size, 1/m³
n_{d2}	number of dispersed particles per unit mixture volume belonging to group 2 with common averaged size, 1/m³
$\dot{n}_{d,col}$	number of collisions per unit time of dispersed particles, 1/s
Re_{cd}	Reynolds number based on relative velocity, continuum properties and size of the dispersed phase, dimensionless
St	$:= \Delta \tau_{cd} / \Delta \tau_{L,c}$ is the *Stokes* number, dimensionless
ΔV_{12}	difference between gas and liquid velocity, m/s
ΔV_{dd}	difference between the velocity of two neighboring droplets, m/s
ΔV_{d1d2}	difference between the velocity of two neighboring droplets with sizes belonging to two different groups, m/s
$\Delta \mathbf{V}_{ml}$	difference between m- and l-velocity vectors, m/s
V'	fluctuation velocity component, m/s
Vol	control volume, m³
v	velocity component in angular direction, m/s
w	axial velocity, m/s
w^*	friction velocity, dimensionless
$\overline{w}$	cross section averaged friction velocity, m/s
w^+	axial velocity, dimensionless
w_c	continuum axial velocity, m/s
$\overline{w}_c$	averaged axial continuum velocity, m/s
w_c^*	continuum axial friction velocity, m/s

Greek

α	volumetric fraction, dimensionless
ε	power dissipated irreversibly due to *turbulent pulsations* in the viscous fluid per unit mass of the fluid (dissipation of the specific turbulent kinetic energy), m²/s³

$\Delta\tau$ time interval, s
$\Delta\tau_{12}$ droplet relaxation time, s
$\Delta\tau_{13}$ particle relaxation time, s
$\Delta\tau_{cd}$ The characteristic time for the response of a single particle, particle relaxation time, s
$\Delta\tau_{d,col}$ time elapsed between two subsequent collisions, s
$\Delta\tau_k$ time constant for the decay of the turbulent kinetic energy, s
$\Delta\tau_e$ time scale of fluctuation of large eddy, s
$\Delta\tau_{ed,\mathrm{int}}$ time required by a particle with relative to the continuum velocity $|\Delta \mathbf{V}_{cd}|$ to cross an eddy with size $\ell_{e,l}$, s
$\Delta\tau_{\mu e,l}$ time scale corresponding to the spatially lowest scale for existence of eddies in field *l* called *inner time scale* or *small time scale* or *Taylor* microscale of turbulence, s
$\Delta\tau_{L,c} := 0.04\, D_h / w_c^*$, time *Lagrangian* scale of turbulence averaged over the cross section, s
$\Delta\tau_{L,cd}$ *Lagrangian* eddy–droplet interaction time, s
$\Delta\tau_{L,cd}(St = 0)$ interaction time for very small particles, s
$\Delta\tau_{L,cd}(St \to \infty)$ interaction time for very large particles, s

Subscripts

c continuous
d dispersed

References

Gosman AD and Ioannides E (Jan. 12-15, 1981) Aspects of computer simulation of liquid-fuelled combustion, AIAA 19th, Aerospace Scientific Meeting, St. Louis, Missouri

Hanratty TJ and Dykhno LA (1977) Physical issues in analyzing gas-liquid annular flows, Experimental Heat Transfer, Fluid Mechanics and Thermodynamics, Edizioni ETS, Eds. Giot M, Mayinger F and Celata GP

Lee MM, Hanratty TJ and Adrian RJ (1989b) An axial viewing photographic technique to study turbulent characteristics of particles, Int. J. Multiphase Flow, vol 15 pp 787-802

Lo S (May 23-27, 2005) Modelling multiphase flow with an Eulerian approach, VKI Lecture Series – Industrial Two-Phase Flow CFD, von Karman Institute

Rusche H (2000) Modeling of interfacial forces at high phase fraction, I. Lift, Project Report III-4, Brite/EuRam BE-4322

Reeks MW (1977) On the dispersion of small particles suspended in an isotropic turbulent field, Int. J. Multiphase Flow, vol 3 pp 319

Reeks MW (1991) On a kinetic equation for transport of particles in turbulent flows, Phys. Fluids A3, pp 446-456

Reeks MW (1992) On the continuum equation for dispersed particles in non-uniform. Phys. Fluids A4, pp 1290-1303

Simonin O (1991) Second-moment prediction of dispersed phase turbulence in particle-laden flows. Proc. 8-th Symposium on Turbulent Shear Flows, pp 741-746

Smoluchowski M (1918) Versuch einer mathematischen Theorie der Koagulationskinetik kolloider Lösungen, Zeitschrift für Physikalische Chemie, Leipzig, Band XCII pp 129-168

Sommerfeld M (1992) Modeling of particle-wall collisions in confined gas-particle flows. Int. J. Multiphase Flow, vol 26 pp 905-926

Somerfeld M and Zivkovic G (1992) Recent advances in numerical simulation of pneumatic conveying through pipe systems, Computational Methods in Applied Sciences, p 201

Staffman PG and Turner JS (1956) J. Fluid Mech., vol 1 p 16

Wolkov EP, Zeichik LI and Pershukov VA (1994) Modelirovanie gorenia twerdogo topliva, Nauka, Moscow, in Russian.

Zaichik LI (1998) Estimation of time between particle collisions in turbulent flow, High Temperature, vol 36 no 3 (translation from Russian)

Zaichik LI and Alipchenkov VM (1999) A statistical model for transport and deposition of high inertia colliding particles in isotropic turbulence, Int. J. Heat Fluid Flow, vol 22 pp 365-371

7 Two group *k-eps* models

7.1 Single phase flow

Hanjalic, Launder and *Schiestel* (1976) proposed to divide the turbulence structures conditionally into two groups. The first one describing the large scale motion and the second describing the transition scale motion leading to dissipation. Using the hypothesis of equilibrium between the transition scale motion and the small scale motion being dissipated the authors derived the following formalism.

$$\frac{\partial}{\partial \tau}(\rho k_p) + \nabla \cdot \left[\rho \left(\mathbf{V} k_p - v_p^k \nabla k_p\right)\right] = \rho \left(\overline{v' P_k} - \varepsilon_p\right), \qquad (7.1)$$

$$\frac{\partial}{\partial \tau}(\rho k_T) + \nabla \cdot \left(\rho \mathbf{V} k_T - v_T^k \nabla k_T\right) = \rho \left(\varepsilon_p - \varepsilon_T\right), \qquad (7.2)$$

$$\frac{\partial}{\partial \tau}(\rho \varepsilon_p) + \nabla \cdot \left[\rho \left(\mathbf{V} \varepsilon_p - v_p^\varepsilon \nabla \varepsilon_p\right)\right] = \rho \left[\frac{\varepsilon_p}{\kappa}\left(c_{\varepsilon p1} \overline{v' P_k} - c_{\varepsilon p2} \varepsilon_p\right)\right], \qquad (7.3)$$

$$\frac{\partial}{\partial \tau}(\rho \varepsilon_T) + \nabla \cdot \left[\rho \left(\mathbf{V} \varepsilon_T - v_T^\varepsilon \nabla \varepsilon_T\right)\right] = \rho \left[\frac{\varepsilon_T}{\kappa}\left(c_{\varepsilon T1} \varepsilon_p - c_{\varepsilon T2} \varepsilon_T\right)\right], \qquad (7.4)$$

With the empirical constants

$$c_{\varepsilon p1} = 2.2, \qquad (7.5)$$

$$c_{\varepsilon p2} = 1.8 - 0.3\left(\frac{k_p}{k_T} - 1\right) \bigg/ \left(\frac{k_p}{k_T} + 1\right), \qquad (7.6)$$

$$c_{\varepsilon T1} = 1.08 \varepsilon_p / \varepsilon_T, \qquad (7.7)$$

$$c_{\varepsilon T2} = 1.15. \qquad (7.8)$$

In this formalism the deformation of the mean velocity field $\overline{\rho v' P_k}$ is the only source for large scale fluctuations. The dissipation of the large scale motion ε_p is simultaneously a source for the transition scale motion. ε_T is the irreversible friction dissipation of the transition eddies, which has to appear in the energy conservation equation.

7.2 Two-phase flow

Following the idea of *Hanjalic* et al. (1976), *Lopez de Bertodano* et al. (1994) proposed a two group model for bubbly flow in which the source, the extra k- equation for small scale eddies, was written as relaxation term

$$\gamma_v \alpha_c \rho_c G_{k,c} \approx \gamma_v \alpha_d \rho_c \frac{1}{\Delta \tau_{cd}} \left(k_{c,d\infty} - k_{c,d} \right), \qquad (7.9)$$

where

$$k_{c,d\infty} = \frac{1}{2} c_c^{vm} \left| \Delta \mathbf{V}_{cd} \right|^2 \qquad (7.10)$$

is the turbulence kinetic energy association with the fluctuation of the so-called added mass of the continuum. For the relaxation time constant the expression

$$\Delta \tau_{cd} = D_d / \left| \Delta \mathbf{V}_{cd} \right|. \qquad (7.11)$$

was used. *Haynes* et al. (2006) continued working on such formalism for two-phase bubbly flow. Following these ideas and extending them to porous body the two group equation for each field describing the turbulence will be

$$\frac{\partial}{\partial \tau}\left(\alpha_l \rho_l k_{pl}\gamma_v\right) + \nabla \cdot \left[\alpha_l \rho_l \left(\mathbf{V}_l k_{pl} - v_{pl}^k \nabla k_{pl}\right)\gamma\right]$$

$$= \alpha_l \rho_l \gamma_v \left(\overline{v_l' P_{k,l}} - \varepsilon_{pl} + G_{k,l} + P_{k\mu,l} + P_{kw,l}\right), \qquad (7.14)$$

$$\frac{\partial}{\partial \tau}\left(\alpha_l \rho_l k_{Tl}\gamma_v\right) + \nabla \cdot \left[\alpha_l \rho_l \left(\mathbf{V}_l k_{Tl} - v_{Tl}^k \nabla k_{Tl}\right)\gamma\right] = \alpha_l \rho_l \gamma_v \left(\varepsilon_{pl} - \varepsilon_{Tl} + G_{k\omega,l}\right), \qquad (7.15)$$

$$\frac{\partial}{\partial t}\left(\alpha_l \rho_l \varepsilon_{pl} \gamma_v\right) + \nabla \cdot \left[\alpha_l \rho_l \left(\mathbf{V}_l \varepsilon_{pl} - v_{pl}^{\varepsilon} \nabla \varepsilon_{pl}\right) \gamma\right]$$

$$= \alpha_l \rho_l \gamma_v \left[\frac{\varepsilon_{pl}}{\kappa_{pl}} \left(c_{\varepsilon p1} v_l' \overline{P_{k,l}} - c_{\varepsilon p2} \varepsilon_{pl} + c_{\varepsilon p3} G_{k,l}\right) + P_{\varepsilon w,l}\right], \quad (7.16)$$

$$\frac{\partial}{\partial t}\left(\alpha_l \rho_l \varepsilon_{Tl} \gamma_v\right) + \nabla \cdot \left[\alpha_l \rho_l \left(\mathbf{V}_l \varepsilon_{Tl} - v_{Tl}^{\varepsilon} \nabla \varepsilon_{Tl}\right) \gamma\right]$$

$$= \alpha_l \rho_l \gamma_v \frac{\varepsilon_{Tl}}{\kappa_{Tl}} \left(c_{\varepsilon T1} \varepsilon_{pl} - c_{\varepsilon T2} \varepsilon_{Tl} + c_{\varepsilon T3} G_{k\omega,l}\right). \quad (7.17)$$

A new source term $G_{k\omega,l}$ is introduced here to model the fine eddy generation in the bubble wake. It is set by *Haynes* et al. (2006) to be function of the type

$$G_{k\omega,l} = f\left(\sqrt{k_{Tl}}, \text{ inter bubble mean distance}\right), \quad (7.18)$$

but not explicitly provided by the authors. The modeling constants $c_{\varepsilon p3}$ and $c_{\varepsilon T3}$ are also not reported. Using

$$v_l' \approx c_\eta \frac{k_{pl}\left(k_{pl} + k_{Tl}\right)}{\varepsilon_{pl}}, \quad (7.19)$$

Haynes et al. (2006), the demonstrated improvement regarding void profiles in bubbly flow is considered to come from this type of splitting. In fact other authors also obtained appropriate void profiles in simulation of similar experiments without splitting. Note that such models are at the very beginning of their development for bubbly flows.

Nomenclature

Latin

c_{cd}^{vm} coefficient for the virtual mass force or added mass force acting on a dispersed particle, dimensionless

$c_{\varepsilon p1}, c_{\varepsilon p2}, c_{\varepsilon T1}, c_{\varepsilon T2}$ modeling constants for single phase flow k-eps model, dimensionless

k_p in sense of two group theory: specific turbulent kinetic energy of the large scale motion group, m²/s²

k_T in sense of two group theory: specific turbulent kinetic energy of the transition scale motion group leading to dissipation, m²/s²

$k_{c,d\infty}$ turbulence kinetic energy association with the fluctuation of the so-called added mass of the continuum, m²/s²

P_k production of the turbulent kinetic energy per unit mass, W/kg

$\overline{P_{k,l}}$ production of the turbulent kinetic energy per unit mass of the velocity field l due to deformation of the velocity field l, W/kg

P_{kw} irreversibly dissipated power per unit flow mass outside the viscous fluid due to turbulent pulsations equal to production of turbulent kinetic energy per unit mass of the flow, W/kg (m²/s³)

$P_{kw,l}$ production of turbulent kinetic energy per unit mass of the field l due to friction with the wall, W/kg

$P_{k\mu,l}$ production of turbulent kinetic energy per unit mass of the field l due to friction evaporation or condensation, W/kg

P_ε production of the dissipation of the turbulent kinetic energy per unit mass, W/kg

$P_{\varepsilon w}$ production of the dissipation of the turbulent kinetic energy per unit mass due to friction with the wall, W/kg

$P_{\varepsilon w,l}$ production of the dissipation of the turbulent kinetic energy per unit mass of the field l due to friction with the wall, W/kg

V velocity vector, m/s

Greek

α volumetric fraction, dimensionless

γ surface permeability defined as flow cross section divided by the cross section of the control volume (usually the three main directional components are used), dimensionless

γ_v volumetric porosity defined as the flow volume divided by the considered control volume, dimensionless

ε_p in sense of two group theory: dissipation of the large scale motion group, m²/s²

ε_T in sense of two group theory: irreversible friction dissipation of the transition eddies, m²/s²

v_l^k total cinematic diffusivity of the turbulent kinetic energy, m²/s

v_l^ε total cinematic diffusivity of the dissipation of the turbulent kinetic energy, m²/s

η dynamic viscosity, kg/(ms)
ρ density, kg/m³
τ time, s
$\Delta\tau_{cd}$ relaxation time constant, s

References

Hanjalic K, Launder BE and Schiestel R (1976) Multiple time scale concepts in turbulent transport modelling, Second symposium on turbulence shear flows, pp 36-49

Haynes P-A, Péturaud P, Montout M and Hervieu E (July 17-20, 2006) Strategy for the development of a DNB local predictive approach based on NEPTUNE CFD software, Proceedings of ICONE14, International Conference on Nuclear Engineering, Miami, Florida, USA , ICONE14-89678

Lopez de Bertodano M, Lahey RT Jr. and Jones OC (1994) Phase distribution of bubbly two-phase flow in vertical ducts, Int. J. Multiphase Flow, vol 20 no 5 pp 805-818

8 Set of benchmarks for verification of *k-eps* models in system computer codes

The emphasis in this chapter is on averaged turbulence modeling in rod-bundles on an intermediate scale that is finer than the sub channel scale but much larger than the scale required for direct numerical simulation. Thirteen benchmarks based on analytical solutions and experimental data are presented and compared with the prediction of the IVA computer code. For the first time distribution of the averaged turbulence structure in boiling bundles is presented. It is demonstrated that k-eps models with two-way coupling possess substantial potential for increasing the accuracy of the description of multi-phase flow problems in bundles especially the effect of the space grids.

8.1 Introduction

We have already concluded that *Euler*-methods with $k-\varepsilon$ models for the continuums and two-way coupling give the most promising framework and have introduced in the IVA three-fluid multi-component model, a set of $k-\varepsilon$ equations for each velocity field, *Kolev 2000*. Later the model was extended to multi-block model in boundary fitted coordinates, *Kolev 2003-2005*. So the following equations are solved as a part of IVA-solution algorithm:

$$\frac{\partial}{\partial \tau}(\alpha_l \rho_l k_l \gamma_v) + \nabla \cdot \left[\alpha_l \rho_l \left(\mathbf{V}_l k_l - v_l^k \nabla k_l \right) \gamma \right]$$

$$= \alpha_l \rho_l \gamma_v \left(\overline{v_l' P_{kl}} - \varepsilon_l + G_{k,l} + P_{k\mu,l} + P_{kw,l} \right), \tag{8.1}$$

$$\frac{\partial}{\partial \tau}(\alpha_l \rho_l \varepsilon_l \gamma_v) + \nabla \cdot \left[\alpha_l \rho_l \left(\mathbf{V}_l \varepsilon_l - v_l^\varepsilon \nabla \varepsilon_l \right) \gamma \right]$$

$$= \alpha_l \rho_l \gamma_v \left[\frac{\varepsilon_l}{\kappa_l} \left(c_{\varepsilon 1} \overline{v_l' P_{kl}} - c_{\varepsilon 2} \varepsilon_l + c_{\varepsilon 3} G_{kl} \right) + P_{\varepsilon w, l} \right], \tag{8.2}$$

where $v_l^k = v_l + \frac{v_l^t}{\text{Pr}_{kl}^t}$, $v_l^\varepsilon = v_l + \frac{v_l^t}{\text{Pr}_{\varepsilon l}^t}$. By computing the source terms we distinguish between sources for pool flow or sources for flow in porous structures. Either one or the other set is implied for each computational cell. We are going to check the performance of the model starting from simple cases and increasing the complexity.

Before verifying *k-eps* models for multi-phase flow, benchmarks for single phase flows have to be reproduced. In what follows we give eleven examples.

8.2 Single phase cases

Problem 1: 2D-steady state developed single phase incompressible flow in a circular pipe.

In this case having in mind that the production of the turbulent kinetic energy is simply $P_k = v' \frac{4}{3} \left(\frac{\partial \overline{w}}{\partial r} \right)^2$ the system of non linear ordinary differential equations simplifies to

$$\frac{\partial \overline{w}}{\partial \tau} = \frac{1}{\rho} \left| \frac{\partial p}{\partial x} \right| + \frac{1}{r} \frac{\partial}{\partial r} \left[(v + v') r \frac{\partial \overline{w}}{\partial r} \right], \tag{8.3}$$

$$\frac{\partial k}{\partial \tau} = P_k - \varepsilon + \frac{1}{r} \frac{\partial}{\partial r} \left[(v + v') r \frac{\partial k}{\partial r} \right], \tag{8.4}$$

$$\frac{\partial \varepsilon}{\partial \tau} = \frac{\varepsilon}{k} (c_{\varepsilon 1} P_k - c_{\varepsilon 2} \varepsilon) + \frac{1}{r} \frac{\partial}{\partial r} \left[(v + v') r \frac{\partial \varepsilon}{\partial r} \right], \tag{8.5}$$

with $v' = c_\eta k^2 / \varepsilon$. We keep the time derivatives because solving the transient problem at constant boundary conditions results naturally in a stable relaxation method. Knowing the wall share stress from macroscopic correlation the structure of k and ε can be reconstructed by solving the above equations.

Practical relevance: This case is useful to check the performance of codes describing more complex geometry or to compute the length scale of the turbulent eddies. In dispersed flow with fine droplets or in bubble flow this length scale is an indication of the maximum possible size of the dispersed particles. This approach is useful for estimating the moisture droplet size in industrial steam flows (if the prehistory is not delivering small sizes) in order to prescribe the requirements for separators.

8.2 Single phase cases

Problem 2: The decay constant for single phase flow.

Bertodano et al. (1994) considered a single phase homogeneous turbulence for incompressible flow and found some useful relations given below. In this case the $k-\varepsilon$ equations reduces to $\dfrac{Dk}{D\tau} = -\varepsilon$, $\dfrac{D\varepsilon}{D\tau} = -c_{\varepsilon 2}\dfrac{\varepsilon}{k}\varepsilon$ with analytical solution over the time $\Delta\tau$

$$\frac{k}{k_0} = \left[\frac{1}{1+(c_{\varepsilon 2}-1)\dfrac{\Delta\tau}{k_0/\varepsilon_0}}\right]^{\frac{1}{c_{\varepsilon 2}-1}} \approx \frac{1}{1+\dfrac{\tau}{k_0/\varepsilon_0}} \approx e^{\frac{\Delta\tau}{k_0/\varepsilon_0}}, \qquad (8.6)$$

which illustrate that the turbulence decays with a time constant

$$\boxed{\Delta\tau_k = k_0/\varepsilon_0 \,.} \qquad (8.7)$$

Conclusion: Comparing with

$$\Delta\tau_{e,l} = \ell_{e,l}/V' = \sqrt{3c_\eta/2}\, k_l/\varepsilon_l = 0.37 k_l/\varepsilon_l$$

we realize that the time scale of the fluctuations of a large eddies is about 1/3 of the turbulence decay time constant. Eq. (8.6) is valid along the characteristic line defined by the continuum velocity.

Practical relevance: The *Bertodano's* et al. Eq. (8.6) is of fundamental importance. It removes the arbitrariness applied by some authors for describing the decay after singularities by purely empirical constants. Is simply says that one has to know k_0 and ε_0 behind the singularity and the decay constant is then uniquely defined. Some examples of works with avoidable empiricism are given here: *Knabe* and *Wehle* (1995) postulated empirical decay coefficients that are generated by comparison with dry out data. Therefore their model incorporating this element is always associated with a specific experimental geometry and boundary conditions. *Nagayoshi* and *Nishida* (1998) performed measurements on a typical sub-channel for BWR-rod-bundles without and with ferrule spacers with 0.5, 1 and 1.5mm. They reported increase of the lateral velocity fluctuation depending on the blockage ratio $1-\gamma_z$, with subsequent decay

$$\frac{|u'|}{|u'_0|} = 1 + 6.5\left(1-\gamma_{z,spacer}\right)^2 e^{-2.7\left[\left(z-z_{grid\_end}\right)/D_h\right]} \qquad (8.8)$$

for *Reynolds* numbers 0.5 to 1.2×10^5. The data indicated that the turbulence is reaching the state after 10 hydraulic diameters. The above empirical relation can be replaced by analytical one.

Problem 3: 1D-Decay of turbulence in a pipe flow

Consider 1D-single phase homogeneous turbulence for incompressible flow with constant velocity. In this case the $k - \varepsilon$ equations reduces to

$$w \frac{dk}{dz} = P_k - \varepsilon, \tag{8.9}$$

$$w \frac{d\varepsilon}{dz} = P_\varepsilon - c_{\varepsilon 2} \frac{\varepsilon^2}{k}. \tag{8.10}$$

Using the first order donor-cell discretization for the convective terms we obtain

$$k = k_{k-1} + (P_k - \varepsilon) \frac{\Delta z}{w}, \tag{8.11}$$

$$\varepsilon = \varepsilon_{k-1} + \left(P_\varepsilon - c_{\varepsilon 2} \frac{\varepsilon^2}{k} \right) \frac{\Delta z}{w}, \tag{8.12}$$

which is a system of nonlinear algebraic equations with respect to the unknowns. Excluding the kinetic energy the quadratic equation

$$a\varepsilon^2 + b\varepsilon + c = 0, \tag{8.13}$$

where

$$a = c_{\varepsilon 2} - 1, \tag{8.14}$$

$$b = P_k + \frac{\Delta z}{w} P_\varepsilon + \frac{w}{\Delta z} k_{k-1} + \varepsilon_{k-1}, \tag{8.15}$$

$$c = -\left(P_\varepsilon + \frac{w}{\Delta z} \varepsilon_{k-1} \right) \left(k_{k-1} + \frac{\Delta z}{w} P_k \right), \tag{8.16}$$

is solved with respect to the dissipation. Then the kinetic energy is computed from the *k*-equation. Using the following characteristics of the flow $D_h = 0.1m$,

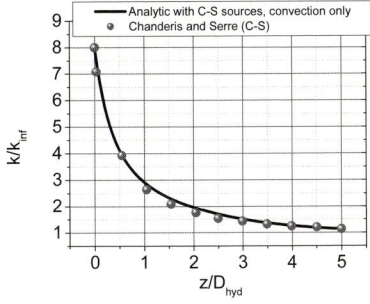 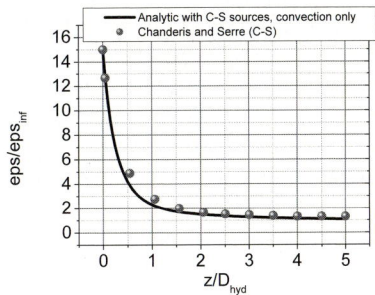

Fig. 8.1. Test problem 3: Decay of turbulence in a pipe liquid flow. Comparison between the 1D-semi analytical solution with a DNS simulation reported by *Chandesris* et al. (2005). a) Turbulent kinetic energy as a function of the axial coordinate; b) Dissipation of the turbulent kinetic energy as a function of the axial coordinate

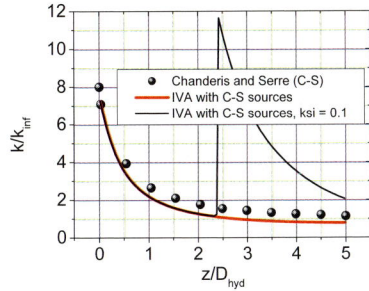 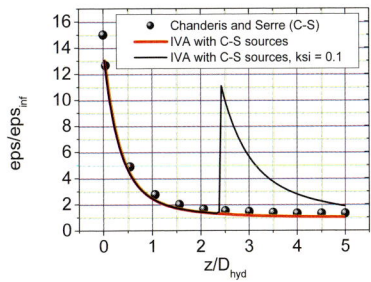

Fig. 8.2. Test problem 4, 5: Decay of turbulence in a pipe liquid flow. Comparison of the IVA computer code solution, *Kolev* (2007a, b), with a DNS simulation reported by *Chandesris* et al. (2005). a) Turbulent kinetic energy as a function of the axial coordinate; b) Dissipation of the turbulent kinetic energy as a function of the axial coordinate. The curves with the jump are obtained by introducing a hydraulic resistance at the middle of the channel

$\bar{w} = 1 m/s$, $Re = 10^5$, $y^+_{lim} = 8$, $c_k = 3.06$, $c_{\varepsilon 2} = 1.92$, $\lambda_{fr} = 1.78 \times 10^{-2}$, $\ell_e = 0.1606 m$, $k_\infty = 4.4945 \times 10^{-3} \ m^2/s^2$, $P_{kw} = 5.5455 \times 10^{-2} \ m^2/s^3$, $\varepsilon_\infty = P_{kw} = 5.5455 \times 10^{-2} \ m^2/s^3$, $P_{\varepsilon w} = 1.3137 \ m^2/s^4$, $k_0 = 8 k_\infty = 3.5956 \times 10^{-2} \ m^2/s^2$, $\varepsilon_0 = 15 \varepsilon_\infty = 0.8318 \ m^2/s^3$, the integration of the system over 5 diameters (100 computational cells) gives the results presented in Fig. 8.1. Figure 8.1 contains direct numerical simulation results reported by *Chandesris* et al. (2005) for these conditions. Obviously, in this case convection is predominant and the diffusion is not important.

Problem 4: Introduce the sources as derived by Chandesris et al. (2005) in IVA computer code and repeat the computation to problem 1.

Now we repeat the same computation using the computer code IVA having complete *k-eps* models. The boundary conditions are the same.

Practical relevance: From the comparison with the previous computation we learn that the resolution of the non-linearities of the source terms in a single time step in the numerical solution method increases the accuracy. This is not done in the IVA computation.

Problem 5: Introduce the sources for distributed hydraulic resistance coefficients in IVA computer code and repeat the computation to problem 4 with and without singularity.

After introducing the sources for the *k-eps* equation for each particular computational cell we compute the test case with and without $\zeta_{fr}\left(2.5D_{hyd}\right)=0.1$ assuming zero viscous dissipation, $\eta_{vis}=0$. The results are presented in Fig. 8.2. We see that the turbulent kinetic energy jumps after the singularity. Then the decay follows the local maximum. Similar is the behavior of the dissipation.

Problem 6: Repeat with IVA the computation to problem 2 without singularity using instead of water gas flow at the same Reynolds number with $\bar{w}=25.56 m/s$. Compare the solution with the analytical solution for convection only.

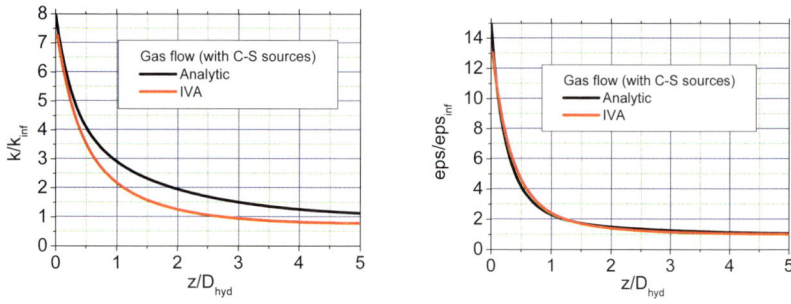

Fig. 8.3. Test problem 4: Decay of turbulence in a pipe gas flow. Comparison of the IVA computer code solution, *Kolev* (2007a, b), with a analytical solution. a) Turbulent kinetic energy as a function of the axial coordinate; b) Dissipation of the turbulent kinetic energy as a function of the axial coordinate

The result of the computation is presented in Fig. 8.3. We see very good comparison for the dissipation of the turbulent kinetic energy. The turbulent kinetic energy is predicted slightly slower. One should keep in mind that the analytical example is solved for incompressible flow. The real compressibility taken into account in

IVA computer code predicts increasing velocity with decreasing pressure. This is the reason for the differences of the kinetic energies.

Problem 7. Given the bundles in Fig. 8.4 with a grid with irreversible pressure loss coefficient $\zeta_{fr} = 0.96$, compute the axial distribution of the averaged turbulent kinetic energy of the liquid and compare it with the measurements reported in Serre and Bestion (2005).

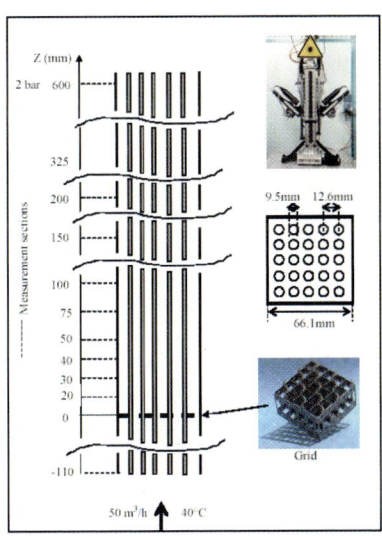

Fig. 8.4. Description of the Agate experiment, taken from *Serre* and *Bestion* (2005)

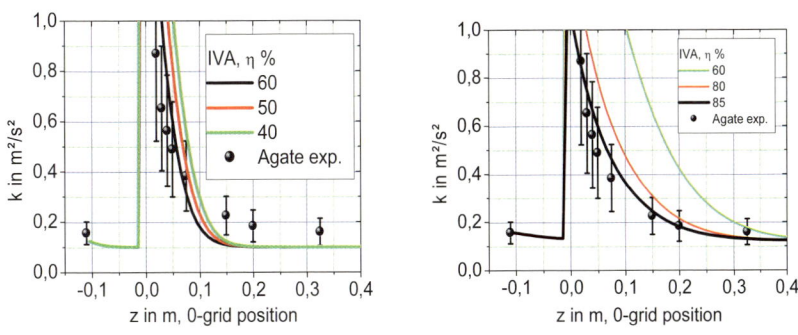

Fig. 8.5. Turbulent kinetic energy as a function of the axial coordinate: a) $y_{lim}^+ = 8$, $c_k = 0.0306$; b) $y_{lim}^+ = 16$, $c_k = 0.0368$

Figure 8.5 contains the obtained computational results for two different sets of the modeling constants as proposed by *Chandesris* et al. (2005) for channels and rod bundles. We vary the percentage of the turbulent energy that is directly dissipated after its generation at the singularity, η_{vis}. We realize that (a) using the modeling constants for rod bundles predicts the experimentally observed decay better, and (b), probably only about 15% of the loosed mechanical energy in the grid region goes for generation of the turbulence.

Practical relevance: Heat transfer in rod bundles in variety of engineering facilities is dependent also on the local singularities that produce turbulence. Therefore understanding these processes allows optimization of technical facilities in this field.

Problem 8. Given a channel with cross section, as given in Fig. 8.6, 2.175m from the entrance a 75mm-high grid with thickness 0.5, 1 and 1.5mm and form given in Fig. 8.6 is mounted. Compute the axial distribution of the averaged turbulent kinetic energy of the liquid considering the generation due to increased hydraulic diameter and compare it with the measurements reported in Nagayoshi and Nishida (1998).

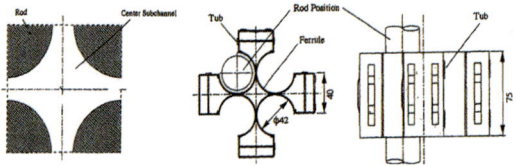

Fig. 8.6. Experimental sub-channel, grid design

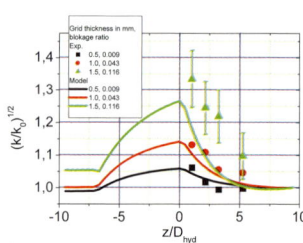

Fig. 8.7. The cross section averaged fluctuation velocity divided by the background velocity fluctuation as a function of the axial coordinate: $y_{\lim}^+ = 16$, $c_k = 0.0368$

We perform the computation taking into account only the change of the hydraulic diameter in the region of the grid without any additional irreversible friction coefficient. The results are presented in Fig. 8.7. Obviously for small blockage ratios < 0.009 the additional turbulence generation is due to the increase of the friction surfaces. For blockage ratios larger than 0.04 the additional weak formation after the blockage have to be taken into account. The decay characteristics are properly predicted.

Problem 9. Given a channel with cross section as given in Fig. 8.8, 0.6m from the entrance, a 40mm-high grid with unknown thickness having mixing vanes and form given in Fig. 8.8 is mounted. The bundle is characterized by 68mm square housing, D=9.5mm rod diameters, W/D=1.4263, P/D=1.326, resulting in hydraulic diameter of 11.21mm. Water at atmospheric pressure and 25°C is pumped into the entrance with 5m/s corresponding to a Reynolds number based on the hydraulic diameter of 62500. At this Reynolds number the measured irreversible resistance coefficient was around 1. Compute the axial distribution of the averaged turbulent kinetic energy of the liquid considering the generation due to increased hydraulic diameter and compare it with the measurements reported in Yang and Chung (1998).

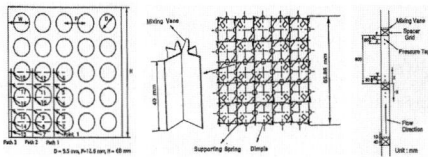

Fig. 8.8. 5x5 rod bundle with vane spacer grids experimental geometry by *Yang* and *Chung* (1998)

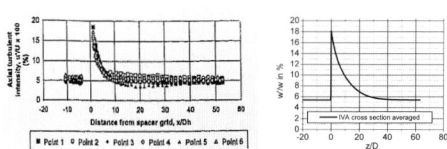

Fig. 8.9. Relative axial fluctuation velocity as a function of the axial coordinate: a) measured at different points; b) computed by IVA – cross section averaged, homogeneous turbulence. Constitutive constants: *Chandesris* et al. (2005) for rod bundles $y^+_{\lim} = 16$, $c_k = 0.0368$ and $\eta_{vis} = 0.85$

We use a discretization size of 1mm for this computation and impose the irreversible resistance coefficient at the exit of the spacer grid. From the computed turbulent kinetic energy the fluctuation velocity is computed assuming homogeneous turbulence. Figure 8.9 a) gives the measured relative axial component of the turbulent pulsations along a path 1 from Fig. 8.8. Figure 8.9 b) gives the computed relative component of the turbulent pulsations. Having in mind that the real turbulent structure is probably heterogeneous and not evenly distributed across the bundle we conclude that the computed averaged structure along the axial coordinate represents well the reality. Figure 8.10 gives the magnitude of the kinetic energy of the turbulence and its dissipation.

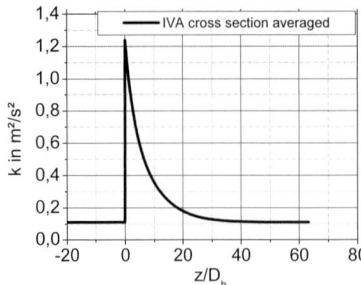 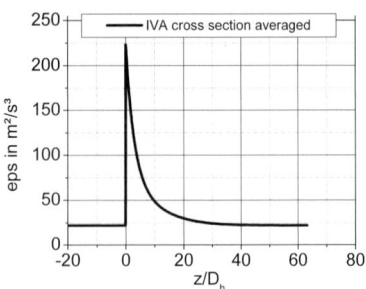

Fig. 8.10. Cross section averaged turbulent properties as a function of the axial distance: a) specific turbulent kinetic energy; b) specific dissipation of the turbulent kinetic energy

Conclusion: Knowing the irreversible resistance coefficient and imposing it as a singularity at the exit of the grid within a k-eps framework provide the appropriate description of the cross section averaged axial structure of turbulence in bundles. Comparing both experiments those by *Serre* and *Bestion* (2005) and those by *Yang* and *Chung* (1998) we realize that only about 15% of the loosed mechanical energy in the grid region goes for generation of the turbulence and that the modeling constants $y^+_{\lim} = 16$, $c_k = 0.0368$ work properly.

Problem 10: Given a 5x5 rod PWR bundle with length of 3m and 5 FOCUS spacer grids with 40mm high, consider a flow of 601.45K-water at 165.5bar with inlet velocity of 4.573m/s. Compute the steady state turbulent kinetic energy and its dissipation. Use them as initial conditions and compute the axial distribution of the turbulent characteristics through and after the spacer grid.

First we compute the distribution along the bundle without grid and find out that the steady state turbulent kinetic energy and its dissipation are $k_\infty = 0.074716 \, m^2/s^2$, $\varepsilon_\infty = 20.985 \, m^2/s^3$, respectively. Then we use them as initial conditions and compute the flow over a spacer grid and behind by using two geometrical models: (a) with changes of the cross section and of the hydraulic diameter over the spacer only, without irreversible friction coefficient; (b) no changes of the cross section and of the hydraulic diameter over the spacer with irreversible friction coefficient posed in the cell after the grid exit. The results are presented in Figs. 8.11a, b and c.

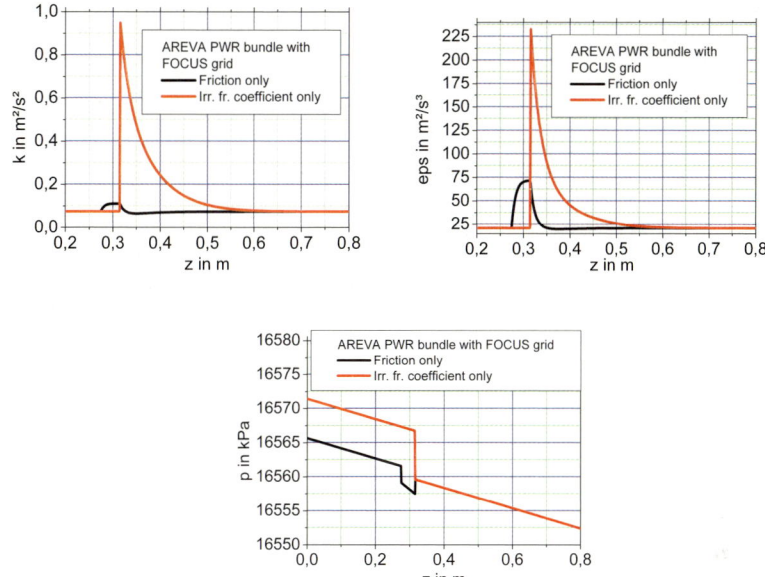

Fig. 8.11. Axial distribution of the turbulent characteristics over a FOCUS grid: a) Turbulent kinetic energy; b) Dissipation of the turbulent kinetic energy; c) Pressure drop

First, we realize that the contribution of the friction inside the grid to the generation of the turbulence is small compared to the blockage and swirling effects. Therefore from the point of view of turbulence generation the second approach can be used in large scale simulations. But one should not forget that the reduction of the cross section leads to increase in the velocity resulting in additional droplet fragmentation which has some influence on the deposition.

Second, in non boiling flows there is no accumulative effect because the next grid is far enough. Only in case of boiling flow the effect can be accumulative as those demonstrated in Fig. 8.14. The stronger the mixture expansion is, the stronger this effect.

Third, the microscopic swirling effect is not allowed to be lumped as a source with the turbulence generation because it will decay within 20-30 diameters behind the grid. Therefore even in a single cell, the swirling effect has to be modeled by macroscopic momentum redistribution. The multi-block approach of IVA in boundary fitted coordinates is of course the better choice for this application but the resources for this study are very limited.

Problem 11: Given a 8x8 rod BWR bundle with length of 3.708m and 7 ferrule spacer grids with 31mm high, consider a flow of 557.95K-water at 71.5bar with inlet velocity of 2.709m/s. Compute the steady state turbulent kinetic energy and

its dissipation. Use them as initial conditions and compute the axial distribution of the turbulent characteristics through and after the spacer grid.

First we compute the distribution along the bundle without grid and find out that the steady state turbulent kinetic energy and its dissipation are $k_\infty = 0.026756\, m^2/s^2$, $\varepsilon_\infty = 3.3525\, m^2/s^3$, respectively. Then we use them as initial conditions and compute the flow over a spacer grid and behind by using two geometrical models: (a) with changes of the cross section and of the hydraulic diameter over the spacer only, without irreversible friction coefficient; (b) no changes of the cross section and of the hydraulic diameter over the spacer with irreversible friction coefficient posed in the cell after the grid exit. The results are presented in Figs. 8.12a, b and c.

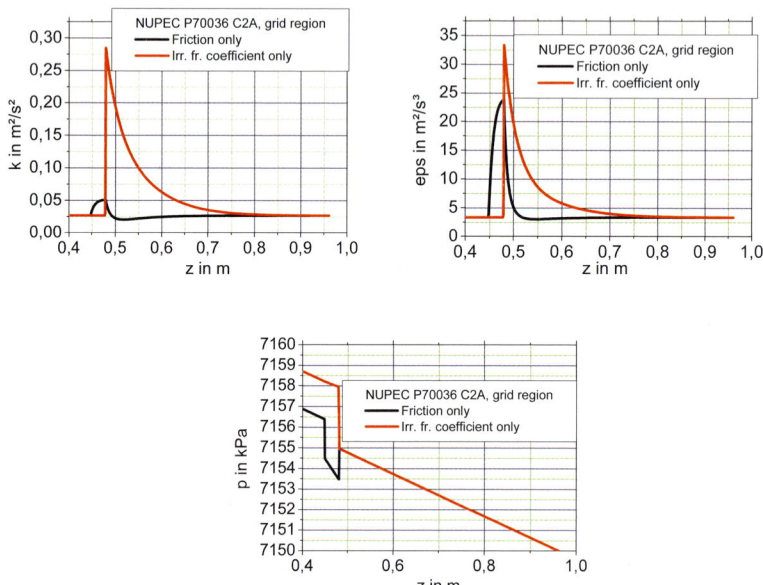

Fig. 8.12. Axial distribution of the turbulent characteristics over a FOCUS grid: a) Turbulent kinetic energy; b) Dissipation of the turbulent kinetic energy; c) Pressure drop

Comparing the turbulence structure generation for the BWR and the PWR cases we see a considerably larger turbulent kinetic energy and its dissipation in the PWR case.

What was concluded for test problem 12 is valid here also.

8.3 Two-phase cases

Problem 12: Given rod bundle for nuclear power plant with the geometry and spatial heat release in the fuel rods specified in OECD/NRC Benchmark (2004). The horizontal cross section of the bundles is illustrated in Fig. 8.11. Under these conditions the flow is boiling and the flow regimes are either liquid only or bubbly flow. Compute the parameters in the bundles including the turbulent kinetic energy and its dissipation in the continuous liquid.

The lateral discretization (18x18x24 cells) used here presented also in Fig. 8.13. The geometry data input for IVA computer code is generated using the software developed by *Roloff-Bock* (2005).

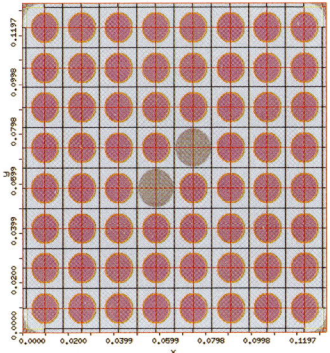

Fig. 8.13. Bundle 1-1, 1 OECD/NRC Benchmark (2004)

The results for a vertical plane crossing the bundle at the middle are presented in Fig. 8.14. The family of curves belongs to each vertical column of cells from one site to the other. We see several interesting elements of the large scale averaged turbulence of the flow:

a) The distance between the spacer grids influences turbulence level. Smaller distance increases the turbulence level. Distance larger than the compete decay distance do not increase the averaged level of turbulence.
b) The boiling in the upper half of the bundle increases also the liquid velocities and therefore the production of turbulence in the wall. In addition, the bubbles increase the production of turbulence due to their relative velocity to the liquid;
c) Comparing figures 8.14 with 8.1 we realize that in order to obtain smooth profiles as those in Fig. 8.1 the resolution in this case in the axial direction have to be substantially increased.

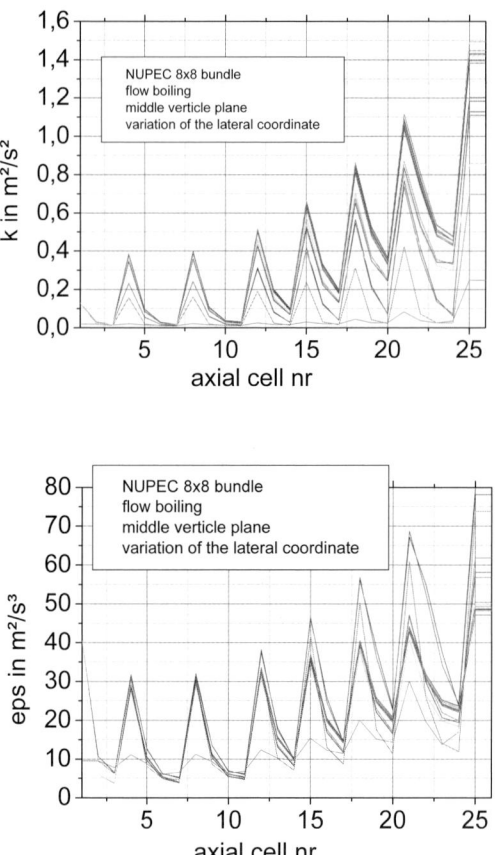

Fig. 8.14. Test problem 5: Turbulence of boiling liquid in rod bundle computed with IVA computer code, *Kolev* (2007a, b). a) Turbulent kinetic energy as a function of the axial coordinate; b) Dissipation of the turbulent kinetic energy as a function of the axial coordinate

Practical relevance: As I obtained this result in 2005 and reported it in Kolev (2006) I did not know any other boiling flow simulation in rod-bundles delivering the large scale averaged level of turbulence. Improving the capabilities in this field opens the door to better prediction of two important safety relevant phenomena in the nuclear power plant: (a) the particles (bubble or droplets) dispersion can be better predicted; (b) the deposition of droplet influencing the dry out can be better predicted; (c) if such methods for prediction of the departure from the nucleate boiling (DNB) could be available that take into account the level of the local liquid turbulence the accuracy of the DNB prediction will increase.

Problem 13: Consider the Bennet et al. (1967) test nr. 5253. Given a vertical pipe with 0.01262m inner diameter and 5.5626m length, the pipe is uniformly heated with 199kW. The inlet water flow happens from the bottom at 68.93bar and 538.90K. Compute the flow parameters inside the pipe. Show the distribution of the turbulent kinetic energy in the continuous gas. In a second computation insert a irreversible friction loss coefficient 0.1 and compare the predictions with the non disturbed flow.

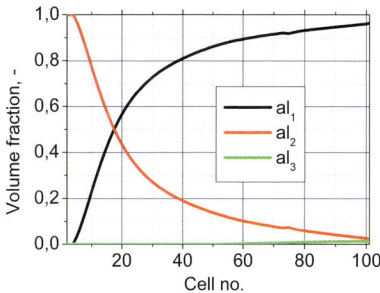
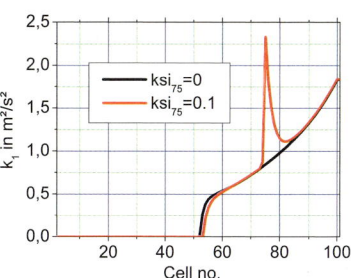

Fig. 8.15. Test problem 6: Turbulence of boiling liquid in pipe computed with IVA computer code, *Kolev* (2007a, b). a) Volumetric fraction of the steam (1) liquid (2) and droplet; b) Turbulent kinetic energy as a function of the axial coordinate with and without singularity

We use again IVA computer code, *Kolev* (2007a, b). Axially 100 equidistant cells are used. The computed local volume fractions of the three fluids are given in Fig. 8.15a. Fig. 8.15b gives the turbulent kinetic energy of the vapor in the film region where the film and the gas are continuous without and with irreversible friction coefficient 0.1 placed in cell 75 with $\eta_{vis} = 0.9$. We see that the singularity creates a jump in the turbulent kinetic energy in the vapor which then dissipates and approaches the undisturbed state. The continuous evaporation feeds turbulent kinetic energy into the vapor which explains the increasing character of the curves.

Conclusions

The emphasis in this section is on averaged turbulence modeling on an intermediate scale that is finer than the sub channel scale but much larger than the scale required for direct numerical simulation.

1. It is identified that the introducing of turbulence modeling of the boiling flow possesses substantial potential for improving our understanding of the CHF phenomena. The recent capabilities of IVA computer code are extensively checked on analytical experiments and benchmarks demonstrating adequate performance in single phase flow. Also the effect of the grids is naturally taken into account

in the prediction of the turbulence characteristics in average. For the first time application to rod bundles demonstrate the effect of the grids on the boiling flow turbulence.

2. More effort is necessary to increase the accuracy of the involved correlation, to derive complete set of source terms for generation of turbulence in all multi-phase flow pattern, to accomplish the right coupling between deposition and local degree of turbulence, and finally to derive appropriate mechanistic criterion for identification of dry out. This is important for the BWR fuel technology. I strongly recommend continuation of this line of research (e.g. to go farther than in Ch. 4.2) theoretically and experimentally.

3. Use of the modeling of the gross turbulence structure for PWR bundles can be made by developing DNB model having as a model element the bulk turbulence of the flow. Such model remains to be developed.

4. In both considered cases (BWR and PWR bundles) the more sophisticated description of the local two-phase turbulence opens the door for better prediction of bubble and droplet dispersion in bundles.

5. The microscopic swirling effect is not allowed to be lumped as a source with the turbulence generation because it will decay within 20-30 hydraulic diameters behind the grid. Therefore even in a single cell, the swirling effect has to be modeled by macroscopic momentum redistribution. The multi-block approach of IVA in boundary fitted coordinates is of course the better choice for this application but the resources for performing this study are very limited.

Nomenclature

Latin

$c_{\varepsilon 1}, c_{\varepsilon 2}, c_{\varepsilon 3}$ constants in the source term for the dissipation turbulent kinetic energy

D_h hydraulic diameter, m

$G_{k,l}$ production of the turbulent kinetic energy due to the buoyancy force deposed into field l, m^2/s^3

k_0 initial turbulent kinetic energy, m^2/s^2

k_l turbulent kinetic energy of field l, m^2/s^2

k_∞ turbulent kinetic energy for steady developed flow, m^2/s^2

ℓ_e effective maximum of the turbulent length scale, m

P_{kl} production of the turbulent kinetic energy deposed into field l, m^2/s^3

P_{kw}	production of the turbulent kinetic energy due to the wall friction, m^2/s^3
$P_{kw,l}$	production of the turbulent kinetic energy due to the wall friction deposed into field l, m^2/s^3
$P_{k\mu,l}$	production of the turbulent kinetic energy due to evaporation or condensation deposed into field l, m^2/s^3
$P_{\varepsilon w}$	production of the dissipation of the turbulent kinetic energy due to the wall friction, m^2/s^4
$P_{\varepsilon w,l}$	production of the dissipation of the turbulent kinetic energy due to the wall friction deposed in the field l, m^2/s^4
$\Pr_{kl}^t$	turbulent Prandtl number describing the diffusion of the turbulent kinetic energy of field l, -
$\Pr_{\varepsilon l}^t$	turbulent Prandtl number describing the diffusion of the dissipation of the turbulent kinetic energy, -
$\dot{q}_w''$	heat flux at the wall, W/m^2
$\dot{q}_{w2}''$	heat flux from the wall into field 2, W/m^2
u'	velocity fluctuation, m/s
u_0'	initial velocity fluctuation, m/s
V_1'	gas velocity fluctuation, m/s
$\mathbf{V}_l$	velocity of field l, m/s
v_l	cinematic viscosity of field l, m^2/s
v_l^k	eddy diffusivity for the turbulent kinetic energy, m^2/s
v_l^ε	eddy diffusivity for the dissipation of the turbulent kinetic energy, m^2/s
v_l^t	turbulent cinematic viscosity of field l, m^2/s
$\overline{w}$	cross section averaged velocity, m/s
w_2	velocity of liquid, m/s
$y_{\lim}^+$	effective thickness of the layer in which part of the turbulence kinetic energy is irreversibly dissipated, dimensionless
z	axial coordinate, m
z_{grid_end}	axial coordinate counted from the end of the grid, m

Greek

α_2	liquid volumetric fraction, -
α_l	volumetric fraction of field l, -
γ	surface permeability (flow cross section divided by the total cross section), -
γ_z	surface permeability in z direction, -

$\gamma_{z,spacer}$ surface permeability in z direction in the region of the spacer, -
γ_v volumetric porosity (flow volume divided by total volume), -
ε_0 initial dissipation of the turbulent kinetic energy, m^2/s^3
ε_l dissipation of the turbulent kinetic energy of field l, m^2/s^3
ε_∞ dissipation of the turbulent kinetic energy for steady developed flow, m^2/s^3
ζ_{spacer} irreversible friction coefficient of the spacer, -
η_{vis} part of the generated turbulent energy that is directly irreversibly dissipated, -
λ_{fr} friction coefficient, -
ρ_l density of field l, kg/m^3
τ time, s
τ_{2w} share stress at the film-wall interface, N/m^2
ζ_{fr} irreversible friction coefficient, -

References

Bennett AW et al. (1967) Studies of burnout in boiling heat transfer, Trans. Instit. Chem. Eng., vol 45 no 8 T319

Bertodano ML, Lahey RT Jr and Jones OC (March 1994) Development of a k-eps model for bubbly two-phase flow, Transaction of the ASME, Journal of Fluids Engineering, vol 116 pp 128-134

Chandesris M, Serre G and Sagaut (2005) A macroscopic turbulence model for flow in porous media suited for channel, pipe and rod bundle flows, 4th Int. Conf. On Computational Heat and Mass Transfer, Paris

Knabe P and Wehle F (Dec. 1995) Prediction of dry out performance for boiling water reactor fuel assemblies based on subchannel analysis with the RINGS code, Nuclear Technology, vol 112 pp 315323

Kolev NI (April 2000) Applied multi-phase flow analysis and its relation to constitutive physics, 8th International Symposium on Computational Fluid Dynamics, ISCFD '99 5 - 10 September 1999 Bremen, Germany, Invited Lecture. *Japan Journal for Computational Fluid Dynamics,* vol 9 no 1 pp 549-561

Kolev NI (2003, 2005) IVA_5M numerical method for analysis of three-fluid multi-component flows in boundary-fitted multi-blocks, Presented in Second M.I.T. Conference on Computational Fluid and Solid Mechanics, (17-20 June 2003) Boston. Computers & Structures, vol 83 (2005) pp 499-523, USA

Kolev NI (2007a) Multiphase Flow Dynamics, Vol. 1 Fundamentals, 3d ed., Springer, Berlin, New York, Tokyo

Kolev NI (2007b) Multiphase Flow Dynamics, Vol. 2 Thermal and mechanical interactions, 3d ed., Springer, Berlin, New York, Tokyo

Kolev NI (26-27 April 2006) IVA Simulations to the OECD/NRC Benchmarks based on NUPEC BWR Full-size Fine-mesh Bundle Tests, 3th Workshop on OECD/NRC

Benchmark based on NUPEC BWR full-size fine-mesh bundle tests (BFBT)-(BFBT- 3), Pisa, Italy

Nagayoshi T and Nishida K (1998) Spacer effect model for subchannel analysis - turbulence intensity enhancement due to spacer, J. of Nucl. Science and Technology vol 35 no 6 pp 399-405

OECD/NRC Benchmark based on NUPEC BWR Full-size Fine-mesh Bundle Tests (BFBT), Assembly Specifications and Benchmark Database, October 4, 2004, Incorporated Administrative Agency, Japan Nuclear Energy Safety Organization, JNES-04N-0015

Roloff-Bock I (2005) 2D-grid generator for heterogeneous porous structures in structured Cartesian coordinates, Framatome ANP, proprietary

Yang SK and Chung MK (1998) Turbulent flow through spacer grids in rod bundles, Transaction of the ASME, J. of Fluid Engineering, vol 120 pp 786-791

9 Simple algebraic models for eddy viscosity in bubbly flow

Simple algebraic models for eddy viscosity still play an important role in the analysis of boiling flows especially in nuclear reactor rod bundles. For this reason I will review this subject in this chapter. I will start with models for single phase flows in bundles in order later to follow the already established methods also for two phase flow. In any case these methods are not as powerful as already described k-eps methods but can be used to improve predictions with existing older computer codes.

Bubbles moving with relative velocity to the liquid create vortices behind them. Eddies with small sizes dissipate quickly. So that part of the generated turbulence energy dissipates into heat. The remaining eddies contribute to shifting the turbulence spectrum of the undisturbed liquid to higher frequencies.

Generally measurements of many authors for bubbly flow e.g. Serizawa et al. (1975) indicated that

$$w'_1 u'_1 \approx w'_2 u'_2 \,. \tag{9.1}$$

This will result in $\rho_1 w'_1 u'_1 \approx \frac{\rho_1}{\rho_2} \rho_2 w'_2 u'_2$ and therefore in $\mathbf{T}'_1 \approx \frac{\rho_1}{\rho_2} \mathbf{T}'_2$.

In upward flows bubbles with small sizes migrate toward the wall, while bubbles with large size tend to collect at the central part.

Keeping bubbles in oscillations with amplitude δr_d and frequency ω_d requires kinetic energy

$$\alpha_d \left(\delta r_d \right)^2 \omega_d^2 / 4 \,, \tag{9.2}$$

Bataille and Lance (1988). Obviously the basic level of turbulence contains this amount of energy.

9.1 Single phase flow in rod bundles

As an example for use of the effective eddy viscosity in single phase flow we consider flow in rod bundles.

9.1.1 Pulsations normal to the wall

The data obtained for single phase fluctuation velocity normal to the wall in pipes can also be used for flow parallel to a rod bundle based on the systematical experimental observations reported by *Rehme* (1992) p. 572: "…The experimental eddy viscosities normal to the wall are nearly independent on the relative gap width and are comparable to the data of circular tubes by *Reichardt* close to the walls…" For the turbulent viscosity normal to the wall *Rehme* reported that the magnitude of the lateral fluctuation velocity is 3.3% of the friction velocity,

$$\frac{V'_{c,y}}{w^*} = \frac{v'_{c,y}}{w^* D_h} = 0.033. \tag{9.3}$$

Using as characteristic length of the lateral turbulence pulsation the hydraulic diameter, defining the eddy viscosity as a product of the fluctuation velocity and the characteristic length (*Prandtl*), $v'_{c,y} = V'_{c,y} D_h$, replacing the friction velocity and rearranging *Rehme* obtained

$$v'_{c,y} = 0.033 \frac{\overline{w_c} D_h}{v_c} v_c \sqrt{\frac{\lambda_{fr}}{8}} = 0.033 v_c \operatorname{Re}_c \sqrt{\frac{\lambda_{fr}}{8}}. \tag{9.4}$$

The coefficient measured by *Rehme* is very close to those measured by other authors: ≈ 0.035, *Hinze* (1955), ≈ 0.04 *Edler* (1959).

Wnek et al. (1975) proposed without any prove to apply the equation

$$v'_{c,y} = const\ v_l \operatorname{Re}_l \sqrt{\frac{\lambda_{fr,l}}{8}} \tag{9.5}$$

to each of the phases based on the gas or liquid *Reynolds* number,

$$\operatorname{Re}_l = D_{h,l} \rho_l |w_l| / \eta'_l. \tag{9.6}$$

$$\lambda_{Rl} = \lambda_{Rl} (\operatorname{Re}_l) \tag{9.7}$$

is the friction factor corresponding to the gas or liquid *Reynolds* number considering each of the phases $l = 1, 2$ flowing in its separated channel with hydraulic diameter $D_{h,l}$ and velocity w_l.

9.1.2 Pulsation through the gap

The *Rehme*'s result is very important for computing the cross intermixing between two parallel channels in rod bundle causing specific spectrum of pulsations for each geometry. The net mass exchange in single phase flow is zero but the effective heat exchange is not zero. *Rehme* generalized the measurements by many authors including his own large data base by the following relation

$$\dot{q}''_{c,ij} = \rho_c c_{p,c} \frac{v'_{c,y}}{\Pr'_c} \frac{T_{c,i} - T_{c,j}}{\Delta S_{ij}}, \qquad (9.8)$$

where the temperatures are the channel averaged fluid temperatures at given elevation and ΔS_{ij} is the distance between the channel axis. The turbulent *Prandtl* number was found to be

$$\Pr'_c = \left(S_{gap}/D_{rod}\right)/0.7, \qquad (9.9)$$

where the gap-to-rod diameter ratio is based either on rod-to-rod or rod-to-wall distance. Three different notations of the above relation are used in the literature. We present also the other two briefly below to make easy for the reader to compare results of different authors to each others.

Introducing the *effective mixing velocity* as follows

$$\dot{q}''_{c,ij} = \rho_c c_{p,c} \frac{v'_{c,y}}{\Pr'_c} \frac{T_{c,i} - T_{c,j}}{\Delta S_{ij}} = \rho_c u'_{c,ij} c_{p,c} \left(T_{c,i} - T_{c,j}\right) \qquad (9.10)$$

results in

$$u'_{c,ij} := \frac{\dot{q}''_{c,ij}}{\rho_c c_{p,c} \left(T_{c,i} - T_{c,j}\right)} = \frac{1}{\Delta S_{ij}} \frac{v'_{c,y}}{\Pr'_c}. \qquad (9.11)$$

In many publications the gap fluctuation *Reynolds* number defined as follows $u'_{c,ij} S_{gap}/v_c$ is used to correlate data. So the *Rehme*'s correlation rewritten in terms of this group is

$$\frac{u'_{c,ij} S_{gap}}{v_c} = 0.0231 \sqrt{\frac{\lambda_{fr}}{8} \frac{D_{rod}}{\Delta S_{ij}}} \operatorname{Re}_c. \qquad (9.12)$$

Petrunik reported in 1973

$$\frac{u'_{c,ij} S_{gap}}{\nu_c} = 0.009 \, \text{Re}_c^{0.827} \, . \tag{9.13}$$

Rogers and *Tahir* reported in 1975

$$\frac{u'_{c,ij} S_{gap}}{\nu_c} = 0.0058 \left(\frac{D_{rod}}{\Delta S_{ij}}\right)^{0.46} \text{Re}_c^{0.9} \quad \text{for bundle geometry,} \tag{9.14}$$

$$\frac{u'_{c,ij} S_{gap}}{\nu_c} = 0.0018 \left(\frac{D_{rod}}{\Delta S_{ij}}\right)^{0.4} \text{Re}_c^{0.9} \quad \text{for simple geometry.} \tag{9.15}$$

The ratio

$$St_{gap} = u'_{c,ij} / \overline{w}_c \tag{9.16}$$

is called *gap Stanton number*. Here $\overline{w}_c$ is the averaged axial velocity. Several authors correlated the gap *Stanton* number with their measurements on bundles. So the *Rehme*'s correlation rewritten in terms of the gap *Stanton* number is

$$St_{gap} = \frac{u'_{c,ij}}{\overline{w}_c} := \frac{\dot{q}''_{c,ij}}{\rho_c c_{p,c} \overline{w}_c (T_{c,i} - T_{c,j})} = \frac{1}{\Delta S_{ij} \overline{w}_c} \frac{v'_{c,y}}{\text{Pr}'_c} = 0.0231 \frac{D_h D_{rod}}{\Delta S_{ij} S_{gap}} \sqrt{\frac{\lambda_{fr}}{8}} \, . \tag{9.17}$$

Several correlations from this type are reported in the literature: *Rogers* and *Rosenhart*, see *Seale* (1981) for rectangular rod array arrangement:

$$St_{gap} = 0.004 \frac{D_h}{\Delta S_{ij}} \text{Re}_c^{-0.1} \, ; \tag{9.18}$$

Wong and *Cao* (1999) derived from experiments with water at 147bar with 3x3 square array rods of 10mm diameter, 13.3mm rod pitch and 1m heated length $St_{gap} = 0.0056 \, \text{Re}_c^{-0.1} D_h / S_{gap}$. The subcooled boiling changed the constant to $St_{gap} = 0.015 \, \text{Re}_c^{-0.1} D_h / S_{gap}$.

Zhukov et al. (1994):

$$St_{gap} = 0.01 \left(1.0744 + 9.1864 \frac{D_h}{\Delta S_{ij}} \right) \text{Re}_c^{-0.1} \tag{9.19}$$

for rectangular arrayed rod bundles;

$$St_{gap} = 0.39 \frac{\frac{2\sqrt{3}}{\pi}\left(\frac{P}{D_{rod}}\right)^2 - 1}{Pe_c \left(\frac{P}{D_{rod}}\right)^{3/2} - 1}, \qquad (9.20)$$

for triangular arrayed rod bundles, $1.1 < P/D_{rod} < 1.35$, $70 < Pe < 1600$ and $Pr_c \ll 1$. Here P is the closest distance between the rod axes and Pe_c is the *Peclet* number. Theoretical expression for bundle is reported also by *Zhukov* in Bogoslovskaya et al. (1996):

$$St_{gap} = 0.01 \left\{ \left[\frac{138}{Pe_c} + \frac{1 - \exp\left(-0.62 \times 10^{-4} \, Re_c \, Pr_c^{1/3}\right)}{8\sqrt{P/D_{rod} - 1}} \right] \frac{D_h}{P} \right.$$

$$\left. + \frac{0.15}{(P/D_{rod} - 1)Re_c^{0.1}} \right\}$$

$$\times \left\{ 1 - \exp\left[-80(P/D_{rod} - 1)\right] \right\} \qquad (9.21)$$

for $1 < P/D_{rod} < 1.32$, $40 < Pe < 1500$ and $0.005 < Pr_c < 0.03$.

Measurements reported by *Baratto, Bailey* and *Tavoularis* (2006) in rod bundles indicate some differences between the gap oscillations characteristics for rod-to-rod gaps and rod-to-wall gaps. The power spectra are similar but the low frequency-oscillations in the rod to wall gap contain more energy. It results in non equal circumferential heat transfer of the rod, *Chang* and *Tavoularis* (2006). The effective velocity of the coherent structure oscillations in the rod to wall gap is reported by *Guellouz* and *Tavoularis* (2000) to become smaller with gap to rod diameter ratio, which is intuitively expected. For air flow at atmospheric conditions, $1.05 < S_{gap}/D_{rod} < 1.25$, $w_c = 10.1$ m/s, $Re_c = 145\,000$ the authors reported the approximation of their data with

$$\frac{u'_{c,ij}}{w_c} = 1.04 \left[1 - \exp\left(10.6 - 10.9 \, S_{gap}/D_{rod}\right)\right], \; S_{gap}/D_{rod} < 1.25.$$

Therefore $1-\exp\left(10.6-10.9 S_{gap}/D_{rod}\right)$ can be used as a dumping factor also for other data as an approximation. This observation has an effect on the local heat transfer on between the rod and the wall, which in accordance with *Guellouz* and *Tavoularis* (1992) also reduces with decreasing S_{gap}/D_{rod} ratio, see their Fig. 15. These authors reported also that in rod bundles for CANDU type reactors the circumferential variation of the local heat transfer for peripheral rods varies with ±5% which is a result of variation of the pulsation velocity components around the peripheral rod as demonstrated in *Ouma* and *Tavoularis* (1991).

9.1.3 Pulsation parallel to the wall

For the turbulent viscosity parallel to the wall *Rehme* reported

$$\frac{v^t_{c,z}}{w^* y_{sym.\,lines}} = \frac{0.0177}{\left(S_{gap}/D_{rod}\right)^{2.42}}. \tag{9.22}$$

$y_{sym.\,lines}$ is the distance from the wall to the symmetry line in the bundles. Replacing the friction velocity and rearranging *results in*

$$v^t_{c,z} = \frac{0.00885}{\left(S_{gap}/D_{rod}\right)^{2.42}} V_c \frac{w_c 2 y_{sym.\,lines}}{V_c} \sqrt{\frac{\lambda_{fr}}{8}}. \tag{9.23}$$

9.2 Two phase flow

There are some attempts in the literature to describe turbulent structure of bubble flows by using simple algebraic turbulence models expressing the turbulent cinematic viscosity of the continuous phase as a function of the local parameters of the flow.

9.2.1 Simple algebraic models

For flow around obstacle *Cook* and *Harlow* (1984) used Eq. (2.47) in the form

$$v^t_{cd} = 0.02\sqrt{2k_c}\,\Delta\ell, \tag{9.24}$$

assuming $\sqrt{2k_c} = 0.3V_c$, where k_c is the turbulent kinetic energy of the continuous field and $\Delta\ell$ is the characteristic geometry size. *Batchelor* (1988) used Eq. (2.47) in the form

$$v'_{cd} = \sqrt{2k_c}D_d = \sqrt{H(\alpha_d)}D_d|\Delta V_{cd}|, \qquad (9.25)$$

where

$$2k_c = H(\alpha_d)\Delta V_{cd}^2, \qquad (9.26)$$

$$\Delta V_{cd} = \sqrt{\Delta u_{cd}^2 + \Delta v_{cd}^2 + \Delta w_{cd}^2}, \qquad (9.27)$$

$$H(\alpha_d) \approx \frac{\alpha_d}{\alpha_{dm}}\left(1 - \frac{\alpha_d}{\alpha_{dm}}\right), \qquad (9.28)$$

and

$$\alpha_{dm} \approx 0.62 \qquad (9.30)$$

is the limit of the closest packing of bubbles. *Zaruba* et al. (2005) confirmed this equation used with the bubble radius instead with the bubble diameter for bubbly flow at atmospheric conditions and $0.002 < \alpha_1 w_1 < 0.004 m/s$. The eddy viscosity takes values in this region between 3.8×10^{-5} and $5 \times 10^{-5} m^2/s$.

Other algebraic model for dispersed flows can be derived using the *Prandtl* mixing length hypothesis

$$v'_{cd} = \left(dI_{cd}^d / dM_c\right)\ell_{e,c}, \qquad (9.31)$$

where $\ell_{e,c}$ is the turbulence length scale of large eddies. For a single particle the momentum

$$dI_{cd}^d = f_{cd}^d \Delta\tau \qquad (9.32)$$

is dissipated into the continuum mass forming idealized trace behind the bubble

$$dM_c = \rho_c \Delta V_{cd} \Delta\tau \frac{1}{4}\pi D_d^2 \qquad (9.33)$$

during the time $\Delta\tau$. The drag force exerted by the particle is

$$f_{cd}^d = c_{cd}^d \frac{1}{4}\pi D_d^2 \frac{1}{2}\rho_c \Delta V_{cd}^2. \qquad (9.34)$$

Therefore the impulse of the friction force per unit continuum mass is

$$dI_{cd}^d / dM_c \approx \frac{1}{2}c_{cd}^d \Delta V_{cd}. \qquad (9.35)$$

In accordance with *Peebles* and *Garber* (1953)

$$c_{cd}^d \approx 0.967 D_d / \ell_{e,c} \qquad (9.36)$$

and consequently

$$v_{cd}^t = \left(dI_{cd}^d / dM_c\right)\ell_{e,c} = \frac{1}{2}c_{cd}^d \Delta V_{cd} \ell_{e,c} \approx 0.48 D_d \left|\Delta V_{cd}\right| \qquad (9.37)$$

valid for single particle or *small particle concentrations*. For *large particle concentrations* the friction energy is dissipated into the continuum mass belonging to a single particle

$$dM_c \approx \rho_c \Delta V_{cd} d\tau / n_d^{2/3}. \qquad (9.38)$$

Therefore the turbulent viscosity is

$$v_{cd}^t \approx 0.58 D_d^{7/9} \left|\Delta V_{cd}\right| \alpha_d^{2/3}. \qquad (9.39)$$

For bubbly flow *Lilienbaum* (1983) proposed intuitively the following relationship based on dimensional analysis

$$v_{cd}^t = 0.4\left(gD_d^4 \Delta V_{cd} \alpha_d\right)^{1/3}. \qquad (9.40)$$

All these expressions are compared in Chapter 3 with Eq. (3.23) and the recommendation was made to use Eq. (3.23) based on sound physical scaling.

An ad hoc idea of how to use the algebraic models is the superposition of the *Reynolds* turbulence and a component coming from the bubble–liquid interaction. An example is given below for the lateral effective eddy diffusivity

$$v_{c,y}^t = 0.033 v_c \, \text{Re}_c \sqrt{\frac{\lambda_{fr}}{8}} + 1.22\left[\frac{3}{4}(1-\eta_{vis})c_{cd}^d \frac{\ell_{e,c}}{D_d}\right]^{1/3}\left(\frac{\alpha_d}{\alpha_c}\right)^{1/3}\left|\Delta V_{cd}\right| \ell_{wake}. \qquad (9.41)$$

9.2 Two phase flow

The lateral mass flow rate of bubbles is then

$$(\rho v)_{\alpha_d,y} = -\rho_d D^t_{\alpha,y} \frac{d\alpha_d}{dy}, \qquad (9.42)$$

with the diffusion coefficient being of order of the eddy diffusivity $D^t_{\alpha,y} \approx v^t_{c,y}$. Not enough validation work is available in the literature for this approach. Note that the effective volumetric flow in each cross section due to turbulent volume fraction diffusion is zero,

$$-D^t_{\alpha_d,y} \frac{d\alpha_d}{dy} = D^t_{\alpha_c,y} \frac{d\alpha_c}{dy} \qquad (9.43)$$

and therefore

$$D^t_{\alpha_d,y} = D^t_{\alpha_c,y}. \qquad (9.44)$$

Note also that with such approach we replace force interaction on smaller scale with diffusion transport of volumetric fractions in large scale.

> Therefore one either has to consider the corresponding force components for small scales in the momentum equations and not to allow microscopic volumetric diffusion in the mass conservation equation or has to replace the action of these forces with volumetric diffusion in the mass conservation equation. The simultaneous use of both the formalisms is wrong.

One important remark regarding computing void diffusion in bundles with sub-channel resolution: In the sense of the sub-channel analysis the void diffusion can be expressed as

$$(\rho v)_{\alpha_d,y} = -\rho_d D^t_{\alpha,y} \left(\alpha_{d,i} - \alpha_{d,j} \right) / \Delta S_{ij}. \qquad (9.45)$$

Note that $(\rho v)_{\alpha_c,y} / \rho_c = -(\rho v)_{\alpha_d,y} / \rho_d$. Because ΔS_{ij} is much larger then the diffusion length scale, the turbulent diffusion coefficient has to be additionally subject to sub-grid modeling

$$v^t_{c,y} = f\left(\text{Re}_c, \Delta S_{ij}, S_{gap}, \ldots \right). \qquad (9.46)$$

There is no well founded systematical work in this direction. Instead, the proposal

$$(\rho v)_{\alpha_d,y} = \rho_d \frac{D^{eff}_{\alpha,y}}{S_{gap}} \left[\alpha_{d,i} - \alpha_{d,j} - (\alpha_{d,i} - \alpha_{d,j})_{mech.\ eq.} \right] \qquad (9.47)$$

by *Lahey* et al. in 1972, see also in *Lahey* and *Moody* (1993), is widely used with empirical modeling of the effective diffusion coefficient $D^{eff}_{\alpha,y}$ and of the so-called equilibrium void difference

$$(\alpha_{d,i} - \alpha_{d,j})_{mech.\ eq.} \approx -1.4 \frac{G_i - G_j}{\overline{G}} \overline{\alpha}_d \qquad (9.48)$$

with $\overline{G} = (G_i - G_j)/2$ and $\overline{\alpha}_d = (\alpha_{d,i} - \alpha_{d,j})/2$.

The mass source due to turbulent diffusion in the mass conservation equation of the dispersed phase is then

$$\mu_d^t = \nabla \cdot (\rho_d D_\alpha^t \nabla \alpha_d). \qquad (9.49)$$

The specific power sources due to turbulent diffusion in the energy conservation for the continuum and disperse phase are then

$$\dot{q}_c^{m't} = \nabla \cdot (\rho_c e_c \gamma D_\alpha^t \nabla \alpha_c) - p_c \nabla \cdot (\gamma D_\alpha^t \nabla \alpha_c) \approx \nabla \cdot (\rho_c c_{p,c} T_c \gamma D_\alpha^t \nabla \alpha_c), \qquad (9.50)$$

$$\dot{q}_d^{m't} = \nabla \cdot (\rho_d e_d \gamma D_\alpha^t \nabla \alpha_d) - p_d \nabla \cdot (\gamma D_\alpha^t \nabla \alpha_d) \approx \nabla \cdot (\rho_d c_{p,d} T_d \gamma D_\alpha^t \nabla \alpha_d). \qquad (9.51)$$

Simple steady state algebraic models for developed flows have strong limitation in its validity for special geometry and flow topology. So, for instance, if turbulence is generated in a singular obstacles and then decaying flow-downward the considerable mixing around the singularity can not be taken properly into account by the above method.

9.2.2 Local algebraic models in the framework of the *Boussinesq*'s hypothesis

In the framework of the *Boussinesq* hypothesis several authors proposed different expressions for the effective viscosity based on the local distribution of the undisturbed Reynolds turbulence and additives coming from the presence of bubbles. Unlike the models discussed in the previous sections based on *cross section averaged properties* these models are based on *local properties*. Now we summarize briefly the formalism how to check the validity of such expressions.

Consider upward directed bubbly flow in a vertical pipe. The zero share line for this flow is the axis. In this case the steady state momentum equation for vertical pipe section Δz is

$$\pi R^2 \frac{\Delta p}{\Delta z} - g \int_0^R \left[\alpha_1 \rho_1 + (1-\alpha_1) \rho_2 \right] r dr = 2\pi R \tau_w, \qquad (9.52)$$

and for the cylinder with radius r

$$\pi r^2 \frac{\Delta p}{\Delta z} - g \int_0^r \left[\alpha_1 \rho_1 + (1-\alpha_1) \rho_2 \right] r dr = 2\pi r \tau_2. \qquad (9.53)$$

Eliminating the pressure gradient results in

$$\tau_2 = \left(1 - \frac{y}{R}\right) \tau_w + \tau_{2,grav}(r), \qquad (9.54)$$

where the gravity component of the share stress is

$$\tau_{2,grav}(r) = \frac{g}{2\pi} \left\{ \frac{r}{R^2} \int_0^R \left[\alpha_1 \rho_1 + (1-\alpha_1) \rho_2 \right] r dr - \frac{1}{r} \int_0^r \left[\alpha_1 \rho_1 + (1-\alpha_1) \rho_2 \right] r dr \right\}. \qquad (9.55)$$

If using the *Boussinesq*'s hypothesis in analogy to the single phase flow the effective cinematic viscosity is postulated to be a function of the distance from the wall, the local share stress at the same distance is

$$\frac{\tau_2(y)}{\rho_2} = (1-\alpha_1) v_{2,eff}(y) \frac{dw}{dy}. \qquad (9.56)$$

Introducing the expression for the share stress results in the momentum equation describing the liquid velocity

$$(1-\alpha_1) \rho_2 v_{2,eff}(y) \frac{dw}{dy} = \left(1 - \frac{y}{R}\right) \tau_w + \tau_{2,grav}(r). \qquad (9.57)$$

Therefore, if the local averaged liquid velocity is known postulated expressions for the effective viscosity can be tested by integrating the above equation and comparing with the measured profiles.

Sato and *Sekoguchi* (1975) proposed to consider the turbulence in the liquid as consisting of the turbulence caused by the undisturbed single phase fluid and additive caused by the motion of the bubbles relative to the liquid. For developed steady state pipe flow the share stress will then take the form

$$\frac{\tau_2(y)}{\rho_2} = (1-\alpha_1)(v_2 + v_2' + v_{12}')\frac{dw}{dy}, \tag{9.58}$$

$$v_{12}' = \alpha_1 \frac{D_1}{2} \Delta w_{12}. \tag{9.59}$$

Compare the contribution of the bubbles with Eq. (3.23) and realize the difference. For computation of the eddy viscosity of the undistorted liquid a conventional single phase model is used. Only two experimental cases are used to check the theory. *Tomiyama* et al. (2000) used v_{12}' in their multi-group bubble approach by summing the contribution of each bubble group weighted with its volume fraction. In a later work by *Sekogushi* et al. (1979) and *Sato* et al. (1981) the undisturbed liquid eddy viscosity is modeled using the *Reichardt* (1951) expression

$$\frac{v_2'}{v_2} = 0.4 y^+ \left[1 - \frac{11}{6}\left(\frac{y}{R}\right) + \frac{4}{3}\left(\frac{y}{R}\right)^2 - \frac{1}{3}\left(\frac{y}{R}\right)^3 \right] f_{damp}^t(y^+) \tag{9.60}$$

multiplied by the *van Driest* (1955) damping factor

$$f_{damp}^t(y^+) = \left[1 - \exp\left(-\frac{y^+}{16}\right) \right]^2. \tag{9.61}$$

In the same work the eddy viscosity component caused by the bubbles locally was modified to

$$v_{12}' = 0.6 \alpha_1 D(y) \Delta w_{12} f_{damp}^t(y^+), \tag{9.62}$$

in order to include damping of the bubble influence in close proximity to the wall. The constant was found by comparing with two void-profile sets: At given void profile the liquid velocity profile is compared with the measured one. After some trials the coefficient 0.6 was found. *Liu* and *Bankoff* (1993a and b) performed experiments on a vertical 2.8m long pipe with 38mm inner diameter. Air was injected in water at atmospheric conditions. The averaged void fraction varied up to 0.5. The authors measured all important characteristics of the flow. Using the *Sato* et al. method, the authors reported favorable comparison with the liquid velocity profiles for high mass flows and bad comparison for low mass flows, see p.1059, *Liu* and *Bankoff* (1993a). Note that it is not clear why bubbly induced turbulence has to be dumped in the same way in the proximity of the wall as the liquid bulk turbulence.

Sato et al. (1981) made an intuitive assumption based on their Fig. 4 without proof that at boiling wall, for instance, the local averaged bubble sizes varies with the distance from the wall

$$D(y) = 0 \text{ for } y < 20 \times 10^{-6} m, \qquad (9.63)$$

$$D(y) = 4y\left(1 - \frac{y}{D_1}\right) \text{ for } 20 \times 10^{-6} m < y < D_1/2, \qquad (9.64)$$

$$D(y) = D_1 \text{ for } y \geq D_1/2. \qquad (9.65)$$

The authors considered the region $y < 20 \times 10^{-6} m$ as a viscous layer with no influence on the bubbles, the region $y < D_1$ as boundary layer, and the remaining region as a core region. Although used without proof this idea is interesting. It can be improved by taking into account the bubble generation dynamics at the heated wall as reported by *Kolev* (2007b) and using the ideas from Section 3.2.

Different attempt to take the effect of the wall boiling on the eddy viscosity was reported by *Pu* et al. (2006). The authors proposed to compute the eddy viscosity as follows

$$\frac{v_2^t}{v_2} = 0.001 y^+ \text{ for } y^+ < 5, \textit{Kays} (1994), \qquad (9.66)$$

and

$$\frac{v_2^t}{v_2} = 0.41 y^+ \left(1 - \frac{y}{\delta_{2F}}\right)^{1.5} \left[1 - \exp\left(-\frac{y^+}{25}\right)\right]^2 \left(\frac{\dot{q}''}{G\Delta h}\right)^{0.3} \left(\frac{1 - X_{1eq}}{X_{1eq}}\right)^{0.1} \text{ for } y^+ \geq 5. \qquad (9.67)$$

Here X_{1eq} is the cross section averaged equilibrium vapor mass flow ratio. Note that the asymptotic of this proposal is unphysical: for zero heat flux the eddy viscosity is zero and for no vapor the eddy viscosity is not defined.

In addition *Sato* et al. (1981) postulated that the components of the eddy conductivity in the flow are equal to the components of the eddy viscosities,

$$\frac{\dot{q}''_{w2}}{\rho_2 c_{p2}}\left(1 - \frac{y}{R}\right) = -(1 - \alpha_1)\left(a_2 + a_2^t + a_{12}^t\right)\frac{d\overline{T}}{dy}. \qquad (9.68)$$

Serizawa et al. (1975a) based on the data collected in *Serizawa* et al. (1975a and b) modified the *Sato* and *Sekogushi* et al. (1975) approach based on mixing length, rather on eddy diffusivity,

$$\frac{\tau_{2w}}{\rho_2}\left(1-\frac{y}{R}\right) = (1-\alpha_1)(v_2 + v_2' + v_{12}')\frac{dw}{dy}, \qquad (9.69)$$

$$\frac{\dot{q}_{w2}''}{\rho_2 c_{p2}}\left(1-\frac{y}{R}\right) = -(1-\alpha_1)(a_2 + a_2' + a_{12}')\frac{d\overline{T}}{dy}. \qquad (9.70)$$

The *Reynolds* and the bubble induced turbulence are modeled as follows

$$v_{12}' = 0 \text{ for } y < 20 \times 10^{-6} m, \qquad (9.71)$$

$$v_{12}' = 0.6\alpha_1 D_1 \left|\overline{\Delta w_{12}}\right| f_{dump}^t(y^+) \text{ for } 20 \times 10^{-6} m < y < D_1, \qquad (9.72)$$

$$v_{12}' = 0.06\alpha_1 \ell_{mix} \left|\overline{\Delta w_{12}}\right| f_{dump}^t(y^+) \text{ for } y \geq D_1, \qquad (9.73)$$

$$\ell_{mix} = \ell_{mix,r=0}\left(\frac{y}{R}\right)^m, \; m = 0.9 \text{ for } 0.02 < \alpha_1 < 0.15. \qquad (9.74)$$

Serizawa and *Kataoka* (1980) suggested that

$$\ell_{mix,r=0}, m = f(\alpha_1, D_h). \qquad (9.75)$$

In a later work *Kataoka* and *Serizawa* (1995) proposed to use the mixing length hypothesis in the following form

$$v_2' = \ell_{2,eff} V_2', \qquad (9.75)$$

$$\ell_{2,eff} = \ell_{2,mix} + \ell_{12,mix}, \qquad (9.76)$$

$$V_2' = \sqrt{\frac{2}{3}k_2}, \qquad (9.77)$$

where mixing length for the continuum turbulence is

$$\ell_{2,mix} = 0.4 y f_{dump}^t, \qquad (9.78)$$

with

$$f^t_{dump} = 1 - \exp(-y^+/26), \quad (9.79)$$

and the contribution due to presence of bubbles is

$$\ell_{12,mix} = \alpha_1 \frac{1}{6}\left[D_1 + \left(\frac{4}{3} - \frac{y}{D_1}\right) \Big/ \left(2 - \frac{4}{3}\frac{y}{D_1}\right)\right] \text{ for } 0 \le y \le D_1, \quad (9.80)$$

$$\ell_{12,mix} = \frac{1}{6}\left[D_1 + \alpha_1\left(y - \frac{D_1}{2}\right)\right] \text{ for } D_1 \le y \le \frac{3}{2}D_1, \quad (9.81)$$

$$\ell_{12,mix} = \frac{1}{3}\alpha_1 D_1 \text{ for } \frac{3}{2}D_1 \le y \le \frac{D_h}{2}. \quad (9.82)$$

Then the continuum turbulent kinetic energy is computed using the transport equation of the turbulent kinetic energy of the continuum

$$-\frac{1}{r}\frac{\partial}{\partial r}\left[r\alpha_2\left(\frac{1}{2}V_2 + v'_2\right)\frac{\partial k_2}{\partial r}\right] = \alpha_2\left(\overline{v'_2 P_{k,2}} + G_{k,2} - \varepsilon_2\right), \quad (9.83)$$

with

$$r = D_h/2 - y. \quad (9.84)$$

The generation of the turbulence due to deformation of the velocity field is

$$\alpha_2 \overline{v'_2 P_{k,2}} = \alpha_2 v'_2 \left(\frac{\partial w_2}{\partial r}\right)^2. \quad (9.85)$$

The contribution of the bubbly induced turbulence generation is considered to be only 7.5% of the power required for the relative motion between the bubble and the liquid,

$$\alpha_2 G_{k,2} = 0.075\alpha_1 \frac{3}{4D_1} c^d_{21} \Delta w^3_{12} f^t_{dump}, \quad (9.86)$$

with dumping function reducing this contribution in the vicinity of the wall. The dissipation of the kinetic energy is considered consisting of three components as follows

$$\alpha_2\varepsilon_2 = k_2^{3/2}\left(0.18\alpha_2/\ell_{2,\mathit{eff}} + \alpha_1/D_1\right) + v_2\left(\frac{\partial V_2'}{\partial r}\right)^2. \tag{9.87}$$

The first two are understandable – dissipation due to small scale eddies; the last component resembles the definition of ε_2 itself and is introduced "…to have compensating effect of the numerical error in the proximity of the wall."

Serizawa and *Kataoka* (1980) reported that the ratio of the eddy diffusivity of heat for two-phase flow to that for single-phase flow varying between 7 and 2 is grossly correlated to the *Lockhart-Martinelli* parameter

$$\frac{a_{12}^t}{a_2 + a_2^t} = 1 + 462 X_{tt}^{-1,27} \tag{9.88}$$

$$X_{tt} = \left(\frac{1-X_1}{X_1}\right)^{0.9}\left(\frac{\rho_1}{\rho_2}\right)^{0.5}\left(\frac{\eta_2}{\eta_1}\right)^{0.1}, \tag{9.89}$$

in the limited region $50 < X_{tt} < 250$. Therefore, there is improvement of the pipe wall heat transfer in bubbly flow compared to single phase flow.

Chu and *Jones* (1980) derived from their own measurements and measurements from other authors the following correlations for heat transfer in vertical non-boiling flow (bubble and slug):

$$\frac{h_{2w}D_h}{\lambda_2} = 0.43\left(\frac{\rho_2 w_2 D_h}{\eta_2}\right)^{0.55} \Pr_2^{1/3}\left(\frac{\eta_2}{\eta_{2w}}\right)^{0.14}\left(\frac{p_a}{p}\right)^{0.17} \pm 15\%, \tag{9.90}$$

upflow, $3\times 10^4 < \frac{\rho_2 w_2 D_h}{\eta_2} < 8\times 10^5$,

$$\frac{h_{2w}D_h}{\lambda_2} = 0.47\left(\frac{\rho_2 w_2 D_h}{\eta_2}\right)^{0.55} \Pr_2^{1/3}\left(\frac{\eta_2}{\eta_{2w}}\right)^{0.14}\left(\frac{p_a}{p}\right)^{0.17} \pm 15\%, \tag{9.91}$$

downflow, $4\times 10^4 < \frac{\rho_2 w_2 D_h}{\eta_2} < 2\times 10^5$. Here p is the pressure and p_a is the atmospheric pressure. Here the improvement of the heat transfer due to presence of bubbles is not directly taken into account. I have already provided sound physical estimate of the improvement of the heat transfer due to presence of gas in Chapter 4.

Idealizing the transversal bubble movement as a diffusion process,

$$(\rho v)_1 = -\rho D_{X,12}^t \frac{dX_1}{dr}, \tag{9.92}$$

Serizawa et al. deduced from experimental data within gas qualities varying in the region of $0 < X_1 < 0.04$

$$D^t_{X,12} \approx (1 \div 2.5) \times 10^{-4} \ m^2/s. \quad (9.93)$$

Here ρ is the two-phase mixture density. There are authors describing lateral diffusion of bubbles driven by void gradients

$$(\rho v)_1 = -\rho_1 D^t_{\alpha,12} \frac{d\alpha_1}{dr}, \quad (9.94)$$

where $D^t_{\alpha,12} \approx v'_2 \ell_{2,mix}$. Note that for constant pressure $D^t_{\alpha,12} = D^t_{X,12}$. Expressed in the form of turbulent *Peclet* number

$$Pe^t_{C,12} = \frac{u'_2 D_{hyd}}{D^t_{X,12}} \approx 2, \quad (9.95)$$

the authors found that the intensity of diffusion is closely connected to the turbulence intensity but not with the local void fraction directly. For comparison see the result obtained by *Zun* (1980): $D^t_{\alpha,12} \approx 0.35 \times 10^{-4} \ m^2/s$ obtained for $\alpha_2 w_2 = 0.748 m/s$. *Lilienbaum* (1983) derived from experiments in inclined channels at atmospheric conditions $D^t_{\alpha,12} \approx 1.5 v'_2$. *Mikiyoshi* and *Serizawa* (1986) reported for the bubble induced component

$$u''_2 = 0.01 |\Delta w_{12}| \alpha_1^{0.5} \text{ for } \alpha_1 < 0.013 \text{ and}$$
$$u''_2 = 0.85 \alpha_1^{0.5} \text{ for } 0.013 \leq \alpha_1 \text{ for } \Delta w_{12} = 0.23 m/s.$$

Conclusions: The weaknesses of the *Boussinesq*'s hypothesis applied in the form discussed in this chapter is that

a) the superposition of the single phase and bubble induced turbulence is valid for very small concentrations, $\alpha_1 < 0.005$, as measured by *Lance* and *Bataille* (1991);
b) the fluctuations are not equal in magnitude in all direction as measured by *Wang* et al. (1987) i.e. the isotropy assumption is not valid;
c) the specific contribution of variety of forces acting on the bubbles is smeared.

Nevertheless important order of magnitude limits are collected by the activities reviewed here.

9.2.3 Modification of the boundary layer share due to modification of the bulk turbulence

In analogy to single phase turbulence *Marié* (1987) expressed the idea that the bubbles outside the boundary layer modify the share stress at the wall. In a later work *Moursalli* et al. (1995) performed measurement of the turbulent characteristics in a vertical bubble flow along a vertical plate. They found that the parameter of the logarithmic low

$$w_2^+ = \left(\ln y^+\right)/\kappa_{12} + c, \qquad (9.96)$$

$$c = y_{lt}^+ \left(1 - \frac{\kappa}{\kappa_{12}}\right) + \frac{\kappa}{\kappa_{12}} 5.45, \quad y_{lt}^+ = 11.23, \quad \kappa = 0.41, \qquad (9.97)$$

are functions on the void fraction far from the wall and on the peaking void as given in Table 9.1.

Table 9.1. Coefficients defining the velocity distribution in the boundary layer as a function of the void fraction

$w_{2,far}, m/s$	$\alpha_{1,far}$	$\alpha_{1,peack}$	κ	c
1	0	0	0.41	4.9
1	0.002	0.016	0.48	5.8
1	0.005	0.038	0.56	6
1	0.015	0.068	0.61	6.5

Troshko and *Hassan* (2001) simplified the axial momentum equation to

$$\tau_{2w}/\rho_2 = (1-\alpha_1)\left(v_2' + v_{12}'\right) dw/dy \qquad (9.98)$$

and assumed

$$\alpha_1 = \alpha_{1,peack}, \qquad (9.99)$$

$$v_2' = \kappa y w^*, \qquad (9.100)$$

$$v_{12}' = \kappa_1 \alpha_{1,\max} \Delta w_{12} y. \qquad (9.101)$$

After integration for $y^+ > y_{lt}^+$ the authors obtained

$$\kappa_{12} = \kappa\left(1 + \frac{\kappa_1 \alpha_{1,\max} \Delta w_{12}}{\kappa w^*}\right). \qquad (9.102)$$

The coefficient

$$k_1 = 4.95 e^{-40.7 w^*} \qquad (9.103)$$

was derived from experimental data. *Koncar* et al. (2005) modified this model by setting empirically

$$y_{lt}^+ = 11.23(1-\alpha_1)^3, \qquad (9.104)$$

$$k_1 = 4.95 e^{-10 w^*}. \qquad (9.105)$$

No resulting expressions have been derived for the wall share stress and for the heat transfer at sub-cooled boiling and no comparisons are reported with such data. The expression of such type is proposed to be used in computational fluid dynamics as a boundary layer treatment without resolving the boundary layer.

Nakoryakov et al. (1981) reported data for the share stress in upward bubbly flow in vertical pipes. The share stress is found to be a non monotonic function of the void fraction. None of the existing models represents this behavior.

Nomenclature

Latin

a	$:= \lambda/(\rho c_p)$, thermal diffusivity, m²/s
a_{12}^t	turbulent liquid thermal diffusivity caused by liquid bubble interaction only, m²/s
a_2	molecular liquid thermal diffusivity, m²/s
a_2^t	turbulent liquid thermal diffusivity without liquid bubble interaction, m²/s
c_{cd}^{vm}	coefficient for the virtual mass force or added mass force acting on a dispersed particle, dimensionless
c_{cd}^d	coefficient for the drag force or added mass force acting on a dispersed particle, dimensionless
c_{cd}^t	coefficient in the term describing bubble turbulent diffusion, dimensionless
c_k	geometry dependent constant for computation of the viscous dissipation in the boundary layer, dimensionless
c_p	specific capacity at constant pressure, J/(kgK)
$c_{\text{van Driest}}$	constant

$c_{\varepsilon 1}, c_{\varepsilon 2}, c_{\varepsilon 3}$ empirical coefficients in the source term of the ε-equation, -
c_η empirical constant or function connecting the eddy cinematic diffusivity with the specific turbulent kinetic energy and its dissipation, -
c'_η constant
D diameter, m
D_1 bubble diameter, m
D_{1d} bubble departure diameter, m
$D_{1,max}$ maximum bubble diameter, m
$D_{1,min}$ minimum bubble diameter, m
D_d diameter of dispersed particle, m
D_h hydraulic diameter, m
$D_{h,l}$ hydraulic diameter of the "tunnel" of field l only, m
D_{rod} diameter of the rods in a rod bundle, m
$D^t_{X,12}$ bubble turbulent diffusion coefficient based on mass concentrations, m²/s
$D^t_{\alpha,12}$ bubble turbulent diffusion coefficient based on volumetric fractions, m²/s
f^d_{cd} drag force experienced by the dispersed phase from the surrounding continuum, N/m³
f^t_{dump} damping factor for the bulk turbulence, dimensionless
f_w friction force per unit flow volume, N/m³
$G_{k,l}$ production of turbulent kinetic energy due to bubble relocation in changing pressure field per unit mass of the filed l, W/kg (m²/s³)
g gravitational acceleration, m/s²
h_{2w} heat transfer coefficient between liquid and wall, J/(m²K)
I^d_{cd} momentum of the dispersed phase dissipated into the continuum, Ns
k without subscript: wall roughness, m
k specific turbulent kinetic energy, m²/s²
ℓ length, m
ℓ_{max} maximum length scale, m
ℓ_{mix} mixing length, m
$\ell_{mix,r=0}$ mixing length at the proximity of the wall, m
ℓ_e size of the large eddy, m
$\ell_{\mu e,l}$ lowest spatial scale for existence of eddies in field l called *inner scale* or *small scale* or *Taylor* micro-scale (µ) of turbulence, m
$\Delta \ell$ characteristic size of the geometry, m
M_c continuum mass belonging to a single particle, kg
m constant in Sato et al. model, -

P_k	production of the turbulent kinetic energy per unit mass, W/kg
$\overline{P_{k,l}}$	production of the turbulent kinetic energy per unit mass of the velocity field l due to deformation of the velocity field l, W/kg
$Pe^t_{c,12}$	$:= \dfrac{u'_2 D_{hyd}}{D^t_{X,12}}$, turbulent Peclet number for lateral bubble diffusion due to turbulence, dimensionless
Pe_c	Peclet number for continuum, dimensionless
Pr^t	turbulence Prandtl number, dimensionless
Pr^l	molecular Prandtl number, dimensionless
P	closest distance between the rod axes, m
p	pressure, Pa
p'	fluctuation of the pressure, Pa
p_c	pressure inside the continuum, Pa
p_d	pressure inside the dispersed phase, Pa
Δp	pressure difference, Pa
$\dot{q}''$	heat flux, W/m²
$\dot{q}''^w_c$	heat flux from the wall into the continuum, W/m²
$\dot{q}''_{w2}$	heat flux from the wall into the liquid, W/m²
$\dot{q}^{mt}_l$	energy inserted in the field l being in the control volume per unit time and unit mixture volume due to turbulent exchange with the neighboring control volumes, W/m³
R	radius, m
R_d	radius of the dispersed particle, m
Re	Reynolds number, dimensionless
Re_{cd}	Reynolds number based on relative velocity, continuum properties and size of the dispersed phase, dimensionless
r	radius, m
δr_d	bubbles oscillations amplitude, m
$r*$	radius, dimensionless
S_{gap}	gap size, the smallest distance between two adjacent rods, m
ΔS_{ij}	distance between vertical channel axis, m
St	$:= \Delta\tau_{cd}/\Delta\tau_{L,c}$ Stokes number, dimensionless
St_{gap}	$:= u'_{c,ij}/\overline{w}_c$, *gap* Stanton *number*, dimensionless
T	temperature, K
T_i	temperature of the vertical sub-channel i, K
T_j	temperature of the vertical sub-channel j, K
T_w	wall temperature, K

$\overline{T}$	averaged temperature, K
u	radial velocity component, m/s
u'	fluctuation of the radial velocity, m/s
u^+	radial velocity, dimensionless
u^*	radial friction velocity, m/s
u_1	bubble radial velocity, m/s
u_1'	fluctuation of the bubble radial velocity, m/s
u_2	liquid radial velocity, m/s
u_l	radial velocity of field l, m/s
u_l'	fluctuation of the radial velocity of field l, m/s
$u_{c,ij}'$	$:= \dfrac{\dot{q}_{c,ij}''}{\rho_c c_{p,c}\left(T_{c,i}-T_{c,j}\right)}$, effective mixing velocity, m/s
Δu_{12}	radial velocity difference between gas and liquid, m/s
Δu_{cd}	radial velocity difference between the dispersed and continuous phase, m/s
$\overline{u'}$	rms of the radial velocity fluctuation, m/s
V'	fluctuation of the velocity, m/s
$\mathbf{V}$	velocity vector, m/s
V_{21}^d	difference between liquid and gas velocity, m/s
V_b	bubble departure volume, m³
ΔV_{12}	difference between gas and liquid velocity, m/s
ΔV_{dd}	difference between the velocity of two neighboring droplets, m/s
ΔV_{d1d2}	difference between the velocity of two neighboring droplets with sizes belonging to two different groups, m/s
$\Delta \mathbf{V}_{ml}$	difference between m- and l-velocity vectors, m/s
Vol	control volume, m³
v	velocity component in angular direction, m/s
v'	fluctuation of the velocity component in angular direction, m/s
Δv_{cd}	angular velocity difference between the continuum and dispersed phase, m/s
$\overline{v'}$	time average of the angular velocity fluctuation, m/s
w	axial velocity, m/s
w^*	friction velocity, dimensionless
$\overline{w}$	cross section averaged friction velocity, m/s
w^+	axial velocity, dimensionless
w_1	bubble axial velocity, m/s
w_2	liquid axial velocity, m/s

$w_{2,far}$	liquid velocity far from the wall, m/s
w_2^+	liquid axial velocity, dimensionless
w_1'	fluctuation of the axial bubble velocity, m/s
w_2'	fluctuation of the axial liquid velocity not taking into account the influence of the bubble, m/s
w_2''	fluctuation of the axial liquid velocity not taking caused only by the presence of bubble, m/s
Δw_{12}	local axial velocity difference between bubbles and liquid, m/s
$\overline{\Delta w_{12}}$	cross section averaged axial velocity difference between bubbles and liquid, m/s
$\Delta w_{12\infty}$	steady state axial bubble rise velocity in liquid, m/s
w_c	continuum axial velocity, m/s
$\overline{w}_c$	averaged axial continuum velocity, m/s
w_c^*	continuum axial friction velocity, m/s
w_l	axial velocity of field l, m/s
w_l'	fluctuation of the axial velocity of field l, m/s
$\overline{w}_l$	cross section axial velocity of field l, m/s
Δw_{cd}	axial velocity difference between dispersed and continuous phase, m/s
X_1	gas mass concentration, m/s
X_{tt}	Lockhart-Martinelli parameter, dimensionless
x	x-coordinate, m
y	y-coordinate, distance from the wall, m
y_0	distance between the bubble and the wall, m
y_{lim}	virtual distance from the wall in which almost all the viscous dissipation is lumped, m
y_{lim}^+	virtual distance from the wall in which almost all the viscous dissipation is lumped, dimensionless
$y_{lim,co}^+$	virtual distance from the wall in which almost all the viscous dissipation is lumped for the total mass flow considered as consisting of *continuum* only, dimensionless
y_{lt}^+	viscous boundary layer limit, dimensionless
$y_{sym.\,lines}$	distance from the wall to the symmetry line in the bundles, m
y^+	distance from the wall, dimensionless
z	axial coordinate, m
Δz	finite of the axial distance, m

Greek

α	volumetric fraction, dimensionless
α_{dm}	volume fraction of the dispersed phase corresponding to the maximum packing of the particles, dimensionless
α'	fluctuation of the volume fraction, dimensionless
α^e	surface averaged volume fraction, dimensionless
$\alpha_{1,far}$	bubble void fraction far from the wall, dimensionless
$\alpha_{1,max}$	maximum bubble void fraction far from the wall, dimensionless
$\alpha_{1,p}$	bubble void fraction at the wall, dimensionless
$\alpha_{1,peack}$	maximum peak of bubble void fraction near to the wall, dimensionless
δ	boundary layer with thickness, m
δ_l	= 1 in case of continuous field l; = 0 in case of disperse field l
ε	power dissipated irreversibly due to *turbulent pulsations* in the viscous fluid per unit mass of the fluid (dissipation of the specific turbulent kinetic energy), m²/s³
η	dynamic viscosity, kg/(ms)
η^t	turbulent or eddy dynamic viscosity, kg/(ms)
η^l	molecular dynamic viscosity, kg/(ms)
η_{vis}	part of the mechanical energy directly dissipated into heat after a local singularity and not effectively generating turbulence, dimensionless
θ	angular coordinate, rad
κ	= 0, Cartesian coordinates; = 1, cylindrical coordinates, - or von Karman constant, -
κ_1	constant in the turbulent model of *Troshko* and *Hassan*, -
κ_{12}	constant in the turbulent model of *Troshko* and *Hassan*, -
κ_d	curvature of the dispersed phase, 1/m
ω_d	bubbles oscillations frequency, 1/s
λ	thermal conductivity, W/(mK)
λ_{fr}	friction coefficient, dimensionless
v	cinematic viscosity, m²/s
v^t_l	turbulent or eddy cinematic viscosity of field l, m²/s
$v_{l,eff}$	effective cinematic viscosity of field l, m²/s
v^t_{12}	turbulent or eddy cinematic viscosity of the liquid caused by the bubbles only, m²/s
$v^t_{c,y}$	turbulent or eddy cinematic viscosity of field l in direction y, m²/s
$v^t_{c,z}$	turbulent or eddy cinematic viscosity of field l in direction z, m²/s
v^*_l	$:= v^t_l + \delta_l v_l$, effective cinematic viscosity, m²/s

ρ	density, kg/m³
$\Delta\rho_{21}$	liquid–gas density difference, kg/m³
σ	surface tension, N/m
τ	time, s
τ_l	share stress in field l, N/m²
$\tau_{l,grav}$	share stress in field l due to gravity, N/m²
$\tau_{l,rr}$	stress in field l in r direction in r-plane, N/m²
τ_{2w}	share stress at the wall liquid interface, N/m²
$\tau_{l,\theta\theta}$	stress in field l in the θ-direction in θ-plane, N/m²
τ_c^w	wall share stress caused by the continuum, N/m²
τ_w	wall share stress, N/m²

Subscripts

1	gas
2	liquid
3	droplet
c	continuum
d	disperse
l	field l
m	field m
e	eddy
μ	associated to mass transfer or microscale
r	radial direction
θ	angular direction
z	axial direction
w	wall
∞	steady, developed flow

Superscripts

'	fluctuation component

References

Baratto F, Bailey SCC and Tavoularis (2006) Measurements of frequencies and spatial correlations of coherent structures in rod bundle flows, Nuclear Engineering and Design, vol 236 pp 1830-1837

Bataille J and Lance M (1988) Turbulence in multiphase flows, Proc. Of the first world congress on Experimental Heat Transfer, Fluid Mechanics, and Thermodynamics, held Sept. 4-9, 1988 in Dubrovnik, Yugoslavia, Elsevier, Shah RK, Ganic´EN and Yang KT eds.

Batchelor GK (1988) A new theory of the instability of a uniform fluidized bed, J. Fluid Mechanic v 193 pp 75-110

Bogoslovskaya GP, Sorokin AP, Kirillov PL, Zhukov AV, Ushakov PA and Titov PA (1996) Experimental and theoretical studies into transverse turbulent transfer of momentum and energy in complex-shaped channels, High Temperature -USSR vol 34 pp 903-908

Chang D and Tavoularis S (July 2006) Convective heat transfer in turbulent flow near a gap, Journal of heat transfer, vol 108 pp 701-708

Chu Y-C and Jones BG (1980) Convective heat transfer coefficient studies in upward and downward, vertical, two-phase non-boiling flows, 19th Nat. Heat Transfer Conf., Orlando, Florida, ed. Stein RP AIChE Symposium Series vol 76 no 199 pp 79-90, New York

Cook TL and Harlow FH (July 1984) VORT: A computer code for bubbly two-phase flow, LA-10021-MS, DE84 017076

Edler JW (1959) J. Fluid Mech., vol 5 pp 242-249

Guellouz MS and Tavoularis S (1992) Heat transfer in rod bundle subchannels with varying rod-wall proximity, Nuclear Engineering and Design, vol 132 pp 351-366

Guellouz MS and Tavoularis S (2000) The structure of the turbulent flow in a rectangular channel containing a cylindrical rod – Part 1: Reynolds-averaged measurements, Experimental Thermal and Fluid Science, vol 23 pp 59-93

Hinze JO (1955) Fundamentals of hydrodynamics of splitting in dispersion processes, AIChE Journal, vol 1 pp 284-295

Kataoka I and Serizawa A (1995) Modeling and prediction of bubbly two phase flow, Proc. 2nd Int. Conf. Multiphase Flow, Kyoto, pp MO2 11-16

Kays WM (1994) Turbulent Prandtl number-where are we? J. Heat Transfer, vol 116 pp 284-295

Kolev NI (2007a) Multiphase Flow Dynamics, Vol. 1 Fundamentals, 3d extended ed., Springer, Berlin, New York, Tokyo

Kolev NI (2007b) Multiphase Flow Dynamics, Vol. 2 Thermal and mechanical interactions, 3d extended ed., Springer, Berlin, New York, Tokyo

Koncar B, Mavko B and Hassan YA (Oct. 2-6, 2005) Two-phase wall function for modeling of turbulent boundary layer in subcooled boiling flow, The 11th Int. Top. Meeting on Nuclear Reactor Thermal-Hydraulics (NURETH-11), Avignon, France

Lance M and Bataille J (1991) Turbulence in the liquid phase of a uniform bubbly air-water flow, J. of Fluid Mechanics, vol 22 pp 95-118

Lahey RT Jr, Shiralkar BS, Radcliffe DW and Polomik EE (1972) Out-of pile subchannel measurements in a nine-rod bundle for water at 1000psia, Progress in Heat Transfer, Hetstroni G ed., Pergamon, London, vol 6 pp 345-363

Lahey RT Jr and Moody FJ (1993) The thermal-hydraulic of boiling water nuclear rector, 2nd ed., ANS, La Grange Park pp 168-184

Lilienbaum W (1983) Turbulente Blasenströmung im geneigten Kanal, Technische Mechanik vol 6 Heft 1 S. 68-77

Liu TJ and Bankoff SG (1993a) Structure of air-water bubbly flow in a vertical pipe-I. Liquid mean velocity and turbulence measurements, Int. J. Multiphase Flow 36, no 4 1049-1060

Liu TJ and Bankoff SG (1993b) Structure of air-water bubbly flow in a vertical pipe-II. Void fraction, bubble velocity

Marié JL (1987) Modeling of the skin friction and heat transfer in turbulent two-component bubbly flow in pipes, Int. J. Multiphase Flow, vol 13 no 3 pp 309-325

Mikiyoshi I and Serizawa A (1986) Turbulence in two-phase bubbly flow, Nuclear engineering and design, vol 95 pp 253-267
Moursalli E, Marié JL and Bataille J (1995) An upward turbulent bubbly boundary layer along a vertical flat plate, Int. J. Multiphase Flow, vol 21 no 1 pp 107-117
Nakoryakov VE, Kashinsky ON, Burdukov AP and Odnoral VP (1981) Local characteristics of upward gas-liquid flows, Int. J. Multiphase Flow vol 7 pp 63-81
Ouma BH and Tavoularis S (1991) Turbulence in triangular subchannels of a reactor bundle model, Nuclear Engineering and Design, vol 128 pp 271-287
Peebles FM and Garber JH (1953) Studies on the motion of gas bubbles, Chem. Eng. Sci. vol 49 pp 88-97
Petrunik K (1973) PhD Thesis, Dept. of Chemical Engineering, University of Windsor, Canada
Pu F, Qiu S, Su G and Jia D (July 17-20, 2006) An investigation of flow, heat transfer characteristic of anular flow and critical heat flux in vertical upward round tube, Proceedings of ICONE14, International Conference on Nuclear Engineering, Miami, Florida, USA , ICONE14-89108
Reichardt H (Juli 1951) Vollständige Darstellung der turbulenten Geschwindigkeiten in glaten Leitungen, Z. angew. Math. Mech., Bd. 31 Nr. 7, S. 208-219
Rehme K (1992) The structure of turbulence in rod bundles and the implications on natural mixing between the subchannels, Int. J. Heat Mass Transfer, vol 35 no 2 pp 567-581
Roger JT and Tahir AEE (1975) ASME paper no 75-HT-31
Sato Y and Sekoguchi K (1975) Liquid velocity distribution in two phase bubbly flow, Int. J. Multiphase Flow, vol 2 pp 79-95
Sato Y, Sadatomi M and Sekoguchi K (1981) Momentum and heat transfer in two-phase bubble-flow-I, Theory, Int. J. Multiphase Flow, vol 7 pp 167-177
Seale WJ (1981) The effect of subchannel shape on heat transfer in rod bundles with axial flow, Int. J. Heat Mass Transfer vol 24, pp 768-770
Sekogushi K, Fukui H and Sato Y (1979) Flow characteristics and heat transfer in vertical bubble flow, Two-Phase Flow Dynamics, Japan-U.S. Seminar, eds. Bergles AE and Ishigai S, Hemisphere Publishing Corporation, Washington
Serizawa A, Kataoka I and Michiyoshi I (1975a) Turbulence structure of air-water bubbly flow – I. Measuring techniques, Int. J. Multiphase Flow, Vol 2 pp 221-233
Serizawa A, Kataoka I and Michiyoshi II (1975b) Turbulence structure of air-water bubbly flow – I. Local properties, Int. J. Multiphase Flow, Vol 2 pp 235-246
Serizawa A, Kataoka I and Michiyoshi III (1975c) Turbulence structure of air-water bubbly flow – I. Transport properties, Int. J. Multiphase Flow, Vol 2 pp 247-259
Serizawa A and Kataoka I (1980) Fundamental aspects of the drift velocity in turbulent bubbly flow, Techn. Reports Inst. Atomic Energy, Kyoto Univ., Rept. No 182
Tomiyama A, Shimada N, Abe S and Zun I (June 11-15, 2000) (N+2)-field modeling of dispersed multiphase flow, ASME 2000 Fluids Engineering Division Summer Meeting, Boston, Massachusetts
Troshko AA, Hassan YA (2001) Law of the wall for two-phase turbulent boundary layers, Int. J. Heat Mass Transfer vol 44 no 4 pp 871-875
Troshko AA, Hassan YA (2001) A two-equation turbulence model of turbulent bubbly flows, Int. J. Multiphase Flow vol 27 pp 1965-2000
van Driest ER (1955) On turbulent flow near a wall, Heat Transfer and Fluid Mechanics Institute and bubble size distribution, Int. J. Multiphase Flow 36, no 4 1061-1072
Wang SK, Lee SJ, Jones OC and Lahey RT Jr (1987) 3-D turbulence structure and phase distribution measurements in bubbly two-phase flows, Int. J. Multiphase Flow, vol 13 no 3 pp 327-343

Zaruba A, Prasser H-M and Kreper E (Oct. 2-6, 2005) Experiments on turbulent diffusion of the gaseous phase in rectangular bubble column using image processing, The 11th Int. Top. Meeting on Nuclear Reactor Thermal-Hydraulics (NURETH-11), Avignon, France

Zhukov AV, Kirillov PL, Sorokin AP and Matjukhin NM (1994) Transverse turbulent momentum and energy exchange in the channels of complicated form, Proc. Heat Transfer 1994, Brighton, vol 4, pp 327-332

Zun I (1980) The transferees migration of bubbles influenced by walls in vertical bubbly flow, Int. J. Multiphase Flow, vol 6 pp 583-588

10 Large eddy simulations

10.1 Phenomenology

If we observe turbulence in flows, we distinguish large scale structures that can well be resolved by overlying the picture with computational grid that is economically feasible and small scale eddies smaller than the used grid size that can not be resolved. The large eddies are directly born from what is subjectively called mean flow. Their size is in a way limited by the geometry of the flow boundaries and in a way how they are generated. They are responsible for effective turbulent transport of mass and energy. Due to their interactions with the mean flow and with the other eddies they collide and coalesce to larger eddies or split to smaller eddies: an endless game that fascinate children and make scientist desperate to describe them mathematically because of the enormous complexity of the process. For the same reason a chain of the smaller eddies with all possible sizes is generated. Those eddies which size is smaller than what is called *Kolmogoroff* small scales dissipate their rotation- and fluctuation energy into heat. While the large eddies hardly have the same structure in all directions, the small scale eddies tend to similarity independent on the flow direction – a property named *isotropy*. Exactly this observation lead *Smagorinski* in 1963 to the idea to look for such conservation equations that describe physics that can really be resolved on the used computational grid and separate the remaining physics that have to be resolved by additional modeling. The non resolved part or the so-called filtered part is modeled in such a way that the energy for the unresolved eddies is taken from the resolved mechanical energy. This approach is called Large Scale Simulation and is getting since that time very popular in the single phase fluid mechanics.

Applying this method to multiphase flow dynamics is very new branch of the science and up to now limited to bubbly and droplet flows only. Nevertheless, because it is very promising, we will describe briefly the main ideas behind this modeling technique.

10.2 Filtering – brief introduction

The reader my start with the book by *Sagaut* (1988) to learn methods for single phase incompressible flow.

Consider the point **x** in the space. Around this point at a distance **x-x′** some flow property $f(\mathbf{x}')$ is normally distributed obeying the *Gauss distribution*

$$G(\mathbf{x},\mathbf{x}') = \sqrt{\frac{6}{\pi\Delta^2}} e^{-\frac{6|\mathbf{x}-\mathbf{x}'|^2}{\Delta^2}}. \tag{10.1}$$

Δ is the *resolution scale* at which the property is observed. Now consider a *control volume* $Vol > \Delta^3$ in which at each point $\mathbf{x}'$ the probability of the property to receive the value $f(\mathbf{x}')$ is defined by the *Gauss* function. Other functions are also used in the literature:

$$G(\mathbf{x},\mathbf{x}') = -\frac{1}{\Delta}, \text{ box filter function, } Lilly \text{ (1967)}$$

$$G(\mathbf{x},\mathbf{x}') = \frac{2\sin\left[\pi(\mathbf{x}-\mathbf{x}')/\Delta\right]}{\pi(\mathbf{x}-\mathbf{x}')}, Fourier\text{-space sharp cut off filter function,}$$

truncated *Fourier* expansion with $|k| < \pi/\Delta$, Leonard (1967). If we are interested in smoothing the function $f(\mathbf{x}')$ over *Vol* we have to perform the averaging

$$\bar{f}(\mathbf{x},\Delta) = \frac{1}{Vol} \int_{Vol} f(\mathbf{x}') G(\mathbf{x},\mathbf{x}') d\mathbf{x}'. \tag{10.2}$$

This operation is called *filtering* because the averaged property is obviously a function of our ability to observe the property f with spatial resolution of Δ. We loose information associated with lower scales. The filtering operation is called *appropriate* if (a) the control volume is selected so that it satisfies the so-called *normalization condition*

$$\int_{Vol} G(\mathbf{x},\mathbf{x}') d\mathbf{x}' = 1, \tag{10.3}$$

and (b) it has to be symmetric and with constant filter with Δ in order to commute with differentiation, *Leonard* (1974).

Now consider a *field indicator*

$$\chi_l(\mathbf{x}) = \begin{cases} 1, & \text{if } \mathbf{x} \in l \\ 0, & \text{if } \mathbf{x} \notin l \end{cases}, \tag{10.4}$$

which is nothing else than the volume faction of filed l resolved in the space perfectly. Knowing that the derivative of the volume faction is equal to the unit vector pointing outward of the filed interface, and considering that this interface is defined only at $\mathbf{x}^\sigma$ we have

$$\frac{\nabla \chi_l}{|\nabla \chi_l|} = -\mathbf{n}_l \delta(\mathbf{x} - \mathbf{x}^\sigma), \tag{10.5}$$

where the *Dirac* delta function $\delta(\mathbf{x} - \mathbf{x}^\sigma)$ has value of unity only at the sharp interface $\mathbf{x} = \mathbf{x}^\sigma$ and zero elsewhere,

$$\delta_l = \begin{cases} 1 \text{ for } \mathbf{x} = \mathbf{x}^\sigma \\ 0 \text{ for } \mathbf{x} \neq \mathbf{x}^\sigma \end{cases}. \tag{10.6}$$

The *averaged volumetric fraction* of the field l within *Vol* is defined by

$$\langle \alpha_l \rangle = \frac{1}{Vol} \int_{Vol} \chi_l(\mathbf{x}') d\mathbf{x}'. \tag{10.7}$$

Obviously

$$\sum_{l=1}^{l_{max}} \langle \alpha_l \rangle = 1. \tag{10.8}$$

The *filtered volume fraction* of the field l within *Vol* is

$$\bar{\alpha}_l(\mathbf{x}, \Delta) = \frac{1}{Vol} \int_{Vol} \chi_l(\mathbf{x}') G(\mathbf{x}, \mathbf{x}') d\mathbf{x}'. \tag{10.9}$$

Again this average over *Vol* depends on our ability to identify field l with spatial resolution of Δ. Note that

$$\bar{\alpha}_l(\mathbf{x}, \Delta \to 0) \to \langle \alpha_l \rangle. \tag{10.10}$$

Otherwise,

$$\alpha_l' = \langle \alpha_l \rangle - \bar{\alpha}_l(\mathbf{x}, \Delta \to 0) \neq 0, \tag{10.11}$$

is the error due to non-perfect resolution. If the selection of the size of the control volume is *appropriate* in the sense discussed above, the condition

$$\sum_{l=1}^{l_{max}} \bar{\alpha}_l = 1 \tag{10.12}$$

is fulfilled. Attention, for non appropriate selection of the control volume the above condition is not fulfilled. The resulting system describing the multiphase flow is then not acceptable.

The local volume averaged product of the property of phase *l* and the volume fraction is defined by

$$\langle \alpha_l f_l(\mathbf{x}) \rangle = \frac{1}{Vol} \int_{Vol} \chi_l(\mathbf{x}') f(\mathbf{x}') d\mathbf{x}'. \tag{10.13}$$

The *local volume average of the property f* over *Vol* is then

$$\langle f_l \rangle = \frac{\langle \alpha_l f_l \rangle}{\alpha_l}. \tag{10.14}$$

The intrinsic *filtered* product of the property of phase *l* and the volume fraction is defined by

$$\overline{\alpha_l f_l}(\mathbf{x}, \Delta) = \frac{1}{Vol} \int_{Vol} \chi_l(\mathbf{x}') f(\mathbf{x}') G(\mathbf{x}, \mathbf{x}') d\mathbf{x}'. \tag{10.15}$$

Again this average over *Vol* depends on our ability to observe it with resolution of Δ. The

$$\overline{f_l}(\mathbf{x}, \Delta) = \frac{\overline{\alpha_l f_l}(\mathbf{x}, \Delta)}{\overline{\alpha_l}}. \tag{10.16}$$

is the definition of the filtered property *f* of phase *l*. Again

$$\overline{f_l}(\mathbf{x}, \Delta \to 0) = \langle f_l \rangle, \tag{10.17}$$

otherwise

$$f_l' = \langle f_l \rangle - \overline{f_l}(\mathbf{x}, \Delta) \tag{10.18}$$

is the error due to non-perfect resolution.

Thus, using these ideas and performing local volume averaging of the conservation equation and splitting the flow variables on filtered and remaining part result in a system of PDE's that looks similar to what we know in this work, but the terms possess the physical context as dictated by the way of obtaining large grid scale equations for multiphase flows. It is beyond the scope of this section to present rigorous derivation. We will confine our attention to some practical models.

Before continuing let me note for completeness, that unlike the time averaging of dual products the filtering operation on dual products produces non-zero terms for the non resolved large grid turbulence $\overline{V_l V_l} - \overline{V}_l \overline{V}_l$ which requires additional modelling. The reader will find interesting discussion to this subject in *Leonard* (1974). In most of the applications reported in the literature these terms are neglected.

The success of the large eddy simulation method is based on the differences of the properties of the large and small eddies as summarized in Table 1, and on the possibility to model simpler the small scale eddies.

Table 10.1. Differences between large scale- and small scale eddies, facilitating creation of simple sub-grid scale models, *Troshko* and *Hassan* (2001)

Large eddies	Small scale eddies
Produced by mean flow	Produced by large eddies
Depends on boundaries	Universal
Ordered	Random
Requires deterministic description	Can be modelled
Inhomogeneous	Homogeneous
Anisotropic	Isotropic
Long-lived	Short-lived
Diffusive	Dissipative
Difficult to model	Easier to model

10.3 The extension of the *Amsden* et al. LES model to porous structures

Each numerical discretization of the space is characterized locally by a specific grid size e.g.

$$\ell_{grid} \approx Vol_{cell}^{1/3} = (\Delta x \Delta y \Delta z)^{1/3}. \tag{10.19}$$

Structures of the flow having scale larger than this are already resolved. To extract the loosed information for structures with lower scales additional modeling is necessary. Such modeling has to generate such an effective turbulent viscosity that is diminishing with the scales converging to the internal micro-scales of turbulence e.g. using the *Kolmogorov* (1942) – *Prandtl* (1945) expression Eq. (2.49) in the following form

$$v'_{c,ss} \approx \sqrt{\frac{2}{3}k_{c,ss}}\ell_{ss} \approx c'_\eta \sqrt{k_{c,ss}}\ell_{ss}, \tag{10.20}$$

where $c'_\eta \approx 0.05$ (instead 0.09) is an empirical constant and $\ell_{ss} \approx 2\ell_{grid}$. The turbulence and viscous term in the momentum equation is then

$$\nabla \cdot \left[\alpha_c^e \gamma \left(\rho_c \overline{\mathbf{V}'_{c,ss} \mathbf{V}'_{c,ss}} - \mathbf{T}_{\eta,c} \right) \right] \tag{10.21}$$

in which the viscous stresses are computed in accordance with the *Stokes* hypothesis

$$\mathbf{T}_{\eta,c} = \eta_c \left[2\mathbf{D}_c - \frac{2}{3}(\nabla \cdot \mathbf{V}_c)\mathbf{I} \right], \tag{10.22}$$

and the non resolved small grid component is

$$\nabla \cdot \left[\left(\alpha_c^e \rho_c \overline{\mathbf{V}'_{c,ss} \mathbf{V}'_{c,ss}} \right) \gamma \right]$$

$$= -\begin{pmatrix} \dfrac{\partial}{\partial y}\left[\gamma_y \alpha_c^e \rho_c v'_{c,ss} \left(\dfrac{\partial v_c}{\partial x} + \dfrac{\partial u_c}{\partial y} \right) \right] + \dfrac{\partial}{\partial z}\left[\gamma_z \alpha_c^e \rho_c v'_{c,ss} \left(\dfrac{\partial w_c}{\partial x} + \dfrac{\partial u_c}{\partial z} \right) \right] \\[4pt] \dfrac{\partial}{\partial x}\left[\gamma_x \alpha_c^e \rho_c v'_{c,ss} \left(\dfrac{\partial v_c}{\partial x} + \dfrac{\partial u_c}{\partial y} \right) \right] + \dfrac{\partial}{\partial z}\left[\gamma_z \alpha_c^e \rho_c v'_{c,ss} \left(\dfrac{\partial w_c}{\partial y} + \dfrac{\partial v_c}{\partial z} \right) \right] \\[4pt] \dfrac{\partial}{\partial x}\left[\gamma_x \alpha_c^e \rho_c v'_{c,ss} \left(\dfrac{\partial w_c}{\partial x} + \dfrac{\partial u_c}{\partial z} \right) \right] + \dfrac{\partial}{\partial y}\left[\gamma_y \alpha_c^e \rho_c v'_{c,ss} \left(\dfrac{\partial w_c}{\partial y} + \dfrac{\partial v_c}{\partial z} \right) \right] \end{pmatrix}$$

$$+ \frac{2}{3} \nabla \left(\gamma \alpha_c^e \rho_c k_{c,ss} \right) \tag{10.23}$$

This method will be then applicable for variety of grid sizes. Methods with such characteristics belong to the large eddy simulation methods. *Smagorinski* (1963) was the first to propose such approach for single phase atmospheric flow which as already mentioned is now very popular in the single phase fluid mechanics. The small scale specific kinetic energy of turbulent pulsation per unit mass in the above relation, $k_{c,ss}$, is associated with fluctuations with sizes smaller than ℓ_{ss}.

Interesting generalization of this idea is proposed by *Amsden* et al. (1985) for fuel injection in combustion chambers which we extend here to *porous structure and multiphase flow*. The turbulent kinetic energy associated with scales smaller than ℓ_{ss} is controlled by the conservation equation

$$\frac{\partial}{\partial t}(\alpha_c \rho_c k_{c,ss} \gamma_v) + \nabla \cdot \left[\alpha_c \rho_c \left(\mathbf{V}_c k_{c,ss} - v^k_{c,ss} \nabla k_{c,ss} \right) \gamma \right]$$

$$= \alpha_c \rho_c \gamma_v \left(v'_{c,ss} \overline{P_{k,l}} - \varepsilon_{c,ss} + G_{k,c} + P_{k\mu,c} + P_{kw,c} \right). \tag{10.24}$$

10.3 The extension of the *Amsden* et al. LES model to porous structures

The term $\alpha_c \rho_c \gamma_v \overline{v'_{c,ss} P_{k,c}}$ represents the production of turbulence by the share of the resolved velocity field. Considering the non resolved turbulence as isotropic and splitting the production term as shown in Chapter 2 on diagonal and of-diagonal part

$$\alpha_c \rho_c \gamma_v \overline{v'_{c,ss} P_{k,c}} = \alpha_c^e \gamma \left(\mathbf{T}'_{c,ss} : \nabla \mathbf{V}_c \right) = -\alpha_c^e \rho_c \frac{2}{3} k_c \mathcal{W} \cdot \mathbf{V}_c + \alpha_c^e \rho_c v_c' \tilde{S}_{k,c}^2, \quad (10.25)$$

where

$$\mathcal{W} \cdot \mathbf{V}_c = \gamma_r \frac{\partial u_c}{\partial r} + \gamma_\theta \frac{1}{r^\kappa} \left(\frac{\partial v_c}{\partial \theta} + \kappa u_c \right) + \gamma_z \frac{\partial w_c}{\partial z}, \quad (10.26)$$

and

$$\tilde{S}_{k,l}^2 = \left[\frac{\partial v_c}{\partial r} + \frac{1}{r^\kappa} \left(\frac{\partial u_c}{\partial \theta} - \kappa v_c \right) \right] \left[\gamma_r \frac{\partial v_c}{\partial r} + \gamma_\theta \frac{1}{r^\kappa} \left(\frac{\partial u_c}{\partial \theta} - \kappa v_c \right) \right]$$

$$+ \left[\frac{\partial w_c}{\partial r} + \frac{\partial u_c}{\partial z} \right] \left[\gamma_r \frac{\partial w_c}{\partial r} + \gamma_z \frac{\partial u_c}{\partial z} \right]$$

$$+ \left[\frac{1}{r^\kappa} \frac{\partial w_c}{\partial \theta} + \frac{\partial v_c}{\partial z} \right] \left[\gamma_\theta \frac{1}{r^\kappa} \frac{\partial w_c}{\partial \theta} + \gamma_z \frac{\partial v_c}{\partial z} \right], \quad (10.27)$$

we obtain

$$\frac{\partial}{\partial \tau} \left(\alpha_c \rho_c k_{c,ss} \gamma_v \right) + \nabla \cdot \left[\alpha_c \rho_c \left(\mathbf{V}_c k_{c,ss} - v_{c,ss}^k \nabla k_{c,ss} \right) \gamma \right]$$

$$= -\alpha_c \rho_c \frac{2}{3} k_{c,ss} \mathcal{W} \cdot \mathbf{V}_c + \alpha_c \rho_c \gamma_v \left(v'_{c,ss} \tilde{S}_{k,l}^2 - \varepsilon_{c,ss} + G_{k,c} + P_{k\mu,c} + P_{kw,c} \right). \quad (10.28)$$

Here

$$p'_c = \rho_c \overline{V'^2_{c,ss}} = \rho_c \frac{2}{3} k_{c,ss} \quad (10.29)$$

and the term

$$\alpha_c \rho_c \frac{2}{3} k_{c,ss} \mathcal{W} \cdot \mathbf{V}_c \quad (10.30)$$

is the compressibility term that represents the turbulent analog to mechanical $pdVol$-work.

The dissipation of the sub-grid kinetic energy into heat is computed by using the *Kolmogoroff* (1941, 1949) equation for isotropic turbulence in the form

$$\varepsilon_{l,ss} \approx c_1 k_{l,ss}^{3/2} / \ell_{ss}, \qquad (10.31)$$

where the constant is of order of unity. This is the term that has to appear as a *source in the energy conservation equation*. Note that for steady developed flow the production equals the dissipation,

$$\overline{v'_{l,ss} P_{k,l}} = \varepsilon_{l,ss}, \qquad (10.32)$$

resulting in the original algebraic sub-grid scale model

$$\overline{v'_{c,ss} P_{k,c}} = k_{c,ss}^{3/2} / \ell_{ss}. \qquad (10.33)$$

Using

$$v'_c \approx c'_\eta \sqrt{k_c} \ell_{ss} \qquad (10.34)$$

the above relation can be rewritten as

$$\boxed{v'_{c,ss} = \left(Sm\ell_{ss}\right)^2 \sqrt{\overline{P_{k,c}}}.} \qquad (10.35)$$

The constant

$$Sm = c'^{3/4}_\eta \qquad (10.36)$$

is called *Smagorinsky* constant (originally 0.28). Note that for

$$\nabla \cdot \mathbf{V}_c \approx 0 \qquad (10.37)$$

we have

$$\boxed{v'_{c,ss} \approx \left(Sm\ell_{ss}\right)^2 \sqrt{\tilde{S}^2_{k,c}/\gamma_v}} \qquad (10.38)$$

10.3 The extension of the *Amsden* et al. LES model to porous structures

For free single phase incompressible flow this model is identical to the *Smagorinski* (1963) and *Deardorff* (1971) proposal

$$v_{ss}^t = \left(Sm\ell_{ss}\right)^2 \sqrt{\tilde{S}_k^2}.\tag{10.39}$$

With this expression the turbulent viscosity is simply a function of the discretization size and of the spatial deformation of the resolved velocity field. Theoretical support for the *Smagorinski* approach was provided by *Lily* (1967) $Sm = \frac{1}{\pi}\left(\frac{2}{3\beta}\right)^{3/4} \approx 0.17$, for $\beta = 1.5$. *Milelli* (2002) analysed bubbly flow with only the algebraic part of this model using $Sm = 0.1$ to 0.33. The sub-grid scale *Prandtl* number for heat transfer was in the order of 0.3 to 0.9. *Yamamotto* et al. (2001) used $Sm = 0.1$ which corresponds to $c_\eta' \approx 0.05$.

Equation (10.19) was proposed by *Reynold* (1990) and *Scotti* et al. (1993). Such measure was found to be useful up to aspect ratio 20:1 by *Reynold* (1990). In the proximity of the wall introduction of turbulence suppression is also possible e.g. by using the *van Driest* (1955) damping factor

$$\ell_{ss} \approx \ell_{grid}\left[1 - \exp\left(-\frac{y^+}{16}\right)\right]^2.\tag{10.40}$$

Dean et al. (2001) used for the constant $c_\eta' \approx 0.046$ and $\ell_{ss} \approx \ell_{grid}$ for bubbly flow and added the contribution of the bubbles to the effective turbulent velocity

$$v_l^{td} = 0.6 D_d \alpha_d \left|\Delta V_{cd}\right|.\tag{10.41}$$

The authors used also $v_d^t = v_c^t$ as proposed by *Jakobsen* et al. (1997). *Lakehal* et al. (2002) used for the constant $c_\eta' \approx 0.059$ and $\ell_{ss} \approx \ell_{grid}$ for bubbly flow and added the contribution of the bubbles to the effective turbulent velocity

$$v_l^{td} = 0.12 \ell_{grid} \alpha_d \left|\Delta V_{cd}\right|.\tag{10.42}$$

Observe that now the characteristic small grid mixing length is not associated with the bubble size as by *Dean* et al. (2001) but depends on the grid scale. *Milelli* (2002) used

$$v_l^{td} = 0.126\pi \frac{D_d}{\Delta} \alpha_d V_c \qquad (10.43)$$

as suggested by *Tran* (1977) or

$$v_l^{td} = 0.12 Sm\alpha_d |\Delta V_{cd}|. \qquad (10.44)$$

Valuable experimental data for testing such type of simulation for vertical 2D-bubble columns are presented by *Milelli* (2002) and *Vanga* (2005).

In summary, three LES methods are available:

1. Algebraic method, using the *Smagorinski* relation for small grid scale eddy viscosity, $v'_{c,ss} = (Sm\ell_{ss})^2 \sqrt{P_{k,c}}$, and the not yet well established additive for the contribution of the dispersed phase from the type $v_l^{td} = 0.12 \ell_{grid} \alpha_d |\Delta V_{cd}|$.

2. k-transport equation providing the kinetic energy of the not resolved turbulence and then computing $v'_{c,ss} \approx c'_\eta \sqrt{k_c} \ell_{ss}$. In this case the contribution of the dispersed phase is introduced as a source into the *k*-conservation equation. The irreversible dissipation in this case is modeled by the *Kolmogoroff* equation for isotropic turbulence $\varepsilon_{l,ss} \approx c_1 k_{l,ss}^{3/2}/\ell_{ss}$. This dissipation appears into the energy conservation equation as turbulence energy that is finally dissipated in heat.

3. k-eps transport equations and $v'_{c,ss} \approx c_\eta k_{c,ss}^2/\varepsilon_{c,ss}$. In this case the contribution of the dispersed phase is introduced as a source into the *k-eps* conservation equations.

Nomenclature

Latin

c'_η empirical constant in Kolmogoroff–Prandtl relation, dimensionless

c_1 modeling constant in the Kolmogoroff equation for dissipation of the subgrid kinetic energy into heat for isotropic turbulence, dimensionless

D diameter, m

$dVol$ infinitesimal volume, m³

$G_{k,l}$ production of turbulent kinetic energy due to bubble relocation in changing pressure field per unit mass of the filed *l*, W/kg (m²/s³)

$k_{l,ss}$	specific turbulent kinetic energy of the unresolved turbulence in velocity field l, m²/s²
ℓ_{grid}	characteristic size of the computational grid, m
ℓ_{ss}	scales of unresolved turbulent kinetic energy, m
p	pressure, Pa
p'	pressure pulsation, Pa
$P_{k\mu,l}$	production of turbulent kinetic energy per unit mass of the field l due to friction evaporation or condensation, W/kg
$P_{kw,l}$	irreversibly dissipated power per unit flow mass outside the viscous fluid due to turbulent pulsations equal to production of turbulent kinetic energy per unit mass of the flow, W/kg (m²/s³)
$\overline{P_{k,l}}$	in $\overline{v'_{l,ss} P_{k,l}}$ which is the production of the turbulent kinetic energy per unit mass of the velocity field l due to deformation of the velocity field l, W/kg
$\tilde{S}^2_{k,l}$	of-diagonal part of $\overline{P_{k,l}}$
$\Delta x, \Delta y, \Delta z$	coordinate increments, m
$\mathbf{V}$	velocity vector, m/s
$V'_{l,ss}$	velocity fluctuation component of the unresolved turbulence, m/s
Vol_{cell}	cell volume, m³
x, y, z	coordinates, m

Greek

α	local volume fraction, dimensionless
γ	surface permeability defined as flow cross section divided by the cross section of the control volume (usually the three main directional components are used), dimensionless
γ_v	volumetric porosity defined as the flow volume divided by the considered control volume, dimensionless
$\nabla.$	divergence
∂	partial differential
$\varepsilon_{l,ss}$	power dissipated irreversibly due to the unresolved *turbulent pulsations* in the viscous fluid per unit mass of the fluid (dissipation of the specific turbulent kinetic energy), m²/s³
$v^t_{l,ss}$	effective viscosity characterizing the unresolved turbulence scales, m²/s
$v^k_{l,ss}$	effective diffusivity of the unresolved turbulent kinetic energy, m²/s
v^{td}_l	bubble induced effective viscosity component, m²/s
ρ	density, kg/m³
τ	time, s

Subscripts

l   field *l*
c   continuous
d   discrete
ss small grid scale

References

Amsden AA, Butler TD, O'Rourke PJ and Ramshaw JD (1985) KIVA-A comprehensive model for 2-D and 3-D engine simulations, paper 850554

Dean NG, Solberg T and Hjertager BH (2001) Large eddy simulation of the Gas-Liquid flow in square cross-sectioned bubble column, Chemical Engineering Science, vol 56 pp 6341-6349

Deardorff JW (1970) J. Fluid Mechanics, vol 41 p 453

Deardorff JW (1971) J. Comp. Physics, vol 7 p 120

Jakobsen HA, Grevskott BH and Svendsen HF (1997) Modeling of vertically bubbly driven flow, Industrial and Engineering Chemistry Research, vol 36 pp 4052-4074

Kolmogoroff AN (1941) The local structure of turbulence in incompressible viscous fluid for very large Reynolds numbers, C. R. Acad. Sci. U.S.S.R., vol 30 pp 825-828

Komogoroff AN (1942) Equations of turbulent motion of incompressible fluid, Isv. Akad. Nauk. SSR, Seria fizicheska Vi., no 1-2 pp 56-58

Lakehal D, Smith BL and Milelli M (2002) Large eddy simulation of bubbly turbulent shear flow, J. Turbulence, vol 3 p 25

Leonard A (1974) Energy cascade in large eddy simulations of turbulent fluid flows, Adv. in Geophysics, A18, pp 237-248

Lily DK (1967) In Proc. IBM Scientific Computing Symposium on Environmental Science, pp 195-210. Thomas Watson Research Center, Yorktown Heights, N. Y.

Milelli M (2002) A numerical analysis of confined turbulent bubble plumes, PhD Thesis, ETH No. 14799, Swiss Federal Institute of Technology, Zurich

Prandtl LH (1945) Über ein neues Formelsystem für die ausgebildete Turbulenz, Nachr. Akad. Wiss., Göttingen, Math.-Phys. Klasse p 6

Reynold WC (1990) In: Wether Turbulence? Turbulence of Crossroads (JL Lumley ed.), pp 313-342, Lecture Notes in Physics, vol 357, Springer-Verlag, Berlin.

Sagaut P (1988) Large eddy simulations for incompressible flow, Springer, Berlin

Scotti A, Monereau C and Lilly DK (1993) Phys. Fluids, vol A5 pp 2306-2308

Smagorinski JS (1963) General circulation experiments with the primitive equations. I. Basic experiment, Month. Weather Review, vol 99 pp 99-165

Tran ML (1977) Modélisation instationare de la distribution spatial des phases dans les écoulementsdiphasiques en régime à bules, PhD Thesis, Université Lyon

Troshko AA and Hassan YA (2001) A two-equations turbulence model of turbulent bubbly flow, Int. J. of Multiphase Flow, vol 27 pp 1965-2000

Yamamotto Y, Potthoff M, Tanaka T, Kajishima and Tsui Y (2001) Large-eddy simulation of turbulent gas-particle flow in a vertical channel : effect of considering inter-particle collisions, J. Fluid Mechanics, vol 442 pp 303-334

van Driest ER (1955) On turbulent flow near a wall, Heat Transfer and Fluid Mechanics Institute and bubble size distribution, Int. J. Multiphase Flow 36, no 4 1061-1072

Vanga BNR (May 2005) Experimental investigation and two-fluid model large eddy simulations of the hydrodynamics of re-circulating turbulent flow in rectangular bubble columns, PhD Thesis, Pardue University, US

11 Solubility of O_2, N_2, H_2 and CO_2 in water

11.1 Introduction

Opening of champagne or soda bottle at atmospheric pressure is a simple example of release of gases dissolved previously into liquid at higher pressure. Liquids absorb gases so that the molecules of the gases move among the molecules of the liquid. Optically no visible bubbles are seen in perfect gas–liquid solutions. If the liquid stays for sufficient long time in contact with gas, the gas concentration inside the liquid reaches a maximum. The experience shows this maximum is a function of the temperature of the liquid and of the partial pressure of the gas component in the gas mixture,

$$C_{2i,\infty} = C_{2i,\infty}(p_{1i}, T_2). \qquad (11.1)$$

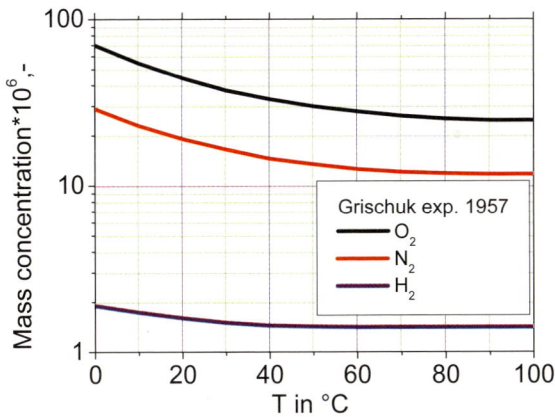

Fig. 11.1. Saturation O_2, N_2 and H_2 concentrations in water being in contact with *pure* gas as a function of the water temperature measured by *Grischuk* (1957). The gas pressure is $10^5 Pa$

This maximum concentration is called *saturation concentration*. Table 11.1 and Figure 11.1 give an example of measurements for the saturation concentrations of

O_2, N_2 and H_2 in water as a function of the water temperature. The pressure of the pure gas is $1\,bar$.

Table 11.1. *Grischuk (1957) data in mg/kg*

T in K	O_2	N_2	H_2
273.15	69.93	29.14	1.92
283.15	54.34	23.06	1.75
293.15	44.33	19.22	1.62
303.15	37.32	16.62	1.52
313.15	33.03	14.63	1.46
323.15	29.89	13.52	1.44
333.15	27.89	12.65	1.43
343.15	26.17	12.15	1.43
353.15	25.17	11.90	1.43
363.15	24.60	11.78	1.43
373.15	24.60	11.78	1.43

If the non condensable gas concentration inside the liquid is larger than the saturation concentration, a gas release starts and continues until the new state of equilibrium is reached. The degassing is visible. Bubbles are generated inside the liquid. The mass of the *generated* non condensable gas per unit time and unit volume of the multi-phase mixture is designated with

$$\mu_{21,i} = f\left[C_{2i} - C_{2i,\infty}\left(p_{1i}, T_2\right)\right] \geq 0 \text{ for } C_{2i} > C_{2i,\infty}\left(p_{1i}, T_2\right). \tag{11.2}$$

The mass of the *solved* non condensable gas per unit time and unit volume of the multi-phase mixture is designated with

$$\mu_{12,i} = f\left[C_{2i} - C_{2i,\infty}\left(p_{1i}, T_2\right)\right] \geq 0 \text{ for } C_{2i} < C_{2i,\infty}\left(p_{1i}, T_2\right). \tag{11.3}$$

For the purpose of correlating experimental data an idealization proposed by *Henry* is used in the literature called *Henry*'s law. The *Henry*'s law says

> "The mass of non condensable gas dissolved in a liquid is proportional to the partial pressure of the gas around the liquid with which the latter is in equilibrium"

Frequently in the chemical thermodynamic literature the solubility data are approximated by the *Henry*'s law in the following form

$$p_{1i} = k_{H,2i}\left(T_2\right) Y_{2i,\infty}. \tag{11.4}$$

Here p_{1i} is the partial pressure of specie i in the gas phase, and $Y_{2i,\infty}$ is the saturation molar concentration of the same specie in the liquid. $k_{H,2i}(T_2)$ is called the *Henry's coefficient*. Later on, we will see that this idealization does not hold for many cases and that there is a pressure dependence on the *Henry*'s coefficient too. Knowing the *Henry*'s coefficient the molar concentration of the saturated solution is then

$$Y_{2i,\infty} = p_{1i}/k_{H,2i}(T_2), \qquad (11.5)$$

and the corresponding mass concentration is

$$C_{2i,\infty} = \frac{Y_{2i,\infty} M_{2i}}{Y_{2i,\infty} M_{2i} + (1-Y_{2i,\infty}) M_{H_2O}}. \qquad (11.6)$$

If the mass concentration of the saturated solution is known the molar concentration is easily computed by

$$Y_{2i,\infty} = \frac{C_{2i}/M_{2i}}{C_{2i,\infty}/M_{2i} + (1-C_{2i,\infty})/M_{H_2O}}. \qquad (11.7)$$

The mass of the dissolved gas in a saturated liquid is then

$$m_{2i,\infty} = \alpha_2 \rho_2 C_{2i,\infty} Vol, \qquad (11.8)$$

where *Vol* is the flow volume, α_2 is the liquid volume fraction and ρ_2 is the liquid density.

The *Grischuk* data from Table 11.1 can be approximated by

$$k_{H,2i}(T_2)(p_{1i}=10^5\,Pa, T_2) = (a_0 + a_1 T_2 + a_2^2 T_2^2 + a_3^3 T_2^3 + a_4^4 T_2^4)10^5\,Pa, \quad (11.9)$$

within 0 and 100°C. The polynomial coefficients and the mean error are given in Table 11.2.

Table 11.2. Coefficients for the approximation of the equilibrium solution of oxygen, nitrogen and hydrogen

	a_0	a_1	a_2	a_3	a_4	error %
O_2	1.2191d6	-16595.37213	78.10548	-0.14912	9.8951d-5	0.30
N_2	9.769d6	-126479.73104	600.04585	-1.23258	9.3109d-4	0.26
H_2	-15.43218d5	0.13585d5	-3.78843d1	3.51564d-2	0	0.46

Note that dn or en means 10^n.

In many literature sources the amount of the dissolved gases in liquids is presented in gas cubic centimeter at standard pressure $p_{norm}=1.0133\times10^5\,Pa$ and temperature T_{norm} per gram of the water (Cc. at S. T. P. per g of water). For the standard temperature different sources use different temperatures e.g. $T_{norm} = 273.15K$ for the so-called *Bunsen absorption coefficient*. In order to obtain the saturation mass concentration one has to multiply such quantities by

$$10^{-3} \rho_{norm} = 10^{-3} \frac{p}{RT} M = \frac{10^{-3} \times 1.0133 \times 10^5}{8314 \times 273.15} M \approx 4.462 \times 10^{-5} M. \quad (11.10)$$

Here the universal gas constant is $R = 8314 \ J/(kg\text{-}mol\ K)$ and M is the mol mass in kg-mole. Having in mind that $M = 28kg$ for nitrogen, $M = 32kg$ for oxygen, $M = 2kg$ for hydrogen and $M = 42kg$ for carbon dioxide we obtain for the multipliers 1.2494×10^{-3}, 1.4279×10^{-3}, 0.08924×10^{-3} and 1.874×10^{-3} respectively.

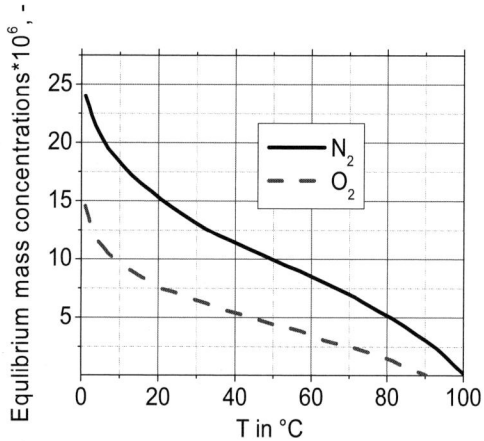

Fig. 11.2. Saturation O_2 and N_2 concentrations in water being in contact with saturated steam–gas mixture as a function of the water temperature. The gas pressure is $10^5 Pa$, Himmelblau and Arends (1959)

Other sources present the data in *mg-mole/l*. In order to obtain the mass concentration one has to multiply such quantities by

$$\frac{10^{-3}}{\rho_2(p, T_2)} M. \quad (11.11)$$

For the case of mixture of a pure gas and steam being in equilibrium with the water the partial pressure of the gas is reduced by the steam component,

$$Y_{2i,\infty} = \frac{p - p'(T_2)}{k_{H,2i}(T_2)}. \quad (11.12)$$

Figure 11.2 shows the solubility of O_2 and N_2 in such case.

Problems:

Problem 1: Given a water pool with air above it at atmospheric pressure. The temperature of the water is 20°C. The air is dry. The water stays long enough in contact with the air so that the surface is saturated with dissolved air. Compute the amount of the dissolved gases into a *kg* of water.

Solution: The air consists of 78.12 vol% nitrogen, 20.96 vol% oxygen and 0.92vol% argon. For practical analysis it is appropriate to assume that the air consists of 0.7812 mole fraction of nitrogen and 0.2188 mole fraction of oxygen. It means that the partial pressure of the nitrogen is 0.7812*bar* and that of the oxygen 0.2188*bar*. Using Eq. (11.9) we obtain 14.8*mg* nitrogen and 9.63*g* oxygen dissolved in 1*kg* of water.

Problem 2: A pump is transporting water from this layer upward. The suction creates pressure lower than one bar. The question is how much of the dissolved gases can be released.

Solution: Evaporation into the bubbles up to the saturation pressure $p'(T_2 = 20°C) = 0.0234bar$ is possible. Then the partial pressures are $p_{N2} = 0.7812(p - 0.0234 \times 10^5)$, $p_{O2} = 0.2188(p - 0.0234 \times 10^5)$. Table 11.3 and Fig. 11.3a give the results. We see the linear dependence due to the validity of the *Henry*'s law at low pressure.

Table 11.3. Released air in water at p < 1*bar* initially saturated at 1*bar*

p in Pa	mgN$_2$/kg	mgO$_2$/kg	$\alpha_{1,hom}$
100000.	0.346	0.225	0.00119
95000.	1.086	0.707	0.00394
90000.	1.825	1.188	0.00698
85000.	2.565	1.670	0.01037
80000.	3.305	2.151	0.01416
75000.	4.044	2.633	0.01844
70000.	4.784	3.115	0.02331
65000.	5.524	3.596	0.02889
60000.	6.263	4.078	0.03537
55000.	7.003	4.559	0.04296
50000.	7.743	5.041	0.05198
45000.	8.482	5.522	0.06288
40000.	9.222	6.004	0.07633
35000.	9.962	6.485	0.09332
30000.	10.701	6.967	0.11548
25000.	11.441	7.449	0.14557
20000.	12.181	7.930	0.18879
15000.	12.920	8.412	0.25614
10000.	13.660	8.893	0.37564
5000.	14.400	9.375	0.64614

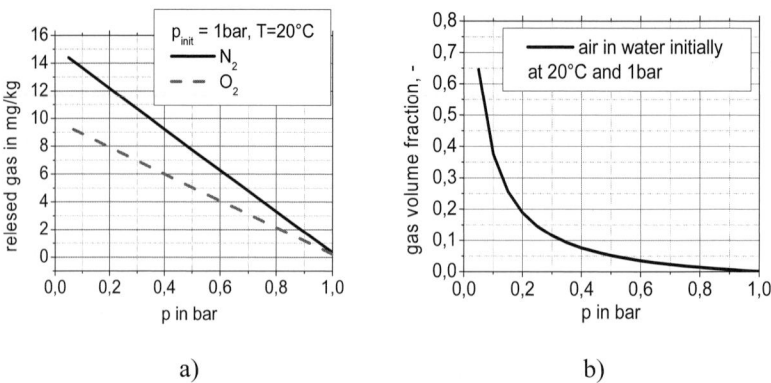

Fig. 11.3. a) Released air from water at p < 1bar initially saturated at 1bar and 20°C. b) gas volume fraction assuming homogeneous two-phase mixture

Problem 3: Compute the volume fraction of Problem 2 assuming that the water and the gas flow with the same velocity and the gases behave as perfect gases. Assume also that the gases have the same temperature as the water.

Solution: The solution is given below:

$$p_{N_2} = Y_{1,N_2} \left[p - p'(T_2 = 20°C) \right],$$

$$p_{O_2} = (1 - Y_{1,N_2}) \left[p - p'(T_2 = 20°C) \right],$$

$$\rho_{N_2} = \frac{p_{N_2} M_{N_2}}{RT_2},$$

$$\rho_{O_2} = \frac{p_{O_2} M_{O_2}}{RT_2},$$

$$\rho = \left[\frac{C_{N_2,\infty} - C_{N_2}}{\rho_{N_2}} + \frac{C_{O_2,\infty} - C_{O_2}}{\rho_{O_2}} + \frac{1 - (C_{N_2,\infty} - C_{N_2}) - (C_{O_2,\infty} - C_{O_2})}{\rho_{H_2O}} \right]^{-1},$$

$$\alpha_1 = \frac{\rho_{H_2O} - \rho}{\rho_{H_2O} - \rho_{N_2} - \rho_{O_2}}.$$

The result is numerically evaluated and presented in Fig. 11.3b. From Fig. 11.3b it is obvious that at very low pressure the gas can occupy considerable amount of the cross section and change totally the flow processes with several consequences.

Problem 4: Given a nuclear reactor core cooled by water at 2.2bar and averaged temperature of 50°C. Due to radiolysis very small part of the water is dissociated into a mixture of [2/3 vol% H_2 i.e. to $Y_{1,H_2} = 2/3$ and 1/3 vol% O_2 resulting in 1/9 mas% H_2 und 8/9 mas% O_2, respectively]. Find the maximum amount of this gas mixture that can be dissolved by the coolant.

Solution: We assume that in a disappearing gas bubble there is a water vapor under partial pressure $p_{H_2O} = p'(T_2 = 50°C) = 0.1235 bar$. The remaining pressure is built by

$$p_{H_2} = Y_{1,H_2} \left[p - p'(T_2 = 50°C) \right],$$

$$p_{O_2} = (1 - Y_{1,H_2}) \left[p - p'(T_2 = 50°C) \right].$$

Under these conditions the maximum of the dissolved gases in accordance with Fig. 11.1 is $2mg$ H_2 and $20.7mg$ O_2 in a $1kg$ of water. Reduction of the pressure under constant temperature leads to mass release as shown in Fig. 11.4a). Computing the volume fraction of the released gases in the same way as in Problem 3 we obtain the result in Fig. 11.4b).

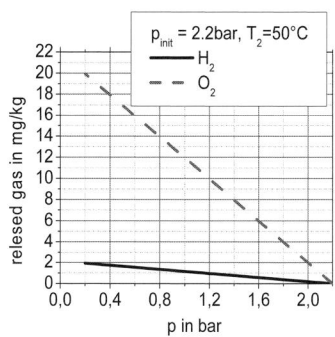
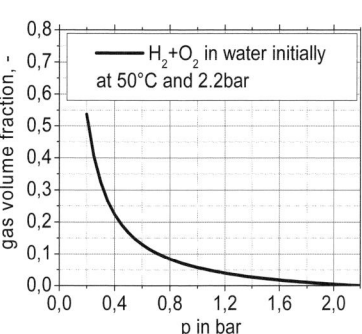

Fig. 11.4. a) Released H_2+O_2 stoichiometric mixture from water at p < 2.2bar initially saturated at 2.2bar and 50°C. b) gas volume fraction assuming homogeneous two-phase mixture

Useful experimental information about the real net production of hydrogen for research water cooled reactors is available in *Dolle* and *Rozenberg* (1977).

Problem 5: Given a volume V_1 filled with water having *known* amount of dissolved gases C_{1,H_2}, C_{1,O_2}, C_{1,N_2} and volume V_2 filled with water having *unknown* amount of dissolved gases C_{2,H_2}, C_{2,O_2}, C_{2,N_2} which has to be estimated. Both volumes have initial pressure p_0 and initial temperature T_0. A third volume V_3 evacuated to a pressure of p_{vac} is connected with the first two volumes. The water enters the evacuated volume violently creating vertices and allowing for generation of so much *active* nucleation centers that the degassing process starts violently amplifying the turbulence. After some finite time $\Delta \tau$ an equilibrium pressure is established equal to p_{final} and temperature T_{gas}. In the third volume a gas volume V_{gas} at its top is measured consisting of Y_{gas,H_2}, Y_{gas,O_2}, Y_{gas,N_2}. Compute C_{2,H_2}, C_{2,O_2}, C_{2,N_2}.

Solution: The mass concentrations are easily estimated from the molar concentrations by $C_{gas,i} = Y_{gas,i} M_i \Big/ \sum_{k=1}^{k_{max}} Y_{gas,k} M_k$, then the mixture gas constant $R_{gas} = \sum_{k=1}^{k_{max}} C_{gas,k} R_k$ and then the mass of the gas $m_{gas} = p_{final} V_{gas} / (R_{gas} T_{gas})$. The mass of the specific components in the gas is then $m_{gas,i} = C_{gas,i} m_{gas}$. The effective pressure of degassing is

$$p_{eff} = p_{eff}(p_{vac}, p_{final}, \Delta \tau, etc.).$$

The final concentration of hydrogen for instance in the first volume is then

$$C_{1,H_2,final} = \min\left[C_{1,H_2}, C_{H_2,sat}(T_0, Y_{gas,H_2} p_{eff}) \right],$$

and in the second volume

$$C_{2,H_2,final} = C_{H_2,sat}(T_0, Y_{gas,H_2} p_{eff}).$$

Applying the mass conservation to the hydrogen content we obtain

$$C_{2,H_2} = C_{H_2,sat} + \frac{m_{gas,H_2}}{\rho_{wasser} V_2} - (C_{1,H_2} - C_{1,H_2,final}) \frac{V_1}{V_2}.$$

Similar is the procedure for computing the other concentrations. It is well known that the degassing and absorption is associated with a hysteresis due to the differences in the initial bubble radius at which the corresponding process starts. The nucleation and the diffusion controlled bubble growth are also associated with characteristic time depending on the level of turbulence and on the bubble-liquid relative velocities. Therefore the process has to be dynamically analyzed by taking into account the separated physical processes which is complicated. Simplification can be done as follows: The gas release starts at p_{vac} reaches the maximum at

$$p_{eff} = p_{vac} + const\left(p_{final} - p_{vac}\right)$$

and is negligible after that. Two limiting cases are of interest: Assuming (a) that the vacuum wave passes the liquid with the velocity of sound for a time much less than $\Delta\tau$ and (b) that the degassing process is instant we obtain $p_{eff} = p_{vac}$. If the gas release is delayed, as it actually is, so that it still continues after the pressure reaches values close to the final pressure, the final concentrations correspond to the saturation concentrations at

$$p_{eff} = p_{final}.$$

This is the more realistic assumption. Setting the first volume to zero results in the method usually used to analyze the dissolved gas content in liquids. In the past the gas temperature was usually reduced to 0°C in order to eliminate the water vapor in the mixture. Modern spectrographs also provide the vapor content in the gas at any temperature and pressure.

11.2 Oxygen in water

Here we collect experimental data reported in the literature and correlate them with analytical expression to facilitate their use in computational analyses. Recommendation for practical use will be given at the end of the chapter together with the error estimate of the correlations.

As already mentioned, oxygen absorption data for atmospheric conditions and water temperatures up to 100°C are reported by *Grischuk* (1957) and given here in Table 11.1. The data are approximated by Eq. (11.9) with mean error of 0.34%. Table 11.4 gives data from earlier measurements at atmospheric pressure for comparison. We see that the *Grischuk* data compare well with the older data. Eq. (11.9) reproduces the data in Table 11.4 with 1.1% mean error. The Eq. (11.9) can be used up to 50bar and up to 160°C with a mean error of 1.7%.

Table 11.4. Oxygen absorption in water in *mg* per *kg* of water as measured by *Morrison* and *Billet* in 1952, *Bohr* and *Bock* in 1891 and *Winkler* in 1891, see *Linke* (1965) p. 1228, compared with *Grischuk* (1957)

°C	W	M&B	B&B	Grischuk (1957)
0	69.82		70.82	69.93
5	61.26		62.69	
10	54.25		55.69	54.34
15	48.83	48.70	49.97	
20	44.26	43.97	45.26	44.33
25	40.41	40.55	41.40	
30	37.26	37.83	38.27	37.32
40	32.99	33.13	33.27	33.03
50	29.84	29.70	29.56	29.89
60	27.84	27.99	26.98	27.89
70	26.13	26.56	25.41	26.17
80	25.13	25.27	24.56	25.17
90	24.56		24.13	24.60
100	24.28		23.99	24.60

Pray et al. (1952) reported measurements at temperatures 298°C to 616.48°C and pressures 9.6 to 21.3*bar* given in Tables 11.4 and 11.5 and Fig. 11.5.

Table 11.5. Oxygen absorption in water in cm^3 at (0°C, 1*atm*) per *g* of water and in *mg* per *kg* of water as measured by *Pray* (1952)

psi	cm^3/g	bar	mg/kg
298.15K (25°C)			
140	0.28	9.65	399
295	0.56	20.34	800
370	0.7	25.51	999
435.93K (162.78°C)			
100	0.15	6.89	214
200	0.31	13.79	443
300	0.46	20.68	657
477.59K (204.44°C)			
100	0.18	6.89	257
150	0.28	10.34	400
533.15K (260°C)			
100	0.64	6.89	914
200	0.91	13.79	1299
300	1.35	20.68	1928
400	1.71	27.58	2442
588.70K (315.55°C)			
100	0.63	6.89	900
200	1.42	13.79	2028
300	2.19	20.68	3127
616.48K (343.33°C)			
104	1,22	7.17	1742
175	1,85	12.07	2641
205	2.29	14.13	3270
280	2.96	19.30	4226
289	2.56	19.93	3655
309	2.99	21.30	4269

Table 11.6. The *Henry* coefficient for oxygen in accordance with *Pray* (1952)

T in K	$k_{H,2,O_2}(T_2)$, Pa
298.15	0.452e10
435.93	0.552e10
477.59	0.458e10
533.15	0.191e10
588.71	0.120e10
616.48	0.084e10

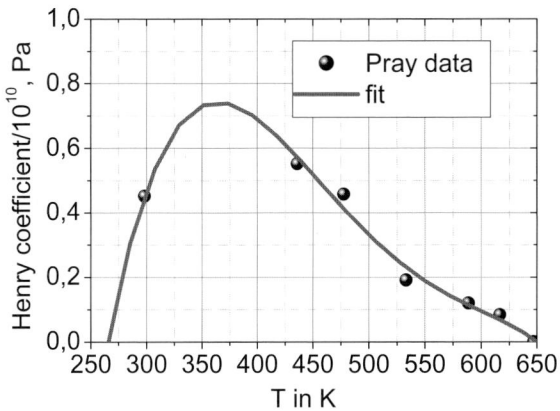

Fig. 11.5. The Henry coefficient for oxygen in accordance with Pray (1952)

We approximated the *Henry* coefficient with

$$k_{H,2i}(T_2) = 1.09\left(a_1 + a_2 T_2 + a_3 T_2^2 + a_4 T^3 + a_5 T^4\right)10^{10}, \qquad (11.12)$$

where $a_{1:5}$ = -23.92186d0, 0.20317d0, -6.06628d-4, 7.78284d-7, -3.67327d-10 as shown in Fig. 11.5. The saturation mole- and mass-concentration is then computed by Eqs. (11.5) and (11.6). The mean error comparing with the *Pray*'s data is only 12.4%. The mean error comparing with all the data given in this section is 11.6%. The mean error comparing with the smoothed data given in Table 11.7 is 15%.

Table 11.7. Values read from smoothed curves drown by *Pray* et al. (1952) and *Frolich* et al. 1933 see *Linke* (1965) p. 1228.

p in bar	6.895	13.39	20.68	27.58	34.47
T in K					
310.9	228.5	471.2	756.8	1028	1271
366.5	142.8	328.4	614.0	842.4	1057
422.0	142.8	371.2	628.2	871.0	1085
477.6	285.6	642.5	885.2	1199	1442
533.15	585.4	1170.8	1585	1999	2856
588.7	1113.7	2113.2	3241	4326	5440

Benson et al. (1979) reported approximations with

$$k_{H,2i}(T_2) = 10^5 \exp(3.71814 + 5596.17/T_2 - 1049668/T_2^2)$$

for $0 \le T_2 - 273.15 \le 100°C$, (11.13)

and

$$k_{H,2i}(T_2) = 10^5 \exp\begin{pmatrix} -4.1741 + 1.3104 \times 10^4/T_2 \\ -3.4170 \times 10^6/T_2^2 + 2.4749 \times 10^8/T_2^3 \end{pmatrix}$$

for $100 < T_2 - 273.15 \le 288°C$. (11.14)

The mean error comparing with the data given in this section is 5.7%.

Ji and Yan (2003) reported the following approximation valid also for higher temperature up to 560.93K and pressures up to 200bar

$$k_{H,2i}(T_2) = 10^5 \exp[a_1 + a_2\bar{p} + a_3\bar{p}^2 + (a_4 + a_5\bar{p} + a_6\bar{p}^2)T_2$$
$$+ (a_7 + a_8\bar{p} + a_9\bar{p}^2)T_2^2 + (a_{10} + a_{11}\bar{p} + a_{12}\bar{p}^2)\ln T_2], \quad (11.15)$$

where $a_{1:12}$ = -2.10973e2, 2.32745e0, -1.19186e-2, -2.02733e-1, 2.45925e-3, -1.21107e-5, 9.77301e-5, -1.43857e-6, 6.84983e-9, 4.79875e1, -5.14296e-1, 2.61610e-3 and $\bar{p} = p/10^5$. The mean error comparing with all the data in this section is 7.1%.

Broden et al. (1978) reported measurements at temperature $\le 150°C$ and pressure $\le 5MPa$ given in Table 11.8. I approximate them here with the correlation

$$C_{O_2} = \frac{M_{O_2}}{\rho_{H_2O}} 10^{-3}\left[a_1 + a_2T_2 + (a_3 + a_4T_2^2 + a_5/T)p + (a_6 + a_7/T)p^2\right], \quad (11.16)$$

where $a_{1:7}$ = -2.545d0, 0.807d-2, -8.414d-5, 2.096d-10, 2.322d-2, 1.027d-12, -3.911d-10. The mol fraction is then computed by Eq. (11.8). The mean error comcomparing with his own data is 3.4% and comparing with all data 19.26%

Table 11.8. Oxygen absorption in water in *mg-mole* per *l* of water and in *kg* per *kg* of water as measured by *Broden* et al. (1978)

T in K	p in Pa	mg-mole/l	mg/kg
325.15	1.e6	9.47	307
350.03	1.e6	8.12	267
374.60	1.e6	7.75	259
398.88	1.e6	7.75	264
424.05	1.e6	8.72	305
325.75	2.e6	18.3	594
350.32	2.e6	15.5	508
374.60	2.e6	14.4	479
398.88	2.e6	15.6	530
424.35	2.e6	17.1	598
326.05	3.e6	27.1	878
350.62	3.e6	22.8	745
374.90	3.e6	21.9	732
399.47	3.e6	23.1	786
424.35	3.e6	26.7	930
326.05	4.e6	35.5	1150
350.62	4.e6	30.3	996
374.60	4.e6	29.1	972
399.17	4.e6	30.6	1043
424.35	4.e6	35.1	1225
326.65	5.e6	43.6	1412
351.22	5.e6	37.6	1234
374.60	5.e6	36.3	1209
400.37	5.e6	38.2	1302
424.95	5.e6	44.4	1550

Tromans (1998) proposed in 1998 an expression derived from the thermodynamic condition of chemical equilibrium valid for temperatures between 273 and 616K and pressures up to 60bar. Multiplying his expression with the constant 0.55 result in the following correlation

$$Y_{O_2,\infty} = 0.55 \frac{22.39 \times 1.4279 \times 10^{-3}}{101325} p$$

$$\times \exp \frac{0.046T_2^2 + 203.35T \ln \frac{T_2}{298} - (299.378 + 0.092T)(T_2 - 298) - 20591}{8.3144T_2},$$

that reproduces all the data we collected here with an accuracy of 5.2% as shown in Fig. 11.6.

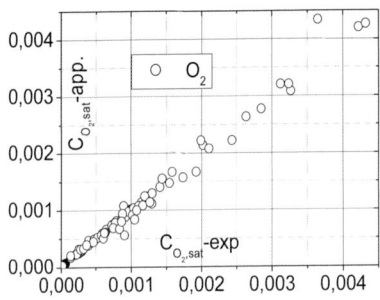

Fig. 11.6. Saturation oxygen concentrations in water: approximated versus measured with the modified *Tromans* equation

Conclusions

Table 11.9. Summary of the approximation correlations, their region of applicability and their error

Author	Approximation	T	p	err. %, own data	err. %, all data
Grischuk (1957)	Eq. (11.9)	0 to 100°C	1e5Pa	0.34	
Pray et al. (1952)	Eq. (11.12)	298°C to 616.48°C	9.6 to 21.3bar	12.4	14.05
Benson et al. (1979)	Eqs. (11.13,11.14)	0 to 288°C			5.7
Ji and *Yan* (2003)	Eq. (11.15)	up to 560.93K	up to 200bar		7.1
Broden et al. (1978)	Eq. (11.16)	$\leq 150°C$	$\leq 5MPa$	3.56	19.26

Table 11.9 gives a summary of the results of this section. For practical use we recommend for pressures around 1*bar* Eq. (11.9). Equations (11.14) or (11.15) are recommended for general applications. The performance of the proposed correlation set is demonstrated in Fig. 11.7.

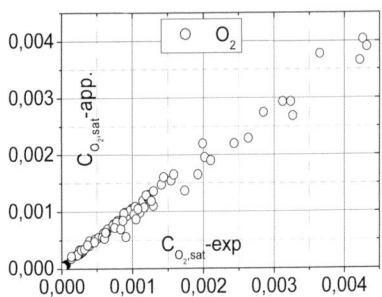

Fig. 11.7. Saturation oxygen concentrations in water: approximated versus measured with the recommended procedure

11.3 Nitrogen water

Here we collect experimental data reported in the literature and correlate them with analytical expression to facilitate their use in computational analyses. Recommendation for practical use will be given at the end of the chapter together with the error estimate of the correlations.

As already mentioned, nitrogen absorption data for atmospheric conditions and water temperatures up to 100°C are reported by *Grischuk* (1957) and given in Table 11.1. The data are approximated by Eq. (11.9) with mean error of 0.26%. Comparison with other data for atmospheric pressure is given in Table 11.10. The agreement between different measurements is within 3% deviation. Equation (11.9) reproduces Table 11.10 within a mean error of 2.4%.

Table 11.10. Nitrogen absorption in water in *mg* per *kg* of water as measured by *Morrison* and *Billet* in 1952, *Bohr* and *Bock* in 1891 and *Winkler* in 1891, see *Himmelblau* and *Arends* (1959), compared with *Grischuk* (1957)

°C	M&B	B&B	W	*Grischuk* (1957)
0		29.86	29.36	29.14
5		26.86	25.99	
10		24.49	23.24	23.06
15	21.80	22.36	20.99	
20	19.98	20.49	19.24	19.22
25	18.55	18.74	17.87	
30	17.20	17.24	16.74	16.62
35	16.03	15.87	15.62	
40	15.18	14.74	14.74	14.63
50	14.06	13.24	14.74	13.52
60	13.56	12.49	12.74	12.65
70	13.22			12.15
80			11.99	11.90
90				11.78
100		12.49	11.87	11.78

Nitrogen absorption in water is measured by *Wiebe* et al. (1933). The results are given in Table 11.11.

Table 11.11. Nitrogen absorption in water in Cc. at (0°C and 1*atm*) per *g* of water and in *g* per *kg* of water as measured by *Wiebe* et al. (1933)

T°C	25		50		75		100	
P atm	cm³/g	g/kg	cm³/g	g/kg	cm³/g	g/kg	cm³/g	g/kg
25	0.383	0.435	0.273	0.341	0.254	0.317	0.266	0.332
50	0.674	0.842	0.533	0.666	0.494	0.617	0.516	0.645
100	1.264	1.579	1.011	1.263	0.946	1.182	0.986	1.232
200	2.257	2.820	1.830	2.286	1.732	2.164	1.822	2.276
300	3.061	3.824	2.534	3.166	2.413	3.015	2.546	3.181
500	4.441	5.548	3.720	4.648	3.583	4.476	3.799	4.746
800	6.134	7.663	5.221	6.523	5.062	6.324	5.365	6.703
1000	7.150	8.933	6.123	7.650	5.934	7.414	6.256	7.816

Within an mean error of 0.3% the *Wiebe* et al. data are approximated here by the following correlation

$$k_{H,2i}(T_2, p) = (a_1 + a_2 T_2 + a_3 T_2^2)10^6 + (a_4 + a_5 T_2 + a_6 T_2^2)p10$$

$$+ (a_7 + a_8 T_2 + a_9 T_2^2)p^2 10^{-4},$$ (11.17)

where $a_{1:9}$ = -133424.80726, 826.81456, -1.17389, -66.79008, 0.49946, $-7.78632\text{d-}4$, 0.02728, $-2.03146\text{d-}4$, $3.28\text{d-}7$. Note the temperature limits of the data.

Other set of data are collected by *Goodman* and *Krase* (1931) and shown in Table 11.12.

Table 11.12. Nitrogen absorption in water in cm^3 at (0°C and 1*atm*) per g of water and in g per kg of water as measured by *Goodman* and *Krase* (1931)

P atm	0°C		25°C		50°C		80°C	
100	1.46	1.82	1.07	1.34	1.003	1.253	0.934	1.167
125	1.76	2.20	1.44	1.80	1.24	1.55	1.15	1.44
200	3.19	3.99	2.76	3.45	2.49	3.11	2.27	2.84
300	3.60	4.50	3.25	4.06	2.99	3.74	2.86	3.57

P atm	100°C		144°C		169°C	
100	0.954	1.192	1.025	1.281	1.08	1.35
125	1.17	1.46	1.30	1.62	1.52	1.90
200	2.25	2.81	2.68	3.35	3.29	4.11
300	2.91	3.63	3.46	4.32	3.83	4.78

With a mean error of 4.8% the *Goodman* and *Krase* (1931) data are reproduced here by

$$C_{2i,\infty} = 1.092 \sum_{i=1}^{3}\left(\sum_{j=1}^{4} a_{ij} T^{j-1}\right) p^{i-1},$$ (11.18)

where the coefficients are given in Table 11.13. These high temperature data are limited for pressures within 100 and 300*atm*.

Table 11.13. Coefficient of Eq. (11.18)

a_{ij}	$j = 1$	$j = 2$	$j = 3$	$j = 4$
$i = 1$	0.10242d+00	- 9.60262d-04	2.90215d-06	- 2.87507d-09
$i = 2$	-9.61364d-09	+ 9.64078d-11	- 3.03074d-13	3.09699d-16
$i = 3$	2.31521d-16	- 2.27411d-18	7.08796d-21	- 7.18264d-24

Other set of data is collected by *Saddington* and *Krase* (1934) and shown in Table 11.14.

Table 11.14. Nitrogen absorption in water in cm^3 at (0°C and 1*atm*) per g of water and in g per kg of water as measured by *Saddington* and *Krase* (1934)

p in atm	t in °C		g/kg
100	65	0.981	1.226
	80	0.977	1.221
	125	1.198	1.497
	180	1.644	2.054
	210	1.817	2.270
	240	2.027	2.532
200	50	1.806	2.256
	80	1.748	2.184
	100	1.825	2.280
	150	2.172	2.714
	200	3.287	4.107
	240	4.378	5.470
300	50	2.572	3.213
	70	2.425	3.030
	105	2.598	3.246
	135	3.126	3.905
	165	3.905	4.879
	230	6.062	7.574

With a mean error of 1.3% the *Saddington* and *Krase* (1934) data are reproduced here by the following correlation

$$k_{H,2i}(T_2, p) = 0.92948 \sum_{i=1}^{4} \left(\sum_{j=1}^{3} a_{ij} p^{j-1} \right) T^{i-1}, \quad (11.19)$$

with the coefficients given in Table 11.15. These high temperature data are limited for pressures within 100 and 300*atm*. With this approximation the data by *Goodman* and *Krase* (1931) are reproduced with 5.2% mean error.

Table 11.15. Coefficients to Eq. (11.19)

a_{ij}	j = 1	j = 2	j = 3
i = 1	-1.16471d11	5610.3d0	- 4.7097d-4
i = 2	1.26395d9	- 71.50575d0	4.21697d-06
i = 3	-3.81683d6	0.254800d0	- 1.21537d-08
i = 4	3565.95174d0	- 2.68626d-4	+ 1.12097d-11

Pray et al. (1952) reported measurements at temperature 298.15K to 588.70K and pressure 10.34 to 40.54*bar* given in Table 11.16.

Table 11.16. Nitrogen absorption in water in cm^3 at (0°C and 1atm) per g of water and in g per kg of water as measured by *Pray* (1952)

Psi		bar	g/kg
298.15K (25°C)			
294	0.28	20.27	0.35
367	0.35	25.30	0.44
588	0.55	40.54	0.69
323.15K (50°C)			
367	0.27	25.30	0.34
384.15K (25°C)			
367	0.25	25.30	0.31
373.15K (111°C)			
367	0.26	25.30	0.32
533.15K (260°C)			
150	0.44	10.34	0.55
400	1.24	27.58	1.55
588.70K (315.55°C)			
150	0.55	10.34	0.69
400	2.32	27.58	2.90

Within 7.7% mean error we approximated them with

$$k_{H,2i}(T_2) = 10^5 \left(a_1 + a_2 T_2 + a_3 T_2^2 + a_4 T^3 + a_5 T^4 \right), \tag{11.20}$$

where $a_{1:5}$ = -4.96679d6, 43617.22615, -135.42537, 0.18072, -8.84973e-5. If the temperature is less than 298.15K then it is set in Eq. (11.20) to 298.15. The mass concentration is then computed by Eqs. (11.5) and (11.6).

Data up to 311°C are collected in *Wiebe* et al. (1933) and given in Table 11.17.

Table 11.17. Nitrogen absorption in water in g per kg

T°C	0	18	25	50	75	100	169	200	260	311
Atm										
10			0.175g						1.243f	2.204f
25			0.435	0.771	0.718	0.752			3.193f	5.511f
50			0.842	1.506	1.396	1.458				
100	1.81a		1.577	2.857	2.673	2.786	3.052a		5.728c,d	
200	3.99a		6.378	5.172	4.895	5.149	9.298a	9.289c	12.37c,d	
300	4.52a		8.650	7.161	6.819	7.195	10.82a		17.13c,d	
500		5.37b	12.55	10.51	10.12	10.74				
800			17.33	14.75	14.30	15.16				
1000		7.72b	20.21	17.30	16.77	17.66				
2000		8.62b								
3000		8.75b								
4000		7.80b								
4500		7.50b								

a *Goodman* and *Krase* 1931; b *Basset* and *Dode* 1936; c *Saddington* and *Krase* 1934; d at 240°C; e at 230°C; f *Pray, Schweikert* and *Minnich* 1952; g *Cassuto* 1904, 1913.

Conclusions:

Table 11.18. Summary of the approximation correlations, their region of applicability and their error

Author	Appr.	T	p	err. %, own data	err. %, all data
Grischuk (1957)	Eq. (11.9)	0 to 100°C	1e5Pa	0.26	24
Pray et al. (1952)	Eq. (11.20)	298.15 to 588.70K	10.34 to 40.54bar	7.7	19.5
Wiebe et al. (1933)	Eq. (11.17)	50 to 100°C	25 to 1000atm	0.3	5.46 for T<100°C
Goodman and Krase (1931)	Eq. (11.18)	0 to 169° C	100 to 300atm	4.84	
Saddington and Krase (1934)	Eq. (11.19)	65 to 240°C	100 to 300atm	1.3	

For up to the atmospheric pressure Eq. (11.9) is the best choice. Eq. (11.9) can also be used up to $100bar$ and temperature < 100°C with an accuracy of about 9.6%. For the region between 100 and 300bar Eq. (11.19) is recommended. For pressures larger than 300bar and temperatures less than 100°C Eq. (11.17) is recommended. For pressure less than 100bar and temperatures higher than 100°C Eq. (11.19) is the best choice due to lack of alternatives. The performance of the recommended procedure is given in Fig. 11.8.

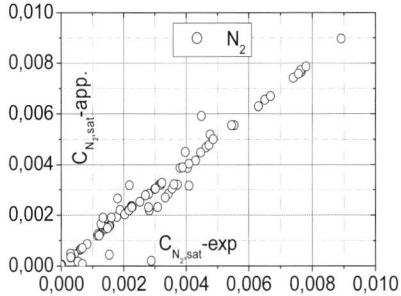

Fig. 11.8. Saturation nitrogen concentrations in water: approximated versus measured with the recommended procedure

11.3 Hydrogen water

Here we collect experimental data reported in the literature and correlate them with analytical expression to facilitate their use in computational analyses. Recommendation for practical use will be given at the end of the chapter together with the error estimate of the correlations.

As already mentioned, hydrogen absorption data for atmospheric conditions and water temperatures up to 100°C are reported by *Grischuk* (1957) and given in Table 11.1. The data are approximated by Eq. (11.9) with mean error of 0.46%.

Pray et al. (1952) reported measurements at temperature 279.04K to 616.48K and pressure 6.895 to 34.47*bar* given in Table 11.19.

Table 11.19. Hydrogen absorption in water in cm^3 at (°C and 1*atm*) per g of water and in *mg* per *kg* of water as measured by *Pray* (1952)

psi		Bar	mg/kg
279.04K (5.89°C)			
300	0.32	20.68	28.56
367	0.44	25.30	39.26
324.82K (51.67°C)			
200	0.33	13.79	29.45
300	0.41	20.68	36.59
350	0.45	24.13	40.16
422.04K (148.89°C)			
100	0.13	6.895	11.60
200	0.28	13.79	24.99
300	0.40	20.68	35.70
375	0.52	25.86	46.40
500	0.70	34.47	62.47
447.04K (173.89°C)			
100	0.15	6.895	13.39
200	0.30	13.79	26.77
300	0.43	20.68	38.37
375	0.56	25.86	49.97
500	0.75	34.47	66.93
472.04K (198.89°C)			
100	0.18	6.895	16.06
200	0.34	13.79	30.34
300	0.52	20.68	46.40
375	0.68	25.86	60.68
497.04K (223.89°C)			
100	0.22	6.895	19.63
200	0.49	13.79	43.73
300	0.75	20.68	66.93
375	0.94	25.86	83.88
500	1.26	34.47	112.4
533.15K (260.00°C)			
100	0.39	6.895	34.80
200	0.91	13.79	81.21
300	1.25	20.68	111.5
588.70K (315.55°C)			
100	0.65	6.895	58.00
200	1.32	13.79	117.8
300	2.01	20.68	179.4
616.48K (343.33°C)			
100	1.40	6.895	124.9
115	1.63	7.929	145.5
120	1.68	8.274	149.9
125	1.74	8.618	155.3

I approximated them within 7.5% error with

$$k_{H,2i}(T_2) = \left(a_1 + a_2 T_2 + a_3 T_2^2 + a_4 T^3 + a_5 T^4 + a_6 T^5\right) 10^5, \quad (11.21)$$

where $a_{1:6}$ = 7.84421d6, −89721.74192, 403.66971, −0.88502, 9.4584d-4, −3.95322d-7. All data are predicted with almost the same accuracy, 14% mean error.

Table 11.20. Hydrogen absorption in water in *kg* per *kg* of water as measured by *Kaltofen* et al. (1986) p. 169

T in K	p in Pa	kg/kg
273.15	25.3e5	48.1e-6
273.15	50.6e5	95.8e-6
273.15	101.3e5	190.6e-6
273.15	303.9e5	543.8e-6
273.15	607.8e5	1015.9e-6
273.15	1013.1e5	1547.9e-6
323.15	25.3e5	37.4e-6
323.15	50.6e5	74.3e-6
323.15	101.3e5	147.7e-6
323.15	303.9e5	426.7e-6
323.15	607.8e5	809.8e-6
323.15	1013.1e5	1275.2e-6
373.15	25.3e5	45.5e-6
373.15	50.6e5	890.0e-6
373.15	101.3e5	175.8e-6
373.15	303.9e5	503.6e-6
373.15	607.8e5	951.8e-6
373.15	1013.1e5	1480.4e-6

Within a mean error of 0.22% the *Kaltofen* et al. data are approximated here by the following correlation

$$k_{H,2i}(T_2,p) = \left(a_1 + a_2 T_2 + a_3 T_2^2\right)10^{10} + \left(a_4 + a_5 T_2\right) p 10^5 + \left(a_6 + a_7 T_2 + a_7 T_2^7\right) p^2, \quad (11.22)$$

where $a_{1:7}$ = -5.56952, 0.0387, - 5.9258d-5, 2.08046d-5, 4.3011d-7, 4.84968d-7, -2.60766d-9, 3.26096d-12. Note the temperature limits of the data.

Conclusions:

Table 11.21 summarizes the results of this section.

11 Solubility of O_2, N_2, H_2 and CO_2 in water

Table 11.21. Summary of the approximation correlations, their region of applicability and their error

Author	Appr.	T	p	err. %, own data	err. %, all data
Grischuk (1957)	Eq. (11.9)	0 to 100°C	1e5Pa	0.46	35.4, 7.8 T<100°C
Pray et al. (1952)	Eq. (11.21)	279.04 to 616.48K	6.895 to 34.47bar	7.5	14
Kaltofen et al. (1986)	Eq. (11.22)	50 to 100°C	25.3 to 1013atm	0.22	5.26 T<100°C

For temperatures below 100°C Eq. (11.22) is recommended. At pressure around 1 bar and temperatures below 100°C Eq. (11.9) is accurate enough. For temperatures higher than 100°C Eq. (11.21) is recommended. The performance of the recommended procedure is demonstrated in Fig. 11.9.

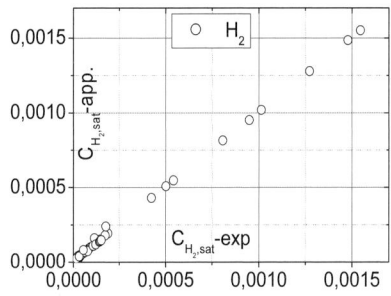

Fig. 11.9. Saturation hydrogen concentrations in water: approximated versus measured with the recommended procedure

11.4 Carbon dioxide–water

Our natural environment contains water and carbon dioxide. Therefore any process associated with the atmospheric and biological phenomena is influenced to some extend by the solubility of the carbon dioxide in the water. Sparking champagne for celebration of important achievements in our life or gas release from a lake in Africa causing thousands of casualties are some of the many examples. Our ecological system is strongly influenced by the dynamics of the appearance and disappearance of CO_2 in the atmosphere. Many industrial processes also require detailed knowledge on solubility. This is the reason why large number of publications is available to this subject. *Carroll, Slupsky* and *Mather* reviewed the state of the art up to 1991 (about 100 papers) and recommended the following correlation valid for $p < 1$MPa and $273 < T < 433$K:

$$k_{H,2i}(T_2,p) = \exp\begin{pmatrix} -6.8346 + 1.2817 \times 10^4/T_2 \\ -3.7668 \times 10^6/T_2^2 + 2.997 \times 10^8/T_2^3 \end{pmatrix} \text{ in } MPa. \quad (11.23)$$

For the enthalpy of solution the authors reported

$$\Delta h_{CO_2 \text{ in Water}} = \frac{1}{R}\frac{\partial}{\partial(1/T_2)}\ln\left[k_{H,2i}(T_2,p)\right]$$

$$= 106.56 - 6.2634 \times 10^4/T_2 + 7.475 \times 10^6/T_2^2 \text{ in } kJ/g\text{-}mol \text{ of } CO_2. \quad (11.24)$$

The specific capacity of the gas dissolved in water is higher than the capacity of the not dissolved gas by

$$\Delta c_p = 6.2634 \times 10^7/T_2^2 - 1.495 \times 10^{10}/T_2^3 \text{ in } J/g\text{-}mol. \quad (11.25)$$

Dodds, *Stutzman* and *Sollami* collected the available data up to 1956 for pressures up to 700atm in the Table 11.22. We learn from this data that up to pressure of about 200bar the solubility decreases with increasing temperature but for higher pressures at about 70°C the solubility starts to increase with increasing temperature.

Table 11.22. Solubility of carbon dioxide in water up to 700atm pressure, *Dodds*, *Stutzman* and *Sollami* (1956).

P in atm	0°C	5	10	12.4	15	18	20	25	31.04
1	3.36	2.8	2.34	2.18	2		1.72	1.49	1.31
	3.36			2.13	2.01		1.57	1.48	
	3.53			2.14	1.96		1.52	1.49	
	3.53							1.48	
								1.49	
5	16.98			10.11	9.01				
	17.1			10.11					
10	31.47			18.94	16.47				
	31.19			18.94					
15	42.83			26.72	23.26				
20	52.31			33.59	29.86				
	52.08			33.59					
25	59.97			39.87	34.63	38.3	31.5		27.84
	59.79			39.93					
	59.89								
30	62.3			45.64	39.87		35.16		
	66.05			45.64					
35	72.1				44.21		38.26		
38	74.34								

11 Solubility of O_2, N_2, H_2 and CO_2 in water

40				47.98				
45				50.23				
50			69.77	53.12	62.87	49.81	53.45	47.41
52			54.32			52.75		
58.1								
67.7								
75			71.32		66.45		61.19	57.57
87.1								
100			72.18		66.7		62.33	59.22
125								
150			75.36		70.18			62.01
200			78.07		72.96			64.35
300			80.62		77.17			
400							75.81	72.20
500							78.01	75.91
600								
700								

P in atm	35	40	50	60	75	100	120
1		1.05	0.87	0.72			
5							
10							
15							
20							
25	25.42	22.81	19.06		13.39	10.54	9.82
30	19.94						
	22.29						
35							
38	33.88			17.24			
	25.11			15.52			
40							
45							
50	43.6	39.95	33.86	19.39	24.71	19.98	18.65
	32.12			21.73			
	41.73						
52							

58.1	38.79			22.91		13.01	
	38.12			24.21			
67.7	46.04			26.61		14.67	
	46.79			27.66		9.29	
75	54.65	50.67	44.23	35.91	33.45	28.05	26.7
	55.65			31.27		16.03	
						13.91	
87.1	55.65			36.01		16.11	
100	57.18	54.59	50.31	42.83	40.46	34.69	33.76
125		56.36	52.55	50.7		39.46	39.46
150	59.91	57.69	54.26		48.25	44.62	43.38
200	62.48	60.34	57.2		52.33	50.43	52.61
300			61.52		57.93	57.97	58.89
400	70.14	68.45	65.35		62.58	63.58	64.39
500	74.57	72.1				67.53	68.94
600			72.1			71.45	73.61
700			75.26		73.79	75.57	78.32

11.5 Diffusion coefficients

The diffusion mass flow rate in kg/(m²s) is proportional to the mass concentration as proposed by *Fick*

$$(\rho w)_{2i} = \rho_2 D_{2i}^* \frac{dC_{2i}}{dz}.$$ (11.26)

Here the D_{2i}^* is the diffusion coefficient for specie i in the liquid 2 in m^2/s. The order of magnitude of the diffusion coefficients for different couples of solute and solvents are given in Table 11.23.

Table 11.23. Order of magnitude of diffusion coefficients

Solute in solvent	Order of magnitude of diffusion coefficients in m^2/s
Gas in gases	1×10^{-5}
Gas in liquids	1×10^{-9}
Gas in polymers and glasses	1×10^{-12}
Gas in solids	1×10^{-14}

Sometimes for gases the pressure dependence is taken into account by the approximation $\rho_2(p,T) D_{2i}^*(p,T) = \rho_2(p=1bar,T) D_{2i}^*(p=1bar,T)$.

For the gases considered here in water the order of magnitude is 10^{-9} m^2/s. While higher liquid temperatures T_2 improve the diffusion of gases, higher liquid dynamic viscosity η_2 reduces it. *Wilke* and *Chang* (1955) proposed the following semi-empirical approximation of the diffusion coefficient:

$$D_{2i}^*\left[\frac{m^2}{s}\right] = 7.4 \times 10^{-15} \frac{(c_{2a} M_2 [g-mole])^{1/2} T[K]}{\eta_2 \left[\frac{kg}{ms}\right] \left(v_{mole,2i} \left[\frac{cm^3}{g-mole}\right]\right)^{0.6}}. \qquad (11.27)$$

c_{2a} is the so-called association parameter. It is given in Table 11.24 for view different liquid. M_2 is the mole mass of the liquid in g per *mole*. The molecular volumes in cm^3/g-mole at the boiling point of some substances are given in Table 11.25.

Table 11.24. Association parameter for different solvents

Solvent	Association parameter c_{2a}
Water	2.6
Methyl alcohol	1.9
Ethyl alcohol	1.5
Benzene	1
Ether	1
Heptane	1

Table 11.25. Molecular volumes in cm^3/g-mole at boiling point

		const								
H2	14.3	1.026e-14	CO	30.7	D2O	20	H2S	32.9	I2	71.5
O2	25.6	7.234e-15	CO2	34	N2O	36.4	COS	51.5		
N2	31.2	6.425e-15	SO2	44.8	NH3	25.8	Cl2	48.4		
Air	29.9	6.591e-15	NO	23.6	H2O	18.9	Br2	53.2		

For water Eq. (11.27) simplifies to

$$D_{2i}^* = const \times T_2/\eta_2 . \qquad (11.28)$$

with constants given in Table 11.24 too.

All data are available for atmospheric pressure only. For high pressure in analogy to gas diffusion one may use the relation proposed by *Reid*, *Sherwood* and *Prausnitz* (1977)

$$\rho(p,T_0) D_i^*(p,T_0) = \rho(p_0,T_0) D_i^*(p_0,T_0), \qquad (11.29)$$

as far as $T_0 < T'(p)$.

Problem: Compute the diffusion coefficients for hydrogen, oxygen and nitrogen in water at atmospheric pressure and 25° temperature. Compare the solutions to each other. Compare the computed oxygen diffusion coefficient with the measured 1.8e-9 m^2/s *Clussler* (1983).

Solution: The water dynamic viscosity is $\eta_2(1atm, 25°C) = 890\text{e-}6 kg/(ms)$. Using Eq. (11.28) we obtain $D^*_{2,H_2} = 3.44\text{e-}09$ m^2/s, $D^*_{2,N_2} = 2.15\text{e-}09$ m^2/s and $D^*_{2,O_2} = 2.42\text{e-}09$ m^2/s. We see that the lighter hydrogen has the largest diffusion coefficient which is expected. The order of magnitude is the same. The computed value for oxygen is 34% higher as reported in *Clussler* (1983).

11.6 Equilibrium solution and dissolution

The solution and dissolution real processes are diffusion controlled processes that take finite time. The instant equilibrium solution and dissolution is an idealization assuming a instant adjustment of the solution gases into the liquid up to the saturation state. This simplification allows computing the mass source term of the gas component inside the liquid field 2 as a function of the change of the partial gas pressure of the component and of the liquid temperature. The total differential of the equilibrium mass concentration is then

$$dC_{2i,\infty}(p_{1i}, T_2) = \left(\frac{\partial C_{2i,\infty}}{\partial p_{1i}}\right)_{T_2} dp_{1i} + \left(\frac{\partial C_{2i,\infty}}{\partial T_2}\right)_{p_{1i,norm}} dT_2 \qquad (11.30)$$

If in the region of interests the linear approximation

$$C_{2i,\infty} \approx C^*_{2i,\infty}(p_{1i,norm}, T_2) \frac{p_{1i}}{p_{1i,norm}} = C^*_{2i,\infty} \frac{p_{1i}}{p_{1i,norm}},$$

is allowed, then

$$dC_{2i,\infty}(p_{1i}, T_2) = \frac{1}{p_{1i,norm}} \left[C^*_{2i,\infty} dp_{1i} + p_{1i} \left(\frac{\partial C^*_{2i,\infty}}{\partial T_2}\right)_{p_{1i,norm}} dT_2 \right] \qquad (11.31)$$

Differentiating $m_{2i,\infty} = \alpha_2 \rho_2 C_{2i,\infty} Vol$ and assuming constant α_2 and ρ_2 we obtain an expression for the equilibrium mass generation per unit volume of the mixture and unit time

$$\mu_{21,i} = \frac{1}{Vol}\frac{dm_{2i,\infty}}{d\tau} \approx \alpha_2 P_2 \left(\frac{\partial C_{2i,\infty}}{\partial p_{1i}}\right)\frac{dp_{1i}}{d\tau} + P_2 \left(\frac{\partial C_{2i,\infty}}{\partial T_2}\right)\frac{dT_2}{d\tau}$$

$$\approx \frac{\alpha_2 P_2}{P_{1i,norm}}\left[C^*_{2i,\infty}\frac{dp_{1i}}{d\tau} + p_{1i}\left(\frac{\partial C^*_{2i,\infty}}{\partial T_2}\right)_{P_{1i,norm}}\frac{dT_2}{d\tau}\right], \quad (11.32)$$

as long as $C_{2i} > C_{2i,\infty} > 0$ and

$$\mu_{12,i} \approx -\frac{\alpha_2 P_2}{P_{1i,norm}}\left[C^*_{2i,\infty}\frac{dp_{1i}}{d\tau} + p_{1i}\left(\frac{\partial C^*_{2i,\infty}}{\partial T_2}\right)_{P_{1i,norm}}\frac{dT_2}{d\tau}\right], \quad (11.33)$$

for $C_{2i} < C_{2i,\infty}$ and $C_{1i,\infty} > 0$.

The next step of sophistication of the mathematical description of this process is to consider the bubble growth as a function of time.

Nomenclature

Latin

a_i, a_{ij}	approximation coefficients
C	mass concentration, dimensionless
$C^*_{2i,\infty}$	$:= C^*_{2i,\infty}\left(p_{1i,norm}, T_2\right)$
C_{2a}	association parameter
$D^*_{2,i}$	diffusion constant for specie i in liquid 2, m²/s
D^*_i	diffusion constant for specie i, m²/s
f	function
$k_{H,2i}$	Henry's coefficient for solubility of specie i in the liquid,
M	mol mass, kg-mole
$m_{2i,\infty}$	mass of the dissolved gas in a saturated liquid, kg
m_{gas}	mass of the gas or gas mixture, kg
$m_{gas,i}$	mass of the specie i in the gas mixture, kg
p	pressure, Pa
p'	saturation pressure, Pa

p_{1i}	partial pressure of specie i in the gas phase, Pa
p_{eff}	effective pressure, Pa
p_i	partial pressure of specie i, Pa
p_{vac}	evacuation pressure, Pa
R	= 8314, universal gas constant, J/(kg-mol K)
R_{gas}	gas constant of the mixture, J/(kgK)
R_k	gas constant for the specie k, J/(kgK)
T	temperature, K
T'	saturation temperature, K
V_1, V_2, V_3, V_{gas}	volumes, m³
Vol	control volume, m³
$v_{mole,2i}$	molar volume of specie i in the liquid,
$Y_{2i,\infty}$	saturation molar concentration of specie i in the liquid, dimensionless
$Y_{1,i}, Y_{gas,i}$	molar concentration of specie i in the gas, dimensionless
$Y_{2,i}$	molar concentration of specie i in the liquid, dimensionless
z	coordinate, m

Greek

α	volume fraction, m³/m³
η	dynamic viscosity, kg/(ms)
$\mu_{12,i}$	mass of the solved non condensable gas per unit time and unit volume of the multi-phase mixture, kg/(sm³)
$\mu_{21,i}$	mass of the generated non condensable gas per unit time and unit volume of the multi-phase mixture, kg/(sm³)
ρ	density, kg/m³
$(\rho w)_{2i}$	diffusion mass flow rate, kg/(m²s)
τ	time, s
$\Delta \tau$	time interval, s

Subscripts

1	field 1, gas
2	field 2, liquid
H_2	hydrogen
O_2	oxygen
N_2	nitrogen
air	air
gas	gas, gas mixture
H_2O	water

i	specie i
∞	saturation concentration
0	initial
final	final state
norm	norm value
hom	for equal velocity of gas and liquid

References

Benson BB, Krause D and Peterson MA (1979) The solubility and isotopic fraction of gases in dilute aqueous solution, I. Oxygen, vol 8 no 9 pp 655-690

Broden A and Simonson R (1978) Solubility of oxygen, Part 1. Solubility of oxygen in water at temperature ≤ 150°C and pressure ≤ 5MPa, Svensk Papperstiding nr 17 pp.541-544

Carroll JJ, Slupsky JD and Mather AE (1991) The solubility of carbon dioxide in water at low pressure, J. Phys. Chem. Ref. Data, vol 20 no 6 pp 1201-1209

Clussler EL (1983) Diffusion mass transfer in fluid systems, Cambridge University Press, Cambridge

Dodds WS, Stutzman LF and Sollami BJ (1956) Carbon dioxide solubility in water, Industrial and Engineering Chemistry, Chemical and Engineering Data Series, vol 1 no 1 pp 92-95

Dolle L and Rozenberg J (1977) Radiolytic yields in water reactor system and influence of dissolved hydrogen and nitrogen, CEA-CONF-4186, 25 Jul 1977. 10 p. Availability: Servece de Documentation, CEN Saclay BP No.2, 91190 Gif-Sur-Yvette, France

Grischuk (1957) Archiv für Energiewirtschaft, vol 11 p 136

Goddman JB and Krase NW (April 1931) Solubility of nitrogen in water at high pressures and temperatures, Industrial and Engineering Chemistry, vol 23 no 4 pp 401-404

Heitmann H-G (1986) Praxis der Kraftwerk-Chemie, Vulkan-Verlag, Essen

Himmelblau und Arends (1959) Chem. Ing. Techn. vol 31 p 791

Hömig HE (1963) Physikalische Grundlagen der Speisewasserchemie, Vulkan Verlag Dr. W. Classen, Essen

Ji X and Yan J (2003) Saturated thermodynamic properties for the air-water system at elevated temperature and pressure, Chemical Engineering Science vol 58 pp 5069-5077

Kaltofen R et al. (1986) Tabelenbuch Chemie, 10th ed., Verlag Harri Deutsch, Thun, Frankfurt/M

Linke WF (1965) Solubilities, vol 2, 4th ed, American Chemical Society, Washington DC

Pray HA, Schweickert CE and Minnich BH (1952) Solubility of hydrogen, oxygen, nitrogen, and helium in water, Industrial and Engineering Chemistry, vol 44 no 5 pp 1146-1151

Reid R, Sherwood TK and Prausnitz JM (1977) Properties of gases and liquids, 3rd ed., McGraw-Hill, New York

Saddington AW and Krase NW (Feb. 1934) Vapour-liquid equilibria in the system nitrogen-water, Journal of the American Chemical Society, vol 56 pp 353-361

Tromans D (1998) Temperature and pressure dependent solubility of oxygen in water: a thermodynamic analysis, Hydrometallurgie vol 48 pp 327-342

Wiebe R, Gaddy VL and Heins C Jr. (March 1933) The solubility of nitrogen in water at 50, 75 and 100° from 25 to 1000 atmospheres, Journal of the American Chemical Society, vol 55 pp 947-953

Wilke CR and Chang P (June 1955) Correlation of diffusion coefficients in dilute solutions, A.I.Ch.E. Journal, vol 1 no 2 pp 264-270

12 Transient solution and dissolution of gasses in liquid flows

Soft drinks and champagne are examples of liquids containing sizable quantities of gases dissolved under pressure. In liquids usually used in technology there are *dissolved inert gases* and also *micro-bubbles*. It is known that at $p = 10^5 Pa$, $T_{2o} = 298.15K$ the amount of dissolved gases and micro-bubbles in the coolant is $\alpha_{1o} \cong 0.005$, for boiling water nuclear reactors, and $\alpha_{1o} \cong 0.001$, for pressurized water reactor, see *Malnes* and *Solberg* (1973). *Brennen* (1995), p. 20, reported that it takes weeks of deaeration to reduce the concentration of air in the water tunnel below $3ppm$ (saturation at atmospheric pressure is about $15ppm$). *Wolf* (1982-1984) reported that it took him about 18 hours per single large scale test to increase the pressure to $\sim 11 MPa$, to warm and mix the water inside the pressure vessel including 5 hours degassing of the water from $8mg\ O_2/l$ to a value of $2mg\ O_2/l$ before each test.

The mass transfer inside the multi-phase flows caused by solution or dissolution of gases is considered as a boundary layer problem. One analyzes the diffusion processes close to the interface leading to the concentration gradient at the surface and then computes the mass flow rate at the interface. As already mentioned the diffusion mass flow rate in kg/s is proportional to the mass concentration gradient as proposed by *Fick* in analogy to the heat transfer

$$(\rho w)_{c,i} = \rho_c D^*_{c,i} \frac{dC_{c,i}}{dz}. \tag{12.1}$$

Here the $D^*_{c,i}$ is the diffusion coefficient for specie i in the continuous liquid c in m^2/s. The order of magnitude of the diffusion coefficients for different couples of solute and solvents are given in Table 11.23. Comparing the diffusion constants in a gas and in a liquid we see a difference of four orders of magnitudes. Therefore if considered together usually the thickness of the gas site boundary layer is neglected and only the limiting process, that is the diffusion in the liquid, is considered.

12.1 Bubbles

Consider a family of mono-disperse bubbles designated with d, moving in a continuum, designated with c, with relative velocity ΔV_{cd}. We are interested in how much mass of the solvent is transferred between the surface of a bubble and the surrounding continuum. We will consider first the steady state mass diffusion problem in a continuum boundary layer, thereafter the transient mass diffusion problem in the bubble and finally we will give some approximate solutions for the average mass transferred per unit time.

The mass transported between the bubble velocity field and the continuum liquid per unit time and unit mixture volume is equal to the product of the interfacial area density a_{cd} and the mass flow rate $(\rho w)_{c,i}^{d\sigma}$ of specie i

$$\mu_{c,i}^{d\sigma} = a_{cd}(\rho w)_{c,i}^{d\sigma}. \tag{12.2}$$

Read the superscripts and the subscripts in the following way: from the interface $d\sigma$ into the continuum c for specie i. The interfacial area density for bubbly flow is

$$a_{cd} = 6\alpha_d / D_d. \tag{12.3}$$

The mass flow rate is defined as

$$(\rho w)_{c,i}^{d\sigma} = \beta_{dc,i}\rho_c\left(C_{c,i}^{d\sigma} - C_{c,i}\right) = \frac{D_{c,i}^* Sh_c}{D_d}\rho_c\left(C_{c,i}^{d\sigma} - C_{c,i}\right). \tag{12.4}$$

Here the interface concentration is $C_{c,i}^{d\sigma}$ and the bulk concentration far from the interface in the continuum is $C_{c,i}$. $\beta_{dc,i}$ is the mass transfer coefficient in dimension of velocity. Usually the dimensionless *Sherwood* number defined as

$$Sh_c = \frac{\beta_{cd,i}D_d}{D_{c,i}^*} = \frac{(\rho w)_{c,i}^{d\sigma}}{\rho_c\left(C_{c,i}^{d\sigma} - C_{c,i}\right)}\frac{D_d}{D_{c,i}^*} \tag{12.5}$$

is used to express theoretical or experimental results describing the diffusion mass transfer at specific surfaces. With these definition the mass source inside the continuum is then

$$\mu_{c,i}^{d\sigma} = a_{cd}(\rho w)_{c,i}^{d\sigma} = a_{cd}\beta_{dc,i}\rho_c\left(C_{c,i}^{d\sigma} - C_{c,i}\right) = 6\alpha_d D_{c,i}^* Sh_c\rho_c\left(C_{c,i}^{d\sigma} - C_{c,i}\right)/D_d^2. \tag{12.6}$$

In excess of component i inside the continuous or inside the disperse phase the surface concentration rapidly reaches the saturation concentration at the corresponding partial pressure and liquid temperature. The pressure in a micro-bubble $\sum_{i=1}^{i_{max}} p_{d,i}$ is greater than the system pressure p due to the surface tension effect by $2\sigma_{cd}/R_d$,

$$\sum_{i=1}^{i_{max}} p_{d,i} = p + \frac{2\sigma_{cd}}{R_d}. \tag{12.7}$$

The partial pressure of the component i is then

$$p_{d,i} = Y_{d,i}\left(p + \frac{2\sigma_{cd}}{R_d}\right), \tag{12.8}$$

and the surface concentration equals the saturation concentration

$$C_{c,i}^{d\sigma} = C_{c,i}^{d,sat}\left(p_{d,i}, T_c\right). \tag{12.9}$$

The inception of the gas release from the liquid happens when

$$C_{c,i}^{d} > C_{c,i}^{d,sat}\left(p_{d,i}, T_c\right). \tag{12.10}$$

Diffusion controlled bubble growth and successive collapse happen with different velocity because of the difference in the initial radius. This is known a *refractory gas release*. The phenomenon is observed by sending acoustic waves through a mixture with micro-bubbles oscillating around a pressure which usually do not lead to gas release. Similar effect can be produced by the turbulent oscillations. *Taylor* (1936) found that the *rms*-values of pressure fluctuation is

$$\sqrt{\overline{p'^2}} = const \rho_c \frac{1}{2}\left(\overline{u_c'^2} + \overline{v_c'^2} + \overline{w_c'^2}\right) = const \rho_c k_c, \tag{12.11}$$

where $1 < const < \sqrt{2}$. *Batchelor* (1953) reported for isotropic turbulence

$$\sqrt{\overline{p'^2}} = 0.583 \rho_c \overline{u_c'^2}. \tag{12.12}$$

In pipe flows we have in the boundary layer $\overline{u_c'^2} \approx (0.01 \text{ to } 0.06)w_c$, see Fig. 7 in *Daily* and *Johnson* (1956). Assuming that the liquid fluctuation velocity is equal to the friction velocity

$$u'_c = w^*_c = \sqrt{\tau_{cw}/\rho_c} = w_c\sqrt{\frac{1}{2}c^d_{cw}} \qquad (12.13)$$

and that the drag coefficient obeys the *Blasius* formula

$$c^d_{cw} = 0.057\,\text{Re}^{1/4}_{cw}, \qquad (12.14)$$

we obtain

$$\sqrt{p'^2} = 0.0166\,\rho_c w^2_c\,\text{Re}^{1/4}_{cw}. \qquad (12.15)$$

Therefore the effective value of the boundary saturation concentration is

$$C^{d\sigma}_{c,i} = C^{d,sat}_{c,i}\left[Y_{d,i}\left(p + \frac{2\sigma_{cd}}{R_d} - \frac{1}{2}0.0166\,\rho_c w^2_c\,\text{Re}^{1/4}_{cw}\right), T_c\right]. \qquad (12.16)$$

The dynamic pressure pulsation correction $\pm 0.0166\,\rho_c V^2_c\,\text{Re}^{1/4}_{cw}$ starts to be important for flow at low system pressure but high velocities. Bubble being at static equilibrium has a pressure

$$p_{gas} + p_{H_2O} = p + \frac{2\sigma_{cd}}{R_d} - \frac{1}{2}0.0166\,\rho_c w^2_c\,\text{Re}^{1/4}_{cw} \qquad (12.17)$$

associated with the bubble radius R_d. Rearranging we have a nonlinear equation with respect to the radius,

$$\frac{3m_{gas}R_{gas}T_c}{4\pi R^3_d} - \frac{2\sigma_{cd}}{R_d} = p - p_{H_2O} - \frac{1}{2}0.0166\,\rho_c V^2_c\,\text{Re}^{1/4}_{cw}. \qquad (12.18)$$

Here m_{gas} is the inert gas mass inside the bubble. The function $f(R_d) = \frac{3m_{gas}R_{gas}T_c}{4\pi R^3_d} - \frac{2\sigma_{cd}}{R_d}$ possess a minimum radius $R_{d,cr}$. This minimum radius satisfying the above equation is then

$$R_{d,cr} = \frac{4\sigma_{cd}}{3\left(p_{H_2O} + \frac{1}{2}0.0166\,\rho_c V^2_c\,\text{Re}^{1/4}_{cw} - p\right)} > 0, \qquad (12.19)$$

which for zero liquid velocity results in the well known expression

$$R_{d,cr} = \frac{4\sigma_{cd}}{3\left(p_{H_2O} - p\right)}.$$ (12.20)

Bubbles having larger sizes are considered unstable and subject to change in their size.

12.1.1 Existence of micro-bubbles in water

Theoretically small gas concentrations in water have to be completely absorbed by molecular diffusion in such a long time so that it is outside of the life of technical facilities and processes. That is the reason why micro-bubbles in treated and untreated waters are very sustainable. A typical example is given in Fig. 12.1.

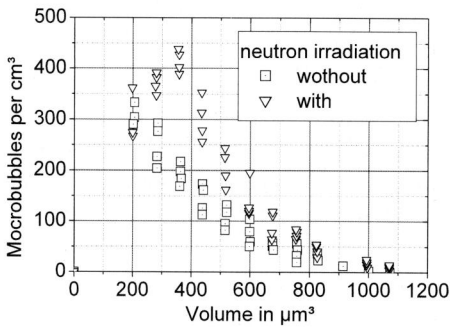

Fig. 12.1. Micro-bubbles in water per unit volume as a function of the single bubble volume for tap water at $1.09 bar$ and $277.6K$ for air volume fraction of 1.53% in accordance with *Hammitt* (1980) p. 76

For untreated tap water about 1.13×10^9 nucleus per m^3 exists. For degassed water they are reduced to 0.911×10^9 with most probable size about $6 \mu m$. The initial bubble diameter is in any case less than the bubble diameter computed after equating the buoyancy force and the surface force

$$D_{1o} < \sqrt{6}\lambda_{RT}.$$ (12.21)

Hammitt reported for untreated tap water about 1.13×10^9 nucleus per m^3 exists with most probable size about $6 \mu m$. For degassed water they are reduced to 0.911×10^9 with most probable size again about $6 \mu m$. *Brennen* (1995) reported that the *free stream nuclei number density* is subject to distribution depending on the nucleation size, e.g. $P_{n_{10}} \approx const \times R_{10}^{-4}$, $R_{10} > 5 \times 10^{-6}$.

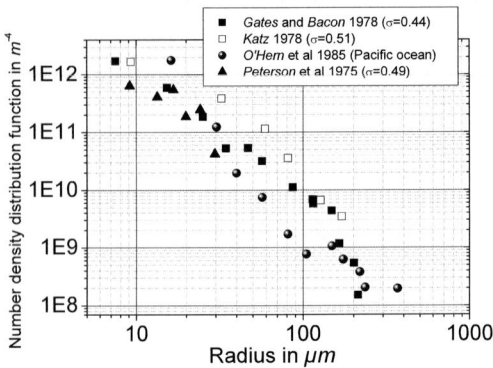

Fig. 12.2. Cavitation nuclei probability number density distribution measured by holography in three different water tunnels. With σ the cavitation number for the particular data is given in *Gates* and *Bacon* (1978), *Katz* (1978), *O'Hern* et al. (1988), *Peterson* et al. (1975)

Figure 12.2 presents the measured cavitation nuclei number density distribution as a function of the nuclei radius reported by different authors. Again the most bubbles are in the region of 7 to 11 µm sizes. *Brennen* (1995) p. 27 reported histograms of nuclei population in treated and untreated tap water and the corresponding cavitation inception number as a function of the Reynolds number. For untreated tap water the 23×10^9 nucleation at sizes of about 3µm are observed. For degassed tap water this number reduces to about 17×10^9 and for filtered water to about 2×10^9.

Conclusions:

1) Tap water has a volumetrically distributed nucleation sites in order of 10^{10} per m^3. Even specially treated water possesses volumetrically distributed nucleation sites of order of 10^9 per cubic meter. The size of the micro-bubbles is in order of 2-15 μm with the largest amount of bubbles having 2 to 10μm size. The resulting volume fraction $\alpha_{1o} = \sum_{all\_sizes} n_{1o} (\pi D_{1o}^3 / 6)$ is so small that it is usually neglected in computational analyses.

2) Micro-bubbles are so small that they follow the liquid motion and fluctuation without slip. By distortion of the solution equilibrium they probably do not contribute much for the gas release.

12.1.2 Heterogeneous nucleation at walls

Simoneau (1981) performed set of experiments with decompression of water containing dissolved nitrogen. He reported that the nucleation happens mainly at the walls of the vessel. That surface roughness plays a role as nucleation seeds are proven experimentally by *Billet* and *Holl* (1979), *Huang* (1984) and *Kuiper* (1979). The question of practical interest is how to describe quantitatively this process. Unfortunately quantitative data for heterogeneous nucleation during gas release from liquids are not available. Let us summarize the data that have to be collected in the future to resolve this problem:

12.1.2.1 Activation of surface crevices

1. Active nucleation site density:

$$n_1'' = \sum_{i=1}^{i_{max}} b_i \left(C_{2,i} - C_{2,i,\infty} \right)^{m_i} . \tag{12.22}$$

The exponent may vary between 2 and 6 in analogy to superheated liquid nucleation at wall. It is not clear whether superposition of the activated nuclei by different dissolved gasses take place or not. Probably all gasses will prefer already active nucleation seeds. It is possible that the coefficient b is a function of the wetting angle between the liquid and the wall material in analogy to the well known dependence from the theory of boiling. The analogy can be placed by asking at which temperature difference across the boundary layer of saturated bulk liquid the evaporation causes the same steam mass flow rate as the gas release at the wall by given driving concentration difference: $\rho_c D_{c,i}^* dC_{c,i}/dz = \left(\lambda_c / \Delta h \right) dT_c/dz$. The answer is $\Delta T_c = \left(\rho_c D_{c,i}^* \Delta h / \lambda_c \right) \Delta C_{c,i}$. The only data to this subject are reported by *Eddington* and *Kenning* (1978) for water at 25°C saturated with nitrogen at atmospheric pressure plus overpressure Δp. The authors relaxed the overpressure and counted the nucleation sites. It was noted that the contact time of the saturated water with the surface before the relaxation reduced the number of activated seeds (10min to about 75%, longer contact time did not change this number). The results for 20min contact time are given below.

Δp in bar	n_1'' in 1/m²
0.55	52×10^4
0.45	40×10^4
0.34	33×10^4
0.27	27×10^4

Therefore for saturated liquid newly wetting the surface, these numbers have to be about 25% higher. Another interesting observation was that over pressurization of the already saturated liquid shortly before the relaxation reduced the number of the activated seeds – an indication that during the contact time surface tensions caused already some minimal gas release to predominant places that are absorbed again.

2. Bubble departure diameter as a function of the buoyancy force and liquid velocity

$$D_{1d} = D_{1d}\left(\Delta\rho_{21}, w_2, etc.\right). \tag{12.23}$$

So for instance for forced bubble share from the wall the departure diameter is expected to be

$$D_{1d} \approx \frac{D_{1d, w_2=1}}{w_2}. \tag{12.24}$$

3. Bubble departure frequency

$$f_{1w} = \frac{1}{\Delta\tau_{1d}} = \left(C_{2,i} - C_{2,i,\infty}, D_{1d}, etc.\right), \tag{12.25}$$

where $\Delta\tau_{1d}$ is the time elapsed for a single bubble until departure. Usually if the mechanism of the diffusion controlled bubble growth is known e.g.

$$D_1 = 2B\tau^{1/2} \tag{12.26}$$

the bubble departure time is easily computed

$$\Delta\tau_{1d} = \frac{1}{4}\left(D_{1d}/B\right)^2. \tag{12.27}$$

Different bubble growth mechanisms leading to different expression for B will be discussed in the next sections. The number of the generated bubbles at the wall per unit time and unit flow volume is then

$$\dot{n}_1''' = \frac{4}{D_h}n_1'' f_{1w}. \tag{12.28}$$

The generated gas mass due to the production of bubbles with departure diameter per unit time and unit mixture volume is then

$$\mu_{1,nucl} = \frac{4}{D_h}n_1'' f_{1w}\frac{\pi D_{1d}^3}{6}\rho_1. \tag{12.29}$$

12.1.2.2 Deposition of the micro-bubbles into the turbulent boundary layer

Another mechanism of transferring micro-bubbles to nuclei capable to grow is the deposition of the micro-bubbles into the boundary layer. What is the difference between bubbles in the turbulent boundary layer and bubbles in the bulk? Micro-bubbles in the boundary layer can be entrapped in vertices close to their center where due to rotation the pressure is extremely low. Bodies in cyclones heavier than the surrounding fluid move outward from the center of the rotation. In contrast, bodies in cyclones lighter than the surrounding fluid move toward the center of the rotation. This mechanism was photographically proven by *Keller* (1979).

Rouse (1953) reported 10 to 13 times larger pressure drop due to vortices in a mixing zone than the rms-pressure fluctuation. Therefore in the boundary layer we can expect order of magnitude stronger gas release as outside the boundary layer due to larger driving concentration differences

$$C_{c,i} - C_{c,i}^{d\sigma} = C_{c,i} - C_{c,i}^{sat}\left[Y_{d,i}\left(p + \frac{2\sigma_{cd}}{R_d} - \frac{1}{2}0.166\rho_c V_c^2 \operatorname{Re}_{cw}^{1/4}\right), T_c\right]. \quad (12.30)$$

The number of the micro-bubbles striking the wall per unit time and unit surface is

$$\dot{n}_d'' = \frac{1}{4}n_d u_c'. \quad (12.31)$$

Here n_d is the number of the micro-bubbles per unit volume and u_c' is the liquid fluctuation velocity that can be considered equal to the micro-bubble fluctuation velocity. The number of micro-bubbles transferred in the turbulent boundary layer per unit flow volume and unit time is then

$$\dot{n}_d''' = \frac{4}{D_h}\dot{n}_d'' = \frac{n_d}{D_h}u_c'. \quad (12.32)$$

There are good reasons that will be discussed in a moment to assume that under given conditions these bubbles are then capable to grow. Assuming that the liquid fluctuation velocity is equal to the friction velocity

$$u_c' = const \times w_c^* = const\sqrt{\tau_{cw}/\rho_c} = const \times w_c\sqrt{\frac{1}{2}c_{cw}^d} \quad (12.33)$$

and that the drag coefficient obeys the *Blasius* formula

$$c_{cw}^d = 0.057 \operatorname{Re}_{cw}^{1/4}, \quad (12.34)$$

we obtain

$$\frac{\dot{n}_d'''}{n_d w_c} = const\, 0.17 \frac{\operatorname{Re}_{cw}^{1/8}}{D_h}. \tag{12.35}$$

From the total micro-bubble flow $w_c n_d$ only the part $const\, 0.17\operatorname{Re}_c^{1/8}/D_h$ is under the boundary layer conditions.

12.1.3 Steady diffusion mass transfer of the solvent across bubble interface

Because of the similarity of the heat conduction and mass diffusion it is possible to use the results obtained for heat transfer coefficients by simply replacing the *Nusselt* number with the *Sherwood* number, the *Peclet* number with the diffusion *Peclet* number and *Prandtl* number with the *Schmidt* number. Table 12.1 contains analytically and experimentally obtained steady state solutions for the *Sherwood* number as a function of the relative velocity and the continuum properties.

Table 12.1. Mass transfer coefficient on the surface of moving solid sphere and liquid droplets

$Sh_c = \dfrac{\beta_{cd,i} D_d}{D_{c,i}^*}$ *Sherwood* number, $Pe_{cd} = \dfrac{D_d |\Delta V_{cd}|}{D_{c,i}^*} = \operatorname{Re}_d Sc_c$ Diffusion *Peclet* number, $\operatorname{Re}_d = \dfrac{D_d \rho_c |\Delta V_{cd}|}{\eta_c}$ *Reynolds* number, $Sc_c = \dfrac{\eta_c}{\rho_c D_{c,i}^*}$ *Schmidt* number

For bubble growth with initial size zero: $D_d = 2B\tau^{1/2}$, $B = \sqrt{D_{c,i}^* Sh_c \left(C_{c,i} - C_{c,i}^{d\sigma}\right)}$

$\operatorname{Re}_d < 1$, potential flow, $\left(\dfrac{\partial R_1}{\partial \tau} \ll \Delta V_{cd}\right)$ no shearing effect in *Navier-Stikes* equation

$Sh_c = 0.99 Pe_{cd}^{1/3}$	**Immobile surface,** *Friedlander* (1961), verified for dissolution of small bubbles *Calderbank* et al. (1962) e.g. $D_d < \sqrt{\sigma_c/(g\Delta\rho_{dc})}$, liquid drops *Ward* et al. (1962), and solid spheres *Aksel'Rud* (1953).

$$Pe_{cd} \gg 9.4\left(\frac{\eta_c}{\eta_d}+1\right)\left(3\frac{\eta_c}{\eta_d}+1\right)^2$$

$$Sh_c = 0.65\left(\frac{\eta_c}{\eta_c+\eta_d}\right)^{1/2} Pe_{cd}^{1/2}$$

Mobile surface, *Hadamard (1911). Freadlander (1957)* come to similar result with a constant 0.61 and verified its equation up to $Re \leq 10$, *Nigmatulin (1978)*

$Pe_{cd} \ll 1$

$$Sh_d = 2 + \frac{9}{16}Pe_{cd} + \frac{9}{64}Pe_{cd}^2 + ...$$

Soo (1969)

$Pe_{cd} \gg 1$

$$Sh_c = \frac{2}{\pi^{1/2}}(cPe_{cd})^{1/2}, \quad c = 1$$

Hunt (1970) for a single-component system

$$Sh_d = \frac{2}{\pi^{1/2}}(cPe_{cd})^{1/2}, \quad c = 0.25\ Sc_c^{-1/3}$$

Isenberg and *Sideman* (1970) for a two-component system

$$Sh_d = \frac{2}{\pi^{1/2}} Pe_{cd}^{1/2}(1-\alpha_d)^{-1/2}$$

Kendouch (1976) for bubbles in a swarm

$$Sh_d = 2 + 0.37 Sc_{cd}^{1/3} Re_d^{3/5}.$$

analog to *Wilson* (1965)

$Re_d < 1$ or $Pe_{cd} < 10^3$ (e.g. for $D_{d\ in\ H2O} < 0.1\ mm$)

$$Sh_c = 2 + \frac{0.65 Pe_{cd}^{1.7}}{1+Pe_{cd}^{1.3}}, \quad \eta_c/\eta_d \cong 0.$$

analog to *Nigmatulin* (1978)

$$Sh_d = 2 + \frac{0.65 Pe_{cd}^{1.7}}{\left[1+(0.84 Pe_{cd}^{1.6})^3\right]^{1/3}(1+Pe_{cd}^{1.2})}.$$

analog to *Brauer* et al. (1976) for bubbles without internal circulation

$Re_d > 1$

$Sh_c = cSc_c^{1/3} Re_d^{1/2}$ $\qquad c = 0.62$, *Calderbank* (1967)
$\qquad\qquad\qquad\qquad\qquad\qquad c = 0.56$ *Froessling* (1938), flow separation at 108°

$$Pe_{cd} \gg \pi \left(\frac{4\eta_d/\eta_c - \left[\rho_d\eta_d/(\rho_c\eta_c)\right]^{1/2}}{1+\left[\rho_d\eta_d/(\rho_c\eta_c)\right]^{1/2}} \frac{1.45}{Re_d^{1/2}} \right)^2$$

$$\times \left(1 + \frac{4.35}{Re_d^{1/2}} \frac{2+3\eta_d/\eta_c}{1+\left[\rho_d\eta_d/(\rho_c\eta_c)\right]^{1/2}} \right)$$

$Sh_c = 1.13\left(1 - \dfrac{2+3\eta_d/\eta_c}{1+\left[\rho_d\eta_d/(\rho_c\eta_c)\right]^{1/2}} \dfrac{1.45}{Re_d^{1/2}}\right)^{1/2} Pe_{cd}^{1/2}$ **Mobile surface,** *Lochiel* (1963).

$100 \leq Re_d \leq 400$ for bubbles $\eta_d/\eta_c \ll 1$

$Sh_c = 1.13\left(1 - \dfrac{2.96}{Re_d^{1/2}}\right)^{1/2} Pe_{cd}^{1/2}$ **Mobile surface,** *Lochiel* (1963).

$Re_d \gg 1$, large bubbles $\eta_d/\eta_c \ll 1$

$Sh_c = 1.13 Pe_{cd}^{1/2}$. **Mobile surface,** *Lochiel* (1963).

$1 < Re_1 < 7 \times 10^4$, $0.6 < Sc_2 < 400$:

$Sh_c = 2 + (0.55 \text{ to } 0.7)Sc_c^{1/3} Re_d^{1/2}$. analog to *Soo* (1969)

Spherical caps

For spherical caps (e.g. 1.8*cm* in diameter with a mushroom like form) with $e_c = width/height$ *Lochiel* (1963) found

$Sh_c = 1.79 \dfrac{(3e_c^2 + 4)}{e_c^2 + 4} Pe_{cd}^{1/2}$.

Rosenberg (1950) and *Takadi* and *Maeda* (1961) proposed to use for all cap bubbles $e_c = 3.5$ resulting in

$Sh_c = 1.28 Pe_{cd}^{1/2}$.

12.1 Bubbles

Problem: Given a nitrogen bubble with initial diameter of D_d in infinite stagnant water at $1 bar$ and $25°C$. The initial concentration of nitrogen in the water is zero. Compute the time required for complete collapse of the bubble for prescribed velocity.

Solution: The change of the bubble size assuming constant bubble density is

$$\frac{dR_d}{d\tau} = -\frac{(\rho w)_{c,i}^{d\sigma}}{\rho_c} \qquad (12.36)$$

Using Eq. (12.4) and rearranging we obtain

$$\frac{dR_d^2}{d\tau} = -D_{c,i}^* Sh_c \left(C_{c,i}^{d\sigma} - C_{c,i} \right). \qquad (12.37)$$

Assuming that the interface concentration is equal to the saturation concentration of the water at the given pressure and temperature, $C_{c,i}^{d\sigma} = C_{c,i}^{sat}(p_c, T_c)$, and that the concentration of the water remains zero we obtain

$$\tau = \frac{R_{d,0}^2}{D_{ci}^* Sh_c C_{c,i}^{sat}(p_c, T_c)}. \qquad (12.38)$$

Bubbles with sizes less than $D_d < \sqrt{\sigma_{cd}/(g \Delta \rho_{dc})} \approx \sqrt{\frac{0.08}{9.91 \times 997}} = 2.85 \times 10^{-3} m$ are considered as stable bubbles with rigid surface. In this case $Sh_c = 0.99 Pe_{cd}^{1/3}$ for $Re_d < 1$ which corresponds to very small velocities $< 3.1 \times 10^{-4} m/s$. In the reality for instance for free rising bubbles in water we have to do with $Re_d \gg 1$ and $\eta_d / \eta_c \approx 0.01 \ll 1$. For this case the appropriate choice from Table 12.1 will be $Sh_c = 1.13 Pe_{cd}^{1/2}$. For bubble diameter of $5mm$ and relative velocity of $0.1m/s$ it takes $84.7h$ to solve completely the bubble. For $1m/s$ the corresponding time is $26.7h$. For comparison note that *Simoneau (1981)* reported that $30l$-$42bar$-$22°C$- water tank aerated with nitrogen bubbles requires 12 days to absorb $0.57cm^3$ in $1g$ water and 28 days to absorb $0.61cm^3$ in $1g$ water which is very close to the complete saturation.

12.1.4 Initial bubble growth in wall boundary layer

Kremeen et al. (1955), *Parkin* and *Kermeen* (1963) reported interesting experiments in which water with $10 m/s$ degases and produces air bubbles attached to a wall with sizes of about $10 \mu m$. The observed growing times are between 1 and $10 ms$. This is much shorter than a pure steady-state molecular diffusion solution could predict. *Van Vingaarden* (1967) reproduced their experiments by modifying a model developed first by *Levich* (1962),

$$\frac{dm_d}{d\tau} = -F(\rho w)_{c,i}^{d\sigma}, \tag{12.39}$$

where

$$(\rho w)_{c,i}^{d\sigma} = -c_{dif} \rho_c \left(C_{c,i,0} - C_{c,i}^{d\sigma} \right) \sqrt{\frac{D_{c,i}^*}{\Delta \tau}} \tag{12.40}$$

with

$$c_{dif} = \frac{2}{\sqrt{\pi}} = 1.13 \tag{12.41}$$

for plane and

$$\Delta \tau = 2 R_d / V_c \tag{12.42}$$

used frequently in the form

$$\begin{aligned}\frac{dm_d}{d\tau} &= -F(\rho w)_{c,i}^{d\sigma} = -4\pi R_d^2 \rho_c \left(C_{c,i,0} - C_{c,i}^{d\sigma} \right) \sqrt{\frac{4 D_{c,i}^*}{\pi \Delta \tau}} \\ &= -\rho_c \left(C_{c,i,0} - C_{c,i}^{d\sigma} \right) \sqrt{32 \pi D_{c,i}^* V_c R_d^{3/2}} \end{aligned} \tag{12.43}$$

by changing the constant

$$c_{dif} = \frac{2^{1/2}}{\pi^{3/4}} = 0.6. \tag{12.44}$$

Equation (12.43) can be rewritten in the form

$$Sh_c \equiv \frac{(\rho w)_{c,i}^{d\sigma} D_d}{\rho_c D_{c,i}^* \left(C_{c,i}^{d\sigma} - C_{c,i,0} \right)} = -c_{dif} \left(Sc_c \operatorname{Re}_c \right)^{1/2} = -c_{dif} Pe_c^{1/2}. \tag{12.45}$$

which is identical with the *Lochiel* (1963) solution from Table 12.1. Note that the diffusion *Peclet* number here is built with the continuum velocity in the pipe.

12.1.3 Transient diffusion mass transfer of the solvent across the bubble interface

12.1.4.1 Molecular diffusion

Consider stagnant bubble with surface concentration $C_{ci}^{d\sigma}$ in continuum with concentration at the beginning of the time $\Delta \tau$, $C_{c,i,0}$. Compute the change of the bubble radius with the time.

The mass transfer is described by the mass conservation equation for specie i in the continuum, similar to the *Fourier* equation, see *Fourier* (1822),

$$\frac{\partial C_{c,i}}{\partial \tau} = D_{c,i}^* \frac{1}{r^2} \frac{\partial}{\partial r}\left(r^2 \frac{\partial C_{c,i}}{\partial r}\right), \qquad (12.46)$$

with the following initial and boundary conditions

$$\tau = 0,\ r = R_d,\ C_{c,i} = C_{c,i}^{d\sigma}, \qquad (12.47)$$

$$\tau = 0,\ r > R_d,\ C_{c,i} = C_{c,i,0}, \qquad (12.48)$$

$$\tau > 0,\ r = R_d,\ C_{c,i} = C_{c,i}^{d\sigma}, \qquad (12.49)$$

$$\tau > 0,\ r = \infty,\ C_{c,i} = C_{c,i,0}. \qquad (12.50)$$

The textbook solution for *thin concentration boundary layer*, *Glasgow* and *Jager* (1959), is used to compute the concentration gradient at the bubble surface and to compute the resulting mass flux as a function of time

$$(\rho w)_{c,i}^{d\sigma} = -\rho_c D_{c,i}^* \left.\frac{\partial C_{c,i}}{\partial r}\right|_{r=R_1} = -\rho_c \left(C_{c,i,0} - C_{c,i}^{d\sigma}\right)\sqrt{\frac{3D_{c,i}^*}{\pi \tau}}$$

$$= -\rho_c \left(C_{c,i,0} - C_{c,i}^{sat}\right)\sqrt{\frac{3D_{ci}^*}{\pi \tau}}\left(1 - \frac{C_{c,i}^{d\sigma} - C_{c,i}^{sat}}{C_{c,i,0} - C_{c,i}^{sat}}\right). \qquad (12.51)$$

For comparison the solution obtained by *Epstein* and *Plesset* (1950) is

$$(\rho w)_{c,i}^{d\sigma} = -\rho_c \left(C_{c,i,0} - C_{c,i}^{d\sigma} \right) \left(\frac{D_{c,i}^*}{R_d} + \sqrt{\frac{D_{c,i}^*}{\pi\tau}} \right). \qquad (12.52)$$

Having minded that

$$\frac{dR_d}{d\tau} = -(\rho w)_{c,i}^{d\sigma} / \rho_c - \frac{R_d}{3\rho_d} \frac{d\rho_d}{d\tau}, \qquad (12.53)$$

we obtain for the change of the bubble radius the following differential equation

$$\frac{dR_d}{d\tau} = \left(C_{c,i,0} - C_{c,i}^{sat} \right) \sqrt{\frac{3D_{c,i}^*}{\pi\tau}} \left(1 - \frac{C_{c,i}^{d\sigma} - C_{c,i}^{sat}}{C_{c,i,0} - C_{c,i}^{sat}} \right) - \frac{R_d}{3\rho_c} \frac{d\rho_c}{d\tau}, \qquad (12.54)$$

or

$$\frac{dR_d}{d\tau} = \frac{B}{2\sqrt{\tau}} \left(1 - \frac{C_{c,i}^{d\sigma} - C_{c,i}^{sat}}{C_{c,i,0} - C_{c,i}^{sat}} \right) - \frac{R_d}{3\rho_c} \frac{d\rho_c}{d\tau}, \qquad (12.55)$$

where

$$B = \left(C_{ci,0} - C_{c,i}^{sat} \right) \sqrt{\frac{12 D_{c,i}^*}{\pi}}. \qquad (12.56)$$

It is interesting to note that comparing this with the expression obtained for the spontaneous evaporation of bubble the dimensionless number corresponding to the Jakob number is simply the concentration difference $C_{c,i,0} - C_{c,i}^{sat}$. For the limiting case of spontaneous flashing of gas for which the surface concentration can be assumed to be the saturation concentration we obtain

$$\frac{dR_d}{d\tau} = \frac{B}{2\sqrt{\tau}} - \frac{R_d}{3\rho_c} \frac{d\rho_c}{d\tau}. \qquad (12.57)$$

In case of pressure change the concentration at the surface also changes. In this case the solution is

$$(\rho w)_{c,i}^{d\sigma} = -\rho_c D_{c,i}^* \left\{ \frac{C_{c,i,0} - C_{c,i,0}^{d\sigma}}{\sqrt{\pi D_{c,i}^* \tau}} - J_{c,i} + \frac{C_{c,i,0} - C_{c,i}^{d\sigma}(\tau)}{R_d} \right\}, \qquad (12.58)$$

see *Churchill* (1958), where

$$J_{c,i} = \frac{1}{R_d\sqrt{\pi}} \int_0^{D_{c,i}^*\tau/R_d^2} \frac{\frac{dC_{c,i}^{d\sigma}}{d\tau'}}{\sqrt{D_{c,i}^*\tau/R_d^2 - \tau'}} d\tau'. \qquad (12.59)$$

Cha and *Henry* (1981) verified this equation on data for carbon dioxide. A bubble had grown from 0.0508mm to about 1.5*mm* for about 40*s*.

Problem: Given a nitrogen bubble with initial diameter of D_d in infinite stagnant water at 1*bar* and 25°C. The initial concentration of nitrogen in the water is zero. Compute the time required for complete collapse of the bubble for prescribed velocity.

Solution: For constant density we have

$$\frac{dR_d}{d\tau} = \frac{B}{2\sqrt{\tau}}, \quad \text{or} \quad R_d = R_{d,0} + B\sqrt{\tau}. \qquad (12.60)$$

The time for complete disappearance of a bubble is then

$$\tau = \frac{\pi}{12 D_{c,i}^*}\left(\frac{R_{d,0}}{C_{c,i}^{sat}(p_c,T_c)}\right)^2. \qquad (12.61)$$

To completely absorb a 5*mm* nitrogen bubble under 1*bar* and 25°*C* in water will take about 78 000 years.

12.1.4.2 Turbulent diffusion

In a bubble flow or in a churn-turbulent flow with considerable turbulence, the bubbles are moving practically with the *same* velocity as the liquid. The mechanism governing the condensation is quite different compared to the mechanism described in the previous sections. In turbulent flows the diffusion is caused mainly *by exchange of turbulent eddies* between the boundary layer and the bulk liquid.

High turbulent Reynolds numbers: At the viscous limit the characteristic time of a turbulent pulsation is $\Delta \tau = (\nu_c / \varepsilon_c)^{1/2}$, see *Kolev* (2004) p. 235 or Ch. 2. Taking this time as scale for small turbulent pulsations (high frequent) that dissipate turbulent kinetic energy we have

$$\Delta \tau = c_t (\nu_c / \varepsilon_c)^{1/2}, \qquad (12.62)$$

where c_t is a constant. The constant may be around $\sqrt{15/2} = 2.7$, see *Taylor* (1935). If it is possible to compute the friction pressure loss between the flow and the structure then it will also be possible to estimate the irreversible dissipation of the turbulent kinetic energy. This can be accomplished either by using e.g. the k-eps models in distributed parameters as described in the previous chapter or by using quasi-steady state models. I give an example for the later case. The dissipated specific kinetic energy of the turbulent pulsations is defined by

$$\alpha_c \rho_c \varepsilon_c = \frac{1}{2} \frac{\rho^2}{\rho_c} |V|^3 \Phi_{co}^2 \left(\frac{\lambda_{fr}}{D_h} + \frac{\zeta}{\Delta x_{eff}} \right) \tag{12.63}$$

Neglecting the viscous dissipation in the boundary layer and assuming that the quasi-steady state dissipation is equal to the generation of the turbulent kinetic energy, and after replacing ε_c from Eq. (12.63) into Eq. (12.62) we obtain

$$\Delta \tau = c_t \frac{\rho_c}{\rho} \left\{ 2\alpha_c v_c \Big/ \left[|V|^3 \Phi_{co}^2 \left(\frac{\lambda_{fr}}{D_h} + \frac{\zeta}{\Delta x_{eff}} \right) \right]^{1/2} \right\} \tag{12.64}$$

The mass flow rate of specie i on the bubble surface can be determined to the accuracy of a constant as

$$(\rho w)_{c,i}^{d\sigma} = -c_{dif} \rho_c \left(C_{c,i,0} - C_{c,i}^{d\sigma} \right) \sqrt{\frac{D_{c,i}^*}{\Delta \tau}}, \tag{12.65}$$

($c_{dif} = \sqrt{\frac{3}{\pi}}$ for sphere, $c_{dif} = \frac{2}{\sqrt{\pi}}$ for plane), where $\Delta \tau$ is the time interval in which the *high frequency eddy* is in contact with the bubble surface. During this time, the mass is transported from the surface to the eddy by molecular diffusion. Thereafter the eddy is transported into the bulk flow again, and its place on the surface is occupied by another one. In this way the mass absorbed from the surface by diffusion is transported from the bubble surface to the turbulent bulk liquid. We substitute $\Delta \tau$ from Eq. (12.64) into Eq. (12.65) and obtain

$$(\rho w)_{c,i}^{d\sigma} = -\frac{c_{dif}}{2^{1/4} \sqrt{c_t}} \rho_c \left(C_{c,i,0} - C_{c,i}^{d\sigma} \right) D_{c,i}^* Sc_c^{1/2} \sqrt{\frac{\rho}{\rho_c}} \frac{1}{(\alpha_c)^{1/4} v_c^{3/4}}$$

$$\times \left[|V|^3 \Phi_{co}^2 \left(\frac{\lambda_{fr}}{D_h} + \frac{\zeta}{\Delta x_{eff}} \right) \right]^{1/4}. \tag{12.66}$$

For one dimensional flow this equation reduces to

$$Sh_{cw} \equiv \frac{(\rho w)_{c,i}^{d\sigma} D_h}{\rho_c D_{c,i}^* \left(C_{c,i}^{d\sigma} - C_{c,i,0}\right)} = \frac{c_{dif}}{2^{1/4}\sqrt{c_t}} Sc_c^{1/2} Re_{cw}^{3/4} \sqrt{\frac{\rho}{\rho_c}} \left(\frac{\Phi_{co}^2}{\alpha_c}\right)^{1/4} \lambda_{fr}^{1/4}, \quad (12.67)$$

where

$$Re_{cw} = \frac{D_h w_c}{\nu_c}. \quad (12.68)$$

The constant can be determined by comparison with a result obtained by *Avdeev* (1986) for one-dimensional flow without local resistance ($\xi = 0$). *Avdeev* used the known relationship for the *friction coefficient of turbulent flow*

$$\lambda_{fr} = 0.184 Re_{cw}^{-0.2}, \quad (12.69)$$

compared the so obtained equation with experimental data for bubble condensation, and estimated the constant in

$$Sh_{cw} = \frac{c_{dif}}{2^{1/4}\sqrt{c_t}} 0.184^{1/4} Sc_c^{1/2} Re_{cw}^{0.7} \sqrt{\frac{\rho}{\rho_c}} (\Phi_{co}^2/\alpha_c)^{1/4} \quad (12.70)$$

as

$$\frac{c_{dif}}{2^{1/4}\sqrt{c_t}} 0.184^{1/4} = 0.228. \quad (12.71)$$

Thus, the so estimated constant can be successfully applied also to three-dimensional flows in porous structures. The final relationship recommended by *Avdeev* for bubble condensation in one-dimensional flow is used as analog to the mass transfer equation

$$Sh_{cw} = 0.228 Sc_c^{1/2} Re_{cw}^{0.7} (\Phi_{co}^2/\alpha_2)^{1/4} \quad (12.72)$$

that describes his own data within ± 30% error band for $D_d/D_{hw} > 80/Re_2^{0.7}$. Note that in the *Avdeev* equation ρ/ρ_c is set to one.

For comparison let us write the relationships obtained by *Hancox* and *Nikol*,

$$Sh_{cw} = 0.4 Sc_c Re_{cw}^{2/3} \quad (12.73)$$

see in *Hughes* et al. (1981), and *Labunsov* (1974),

$$Sh_{cw} = \frac{\lambda_{fr}/8}{1-12\sqrt{\frac{\lambda_{fr}}{8}}} Sc_c Re_{cw} \frac{D_d}{D_{hw}} \cong \frac{0.023}{1-1.82 Re_{cw}^{-0.1}} \frac{D_d}{D_h} Sc_c Re_{cw}^{0.8}. \quad (12.74)$$

We see that in the three equations obtained independently from each other the dependence on Re_{cw} is $Re_{cw}^{0.7 \text{ to } 0.8}$, and the dependence on Sc_c is $Sc_c^{0.5 \text{ to } 1}$. In case of $Re_{cw} \approx 0$ the mechanical energy dissipated behind the bubbles should be taken into account.

Problem: Given a nitrogen bubble with initial diameter of D_d in water at $1 bar$ and $25°C$ flowing with $3 m/s$ in a pipe with diameter of $0.08m$. The initial concentration of nitrogen in the water is zero. Compute the time required for complete collapse of the bubble. Assume that the bubble move with the liquid without relative velocity.

Solution: For constant gas density we have

$$\frac{dR_d}{d\tau} = -\frac{(\rho w)_{c,i}^{d\sigma}}{\rho_c} = -\frac{Sh_{cw} D_{c,i}^*}{D_h} \left(C_{c,i}^{d\sigma} - C_{c,i,0} \right), \quad (12.75)$$

$$\tau = \frac{R_{d0} D_h}{D_{c,i}^* Sh_{cw} C_{c,i}^{sat}(p_c, T_c)}. \quad (12.76)$$

Using the three above introduced correlations we obtain

Avdeev   50min
Hancox and *Nikol*   2min
Labunsov   187min

We realize how unreliable the models are today. The results vary within two orders of magnitude. In any case the absorption in turbulent flows is strongly accelerated compared to the pure molecular diffusion and compared to the steady state diffusion in laminar flows.

Low turbulent Reynolds numbers: The characteristic time of one cycle for large eddies, estimated by dimensional analysis of the turbulent characteristics of the continuous velocity field, is of the order of

$$\Delta \tau_t = l_{ec}/V_c' \quad (12.77)$$

for low frequency pulsations. Replacing the characteristic size of the large eddies in the liquid with

$$l_{ec} = 0.03 D_h \qquad (12.78)$$

and with the characteristic fluctuation velocity equal to the friction velocity for which the *Blasius* equation

$$\lambda_{Rw} = 0.316 Re_{cw}^{-1/4} \qquad (12.79)$$

is valid,

$$V_c' = \sqrt{\frac{\tau_{cw}}{\rho_c}} = \sqrt{\frac{\lambda_{Rw}}{8}} w^2 = w \sqrt{\frac{0.316 Re_{cw}^{-1/4}}{8}} = 0.2 \frac{V_c}{D_h} Re_{cw}^{7/8} \qquad (12.80)$$

Theofanous et al. (1975) obtain for the time constant

$$\Delta \tau_t = \frac{0.15 D_h^2}{V_c Re_{cw}^{7/8}}. \qquad (12.81)$$

Replacing in Eq. (12.65) results in

$$(\rho w)_{c,i}^{d\sigma} = -c_{dif} \rho_c \left(C_{c,i,0} - C_{c,i}^{d\sigma} \right) \sqrt{\frac{D_{c,i}^*}{\Delta \tau}}$$

$$= -\frac{c_{dif}}{\sqrt{0.15}} \rho_c \left(C_{c,i,0} - C_{c,i}^{d\sigma} \right) \frac{1}{D_h} \sqrt{D_{c,i}^* V_c} \, Re_{cw}^{7/16}, \qquad (12.82)$$

or in dimensionless form

$$Sh_c = \frac{c_{dif}}{\sqrt{0.15}} Sc_c^{1/2} Re_{cw}^{7/16} = 2.52 Sc_c^{1/2} Re_{cw}^{7/16}. \qquad (12.83)$$

We see that the exponent of the *Reynolds* number 0.44 is somewhat less than that in the case of strong turbulence, 0.7. Note that *Lamont* and *Scott* (1970) found in their experiments with CO_2 in water 0.52, $\beta_{dc,i} = 3.83 \times 10^{-6} Re_{cw}^{0.52}$ in m/s. *Theofanous* et. (1975) proposed to introduce a correction to Eq. (12.82)

… 12 Transient solution and dissolution of gasses in liquid flows

$$f = f\left(\sqrt{2}\frac{\Delta \tau_d}{\Delta \tau}\right), \qquad (12.84)$$

given in graphical form. The form of the mass transfer coefficient is then

$$\beta_{dc,i} = \frac{(\rho w)_{c,i}^{d\sigma}}{\rho_c \left(C_{c,i}^{d\sigma} - C_{c,i,0}\right)} = 0.63 f\left(\sqrt{2}\frac{\Delta \tau_d}{\Delta \tau}\right)\sqrt{\frac{D_{c,i}^*}{\Delta \tau}}, \qquad (12.85)$$

or in dimensionless form

$$Sh_{cw} = f 1.63 Sc_c^{1/2} \operatorname{Re}_{cw}^{7/16}, \qquad (12.86)$$

where

$$\Delta \tau_d = \frac{D_d}{|\Delta w_{cd}|} \qquad (12.87)$$

is some characteristic time associated with the bubble size and the relative velocity. The correction function takes into account the improving of the mass transfer due to the relative motion of the bubble with respect to surrounding liquid. The correction can be approximated by

$$f = 1.37497 + 6.93268 e^{-\sqrt{2}\frac{\Delta \tau_d}{\Delta \tau}/0.04136} + 2.93661 e^{-\sqrt{2}\frac{\Delta \tau_d}{\Delta \tau}/0.26668}. \qquad (12.88)$$

The correlation is verified by comparison with data in the region of Re_{cw} =1810 to 22400 giving $\beta_{dc,i} = 1\times 10^{-6}$ to 7.4×10^{-6}.

Problem: Given a nitrogen bubble with initial diameter of D_d in water at 1bar and 25°C flowing with 3m/s in a pipe with diameter of 0.08m. The initial concentration of nitrogen in the water is zero. Compute the time required for complete absorption of the bubble. Assume that the bubble is moving with the liquid without relative velocity.

Solution: For constant gas density we have again

$$\tau = \frac{R_{d0} D_h}{D_{ci}^* Sh_{cw} C_{c,i}^{sat}(p_c, T_c)}. \qquad (12.89)$$

Using the above introduced correlations we obtain 136min, a result that is very similar to the prediction using the *Labuntzov*'s correlation in the previous section.

Problem: Given a nitrogen bubble with initial diameter of D_d in water and $25°C$ flowing with $1.38 m/s$ in a pipe with diameter of $0.08m$. The length of the pipe is $48.5m$. The bubble is assumed to drift with the liquid without relative velocity. Therefore the bubble travels at $35.92s$ along the pipe. The entrance pressure is equal to $0.9 bar$ and the exit pressure is $2.2 bar$. The initial concentration of nitrogen in the water corresponds to the saturation concentration at the entrance pressure. Compute the size of the bubble at the exit of the pipe.

Solution: The bubble mass change with the time is

$$\frac{dm_d}{d\tau} = -F(\rho w)_{c,i}^{d\sigma} = -(4\pi)^{1/3} 3^{2/3} \left(\frac{m_d}{\rho_d}\right)^{2/3} (\rho w)_{c,i}^{d\sigma}$$

$$= -(4\pi)^{1/3} 3^{2/3} \left(\frac{m_d R_{gas} T}{p}\right)^{2/3} \frac{Sh_{cw} D_{c,i}^*}{D_h} \rho_c \left[C_{c,i}^{sat}(p, 25°C) - C_{c,i}^{sat}(p_0, 25°C)\right].$$

(12.89)

This is a non linear differential equation which has to be integrated numerically. With some approximation analytical solution can be obtained as follows. Approximating the saturation concentrations as a linear function of pressure starting with the values at $1 bar$ which is very good for such a small concentrations results in

$$\frac{dm_d}{d\tau} \approx -(4\pi)^{1/3} 3^{2/3} \left(\frac{m_d R_{gas} T}{p}\right)^{2/3} \frac{Sh_{cw} D_{c,i}^*}{D_h} \rho_c \frac{C_{c,i}^{sat}(1bar, 25°C)}{10^5} (p - p_0).$$

(12.90)

Replacing the pressure change with the linear function of time

$$p = p_0 + w\frac{dp}{dz}\tau = p_0 + p_\tau \tau,$$

and rearranging we obtain

$$\frac{dm_d}{m_d^{2/3}} \approx -A p_\tau \frac{\tau d\tau}{(p_0 + p_\tau \tau)^{2/3}},$$

(12.91)

where

$$A = (4\pi)^{1/3} 3^{2/3} (R_d T)^{2/3} \frac{Sh_{cw} D_{c,i}^*}{D_h} \rho_c \frac{C_{c,i}^{sat}(1bar, 25°C)}{10^5},$$

(12.92)

can be considered as a constant. The analytical solution is then

$$\frac{m_d}{m_{d0}} = \left\{ 1 - \frac{1}{4} \frac{A}{p_\tau m_{d0}^{1/3}} \left[(p_0 + p_\tau \tau)^{1/3} (-3p_0 + p_\tau \tau) + 3p_0^{4/3} \right] \right\}^3. \qquad (12.93)$$

Obviously, at the end of the pipe the mass ratio is a function also on the initial size of the bubble. This dependence is presented in Fig. 12.3.

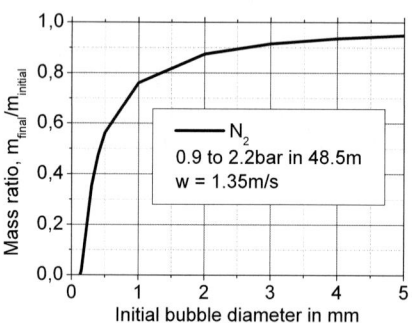

Fig. 12.3. Mass ratio at the end of the 48.5m long pipe with 0.08m-diameter and 1.38m/s water velocity as a function of the initial bubble size

The 5mm bubbles change their mass very little. At the exit of the pipe they still have 97.58% of their mass, which means only 2.4% mass reduction. Only bubbles smaller than 0.2mm will disappear over the considered distance.

Problem: Consider the same situation as in the previous problem. The only difference is that the initial bubble size is 1.18μm. Find the time required for the final dissolution of this bubble. The time required for complete dissociation can be iteratively found by

$$\tau^{n+1} = \frac{1}{p_\tau} \left[3p_0 + \frac{4p_\tau m_{d0}^{1/3} - 3p_0^{4/3}}{A \left(p_0 + p_\tau \tau^n \right)^{1/3}} \right]. \qquad (12.94)$$

Here n and $n+1$ designates successive iteration values of the time. In this particular case it is 0.608s.

Problem: Given are nitrogen micro bubbles with initial diameter of $D_{d0} = 1\mu m$ in water and 25°C flowing with 0.25m/s in a pipe with diameter of 0.025m and 11m length. The bubble number density is about $n_d = 10^6$ per m^3. The concentration of nitrogen is the saturation concentration at $1 bar$. The pressure at the end of the pipe is $0.8 bar$. Compute the release of the gas at the exit of the pipe.

Solution: Using the *Theofanous* et al. correlation we compute the mass transfer coefficient for $1 \mu m$ bubble and assume it constant. Then from above we obtain a mass growth of 18%. The volume fraction is then $\alpha_d = n_d \dfrac{m_d}{m_{d0}} \dfrac{m_{d0}}{\rho_d} = 0.044$.

Conclusions: The speed of solution or dissolution of gases depends strongly on the flow pattern. To completely dissolve bubbles of nitrogen for instants needs thousands of years due to molecular diffusion, tens of hours due to relative motion without turbulence and tens of minutes due to turbulence. Abrupt change of the pressure can cause much stronger mass transfer due to two reasons: a) The driving difference of the concentration in the transient is larger and b) flow induced turbulence in such cases is much stronger.

12.2 Droplets

Consider a family of mono-disperse particles designated with d, moving in a continuum, designated with c, with relative velocity ΔV_{cd}. We are interested in how much mass of the solvent is transferred between the surface of a particle and the surrounding continuum. We will consider first the steady state mass diffusion problem in a continuum boundary layer, thereafter the transient mass diffusion problem in a particle and finally we will give some approximate solutions for the average mass transferred.

12.2.1 Steady state gas site diffusion

The mass transported between the particle velocity field and the continuum per unit time and unit mixture volume for a steady state case is frequently approximated by

$$\mu_{c,i} = a_{cd} \left(\rho w \right)_{c,i}^{d\sigma} = a_{cd} \beta_{dc,i} \rho_c \left(C_{c,i}^{d\sigma} - C_{c,i} \right) = 6 \alpha_d D_{c,i}^* Sh_c \rho_c \left(C_{c,i}^{d\sigma} - C_{c,i} \right) / D_d^2. \quad (12.95)$$

Here $a_{cd} = 6\alpha_d / D_d$ is the interfacial area density, $\beta_{cd,i} = D_{d,i}^* Sh_d / D_d$ is the mass transfer coefficient and Sh_d is the *Sherwood* number defined as given in Table 12.2.

Table 12.2. Mass transfer coefficient on the surface of moving solid sphere and liquid droplets

$Sh_{d,i} = \dfrac{\beta_{cd,i} D_d}{D_{c,i}^*}$ *Sherwood* number, $Pe_{cd} = \dfrac{D_d |\Delta V_{cd}|}{D_{ci}^*} = Re_d Sc_c$ *Diffusion Peclet number*, $Re_d = \dfrac{D_d \rho_c |\Delta V_{cd}|}{\eta_c}$ *Reynolds* number, $Sc_c = \dfrac{\eta_c}{\rho_c D_{c,i}^*}$ *Schmidt number*

$Re_d \ll 1$ Potential flow

$Pe_{cd} \ll 1$

$Sh_d = 2 + \dfrac{1}{2} Pe_{cd} + \dfrac{1}{6} Pe_{cd}^2 + ...$ Soo (1965)

$Pe_{cd} \gg 1$

$Sh_d = 0.98 Pe_{cd}^{1/3}$ $\eta_d/\eta_c = \infty$ Nigmatulin (1978)

$Sh_d = \dfrac{2}{\pi^{1/2}} Pe_{cd}^{1/3}$ Boussinesq (1905), isothermal sphere

$Sh_d = \left(\dfrac{3}{4\pi} \dfrac{1}{1+\eta_d/\eta_c} Pe_{cd} \right)^{1/2}$ Levich (1962)

$Sh_d = 0.922 + 0.991 Pe_{cd}^{1/3}$ Acrivos-Goddard (1965)

$Sh_d = 4.73 + 1.156 Pe_{cd}^{1/2}$ Watt (1972) isothermal sphere

$Re_d < 1$,

$Pe_{cd} < 10^3$

$Sh_d = 2 + \dfrac{\tfrac{1}{3} Pe_{cd}^{0.84}}{1+\tfrac{1}{3} Pe_{cd}^{0.51}}$ $\eta_d/\eta_c = \infty$ Nigmatulin (1978)

$Pe_{cd} < 1$

$Sh_d = 2 + \dfrac{1}{2} Pe_{cd}^{1/3}$ 　　　　　　　　*Acrivos* and *Taylor* (1965) only the resistance in the continuum is taken into account

$1 < Re_d < 7 \times 10^4$, $0.6 < Sc_c < 400$

$Sh_d = 2 + (0.55 \text{ to } 0.7) Re_d^{1/2} Sc_c^{1/3}$ 　　　*Soo* (1965)

Droplets and bubbles 　　　　　　　*Michaelides* (2003)

$0 < Re_d < 1$, $Pe_{cd} > 10$

$Sh_d = \left(\dfrac{0.651}{1 + 0.95\,\eta_d/\eta_c} Pe_{cd}^{1/2} + \dfrac{0.991\,\eta_d/\eta_c}{1 + \eta_d/\eta_c} Pe_{cd}^{1/3} \right)(1.032 + A)$

$+ \dfrac{1.651(0.968 - A)}{1 + 0.95\,\eta_d/\eta_c} + \dfrac{\eta_d/\eta_c}{1 + \eta_d/\eta_c}$, 　$A = \dfrac{0.61 Re_d}{21 + Re_d}$.

$10 \leq Pe_{cd} \leq 1000$, $Re_d > 1$

$Sh_d = \dfrac{2 - \eta_d/\eta_c}{2} Sh_{d,0} + \dfrac{4\,\eta_d/\eta_c}{6 + \eta_d/\eta_c} Sh_{d,2}$ 　for $0 \leq \eta_d/\eta_c \leq 2$

$Sh_d = \dfrac{4}{\eta_d/\eta_c + 2} Sh_{d,2} + \dfrac{\eta_d/\eta_c - 2}{\eta_d/\eta_c + 2} Sh_{d,\infty}$ 　for $2 < \eta_d/\eta_c < \infty$

where

$Sh_{d,0} = 0.651 Pe_{cd}^{1/2}(1.032 + A) + 1.6 - A$ 　for $\eta_d/\eta_c = 0$,

$Sh_{d,\infty} = 0.852 Pe_{cd}^{1/3}(1 + 0.233 Re_d^{0.287}) + 1.3 - 0.182 Re_d^{0.355}$ 　for $\eta_d/\eta_c = \infty$,

$Sh_{d,2} = 0.64 Pe_{cd}^{0.43}(1 + 0.233 Re_d^{0.287}) + 1.41 - 0.15 Re_d^{0.287}$ 　for $\eta_d/\eta_c = 2$.

$100 \leq Re_d \leq 400$ for bubbles $\eta_d/\eta_c \ll 1$

$$Sh_d = 1.13\left(1 - \frac{2.96}{Re_d^{1/2}}\right)^{1/2} Pe_{cd}^{1/2} \qquad \text{Lochiel (1963), \textbf{mobile surface}.}$$

$Re_d \gg 1$, large bubbles $\eta_d / \eta_c \ll 1$

$$Sh_d = 1.13 Pe_{cd}^{1/2}. \qquad \text{Lochiel (1963), \textbf{mobile surface}.}$$

Solid sphere undergoing a step temperature change, *Feng* and *Michaelides* (1986)

$$Sh_d = 2\left\{1 + 2Pe_{cd}^2 \ln(2Pe_{cd}) + Pe_{cd}\left[\frac{1}{2}erf\left(2Pe_{cd}\sqrt{\tau^*}\right) + \frac{\exp(-4Pe_{cd}^2 \tau^*)}{4Pe_c\sqrt{\pi\tau^*}}\right]\right\}$$

$$Sh_d = 2 + 0.6 Re_d^{1/2} Sc_c^{1/3} \qquad \text{\textit{Ranz} and \textit{Marshal} (1952), verified for heat transfer on water droplets}$$

$64 < Re_d < 250$, $0.00023 < D_d < 0.00113m$, $1.03 \times 10^5 < p < 2.03 \times 10^5 Pa$, $2.8 < T_c - T_d < 36K$, $2.7 < \Delta V_{cd} < 11.7 m/s$

$$Sh_d = 2 + 0.738 Re_d^{1/2} Sc_c^{1/3}. \qquad \text{\textit{Lee} and \textit{Ryley} (1968)}$$

For much stronger turbulence in the carrier phase probably the assumption $C_{ci} = const$ over the considered time interval has to be abandoned. If we neglect the convection and the diffusion from the neighboring elementary cells, which is valid for a mass transfer time constant considerably smaller than the flow time constant, the continuum mass conservation can be approximated by

$$\alpha_c \rho_c \frac{dC_{c,i}}{d\tau} = \mu_{c,i} := 6\alpha_d D_{c,i}^* Sh_c \rho_c \left(C_{c,i}^{d\sigma} - C_{c,i}\right)/D_d^2 \qquad (12.96)$$

or

$$\frac{dC_{c,i}}{d\tau} = \frac{C_{c,i}^{d\sigma} - C_{c,i}}{\Delta\tau_c^*}, \qquad (12.97)$$

where

$$\Delta \tau_c^* = \frac{D_d^2}{6 D_{c,i}^* Sh_c} \tag{12.98}$$

approximates the characterizing continuum time constant of the gas absorption or release.

Thus for some effective and not priory known $C_{d,i}^{c\sigma}$ the difference between the interface concentration and the continuum concentration decreases exponentially

$$\overline{C}_{m,c,i}(\Delta\tau) = \frac{C_{c,i} - C_{c,i,0}}{C_{c,i}^{d\sigma} - C_{c,i,0}} = \exp\left(-\frac{\Delta\tau}{\Delta\tau_c^*}\right) \tag{12.99}$$

during the time interval $\Delta\tau$. Therefore, the time-averaged mass, transferred from the surface into the continuum per unit mixture volume and unit time is

$$\mu_{c,i}^{d\sigma} = 6\alpha_d \beta_{c,i} Sh_{c,i} \left(C_{c,i}^{d\sigma} - C_{c,i}\right) f_c / D_d^2 = \omega_c \left(C_{c,i}^{d\sigma} - C_{c,i}\right) \tag{12.100}$$

where

$$f_c = \left[1 - \exp\left(-\frac{\Delta\tau}{\Delta\tau_c^*}\right)\right] \frac{\Delta\tau_c^*}{\Delta\tau}. \tag{12.101}$$

Note that for $\Delta\tau \to 0$, $f_c \to 1$.

12.2.2 Transient diffusion inside the droplet

In the nature the concentration of both media changes with the time. As already mentioned the diffusion inside liquids is much slower than inside the gases and therefore process limiting. We consider next the transient diffusion mass transport inside the droplet associated with the mass transfer at the surface.

Consider droplet with surface mass concentration of specie i, $C_{d,i}^{c\sigma}$ cared by the continuum with the same velocity. The mass concentration of the droplet at the beginning of the time $\Delta\tau$ is $C_{d,i,0}$. Compute the mass release or absorption per unit time and unit mixture volume. The transport of specie i inside the droplet is controlled by the mass conservation equation for specie i

$$\frac{\partial C_{d,i}}{\partial \tau} = D_{d,i}^* \left(\frac{\partial^2 C_{d,i}}{\partial r^2} + \frac{2}{r} \frac{\partial C_{d,i}}{\partial r}\right), \tag{12.102}$$

which is in fact the *Fourier* equation in terms of concentration. Usually the equation is written in terms of the dimensionless concentration

$$\overline{C}_{d,i} = \frac{C_{d,i}(r,\tau) - C_{d,i,0}}{C^{c\sigma}_{d,i} - C_{d,i,0}} \qquad (12.103)$$

for $r = 0$, R_d, $\tau = \tau, \tau + \Delta\tau$ in the following form

$$\frac{\partial \overline{C}_{d,i}}{\partial \tau} = D^*_{d,i}\left(\frac{\partial^2 \overline{C}_{d,i}}{\partial r^2} + \frac{2}{r}\frac{\partial \overline{C}_{d,i}}{\partial r}\right), \qquad (12.104)$$

where $D^*_{d,i}$ is the diffusion constant of specie i inside the droplet. Text book analytical solution of the *Fourier* equation is available for the following boundary conditions:

(a) Droplet initially at uniform concentration: $\overline{C}_{d,i}(r,0) = 0$;

(b) Droplet surface at $r = R_d$ immediately reaches the surface concentration $C^{c\sigma}_{d,i}$, $\overline{C}_{d,i}(R_d,\tau) = 1$;

(c) Symmetry of the concentration profile: $\left.\dfrac{\partial \overline{C}_{d,i}}{\partial r}\right|_{r=0} = 0$.

The solution is represented by the converging *Fourier* series

$$\overline{C}_{d,i}(r,\tau) = 1 - \frac{2}{\pi}\frac{R_d}{r}\sum_{n=1}^{\infty}\frac{(-1)^n}{n}\sin\left(n\pi\frac{r}{R_d}\right)\exp(-n^2\tau/\Delta\tau_d) \qquad (12.105)$$

where $\Delta\tau_d = D_d^2/(4\pi^2 D^*_{d,i})$ is the characteristic time constant of the mass diffusion process. The intrinsic volume-averaged non-dimensional concentration is

$$\overline{C}_{m,d,i} := \frac{C_{m,d,i}(\tau) - C_{d,i,0}}{C^{c\sigma}_{d,i} - C_{d,i,0}} = 1 - \frac{6}{\pi^2}\sum_{n=1}^{\infty}\frac{1}{n^2}\exp(-n^2\tau/\Delta\tau_d). \qquad (12.106)$$

Convection inside the droplet caused by the interfacial shear due to relative velocity can improve the turbulent diffusion coefficient of the droplet $D^*_{d,i}$ by a factor $f \geq 1$. Therefore

$$\bar{C}_{m,d,i} = 1 - \frac{6}{\pi^2} \sum_{n=1}^{\infty} \frac{1}{n^2} \exp\left(-fn^2 \tau / \Delta\tau_d\right). \tag{12.107}$$

Celata et al. (1991) correlated their experimental data for effective thermal conductivity for condensation with the following expression for f,

$$f = 0.53 Pe_d^{*\,0.454}, \tag{12.108}$$

where a special definition of the *Peclet* number is used $Pe_d^* = \dfrac{D_d |\Delta V_{cd}|}{D_{di}^*} \dfrac{\eta_c}{\eta_c + \eta_d}$.

For values of $\bar{C}_{m,d,i}$ greater than 0.95 (long contact times) only the first term of Eq. (12.105) is significant

$$\bar{C}_{m,d,i} \approx 1 - \frac{6}{\pi^2} \exp\left(-f n^2 \tau / \Delta\tau_d\right). \tag{12.109}$$

One alternative solution of Eq. (12.104) can be obtained by the method of *Laplace* transformations which takes the form of a diverging infinite series. For values of $\bar{C}_{m,d,i}$ less than 0.4 (short contact times) the first term of this series,

$$\bar{C}_{m,d,i} = \frac{3\sqrt{\pi}}{D_d \sqrt{D_{d,i}^* \tau}}, \tag{12.110}$$

is the only significant term. This equation applies for short contact times when the concentration gradient of the surface surroundings has not penetrated to the center of the sphere which consequently behaves as a semi-infinite body.

From the mass balance of the specie i inside the droplet velocity field we obtain the time-averaged mass transfer from the surface to the droplets per unit time and unit mixture volume

$$\mu_{d,i}^{\sigma c} = \alpha_d \rho_d \frac{1}{\Delta\tau} \int_0^{\Delta\tau} \frac{dC_{m,d,i}}{d\tau} d\tau = \alpha_d \rho_d \frac{1}{\Delta\tau}\left[C_{m,d,i}(\Delta\tau) - C_{m,d,i}(0)\right]$$

$$= \alpha_d \rho_d \frac{\bar{C}_{m,d,i}(\Delta\tau)}{\Delta\tau}\left(C_{d,i}^{c\sigma} - C_{d,i,0}\right) = \omega_d \left(C_{d,i}^{c\sigma} - C_{d,i,0}\right). \tag{12.111}$$

Note that the mass conservation at the interface dictates

$$\mu_{d,i}^{c\sigma} = -\mu_{c,i}^{d\sigma}, \tag{12.112}$$

which is valid not only for the instantaneous but also for the time averaged values and therefore

$$C_{c,i}^{d\sigma} = \frac{\omega_d C_{d,i,0} + \omega_c C_{c,i}}{\omega_d + \omega_c}. \qquad (12.113)$$

If the gas site diffusion is ignored as a much faster than the liquid site diffusion then $C_{c,i}^{d\sigma} = C_{c,i}^{sat}$ is the best assumption.

12.3 Films

There are different flow pattern leading to a film–gas interface in multi-phase flows. Stratified and annular flows in channels with different geometry, walls of large pools with different orientations and part of the stratified liquid gas configuration in large pools are the flow patterns of practical interest. The problem of the mathematical modeling of interface mass transfer consists of (a) macroscopic predictions of the geometrical sizes and local averaged concentrations, velocities, and pressure and (b) microscopic modeling of the interface mass transfer. The purpose of this section is to review the state of the art for modeling of the interface mass transfer for known geometry.

12.3.1 Geometrical film-gas characteristics

Next we summarize some important geometrical characteristics needed further.

The cross section occupied by gas and liquid in a channel flow is

$$F_1 = \alpha_1 F \qquad (12.114)$$

and

$$F_2 = \alpha_2 F \qquad (12.115)$$

respectively, where F is the channel cross section. Depending on the gas velocity the film structure can be

(a) symmetric, or
(b) asymmetric.

The symmetric film structure is characterized by uniformly distributed film on the wet perimeter in the plane perpendicular to the main flow direction. For this case the film thickness is

$$\delta_{2F} = D_h \left(1 - \sqrt{1-\alpha_2}\right)/2, \qquad (12.116)$$

where D_h is the hydraulic diameter of the channel in a plane perpendicular to the main flow direction. The liquid gas interface per unit flow volume that is called the interfacial area density is

$$a_{12} = \frac{4}{D_h}\sqrt{1-\alpha_2} \ . \tag{12.117}$$

The hydraulic diameter of the gas is

$$D_{h1} = D_h\sqrt{1-\alpha_2} \ . \tag{12.118}$$

Asymmetric film structure results in the case of dominance of the gravitation force. The flow can be characterized by three perimeters, the gas–wall contact, Per_{1w}, the gas–liquid contact, Per_{12}, and the liquid–wall contact, Per_{2w}. To these three perimeters correspond three surface averaged shear stresses, one between the gas and the wall, τ_{1w}, one between gas and liquid, τ_{12}, and one between the liquid and the wall, τ_{2w}. The hydraulic diameters for computation of the friction pressure loss of the both fluids can be defined as

$$D_{h1} = 4\alpha_1 F / (Per_{1w} + Per_{12}), \tag{12.119}$$

$$D_{h1} = 4\alpha_2 F / (Per_{2w} + Per_{12}) . \tag{12.120}$$

Two characteristic *Reynolds* numbers can be defined by using these length scales

$$Re_1 = \frac{\rho_1 V_1 D_{h1}}{\eta_1} = \frac{\alpha_1 \rho_1 V_1}{\eta_1} \frac{4F}{Per_{1w} + Per_{12}}, \tag{12.121}$$

$$Re_2 = \frac{\rho_2 V_2 D_{h2}}{\eta_2} = \frac{\alpha_2 \rho_2 V_2}{\eta_2} \frac{4F}{Per_{2w} + Per_{12}} . \tag{12.122}$$

In the three-dimensional space it is possible that the liquid in a computational cell is identified to occupy the lower part of the cell. In this case the gas–liquid interfacial area density is

$$a_{12} = 1/\Delta z \ . \tag{12.123}$$

The film thickness in this case is

$$\delta_{2F} = \alpha_2 \Delta z , \tag{12.124}$$

where the α_2 is the local liquid volume fraction in the computational cell. In the case of film attached at the vertical wall of radius r of a control volume in cylindrical coordinates

$$\delta_{2F} = r\left(1 - \sqrt{1 - \alpha_2\left[1 - (r_{i-1}/r)^2\right]}\right). \tag{12.125}$$

For the limiting case of $r_{i-1} = 0$ Eq. (12.125) reduces to Eq. (12.116). The interfacial area density in this case is

$$a_{12} = 2\frac{1 - \delta_{2F}/r}{r\left[1 - (r_{i-1}/r)^2\right]} = 2\frac{\sqrt{1 - \alpha_2\left[1 - (r_{i-1}/r)^2\right]}}{r\left[1 - (r_{i-1}/r)^2\right]}. \tag{12.126}$$

For the case of $r_{i-1} = 0$ Eq. (12.126) reduces to Eq. (12.117). For such flow pattern the following film *Reynolds* number is used

$$Re_{2F} = \rho_2 V_2 \delta_{2F}/\eta_2 . \tag{12.127}$$

12.3.2 Liquid side mass transfer due to molecular diffusion

Next we consider the heat transfer from the interface to the bulk liquid. The liquid can be laminar or turbulent. For laminar liquid the heat is transferred from the interface to the liquid due to heat conduction described by the *Fourier* equation, (1822),

$$\frac{\partial C_{2,i}}{\partial \tau} = -D^*_{2,i}\frac{\partial^2 C_{2,i}}{\partial y^2}, \tag{12.128}$$

where the positive y-direction is defined from the interface to the bulk liquid. The textbook solution for the following boundary conditions

$$\tau \geq 0, y = 0, C_{2,i} = C_{2,i}^{1\sigma} \text{ (interface)}, \tag{12.129}$$

$$\tau = 0, y > 0, C_{2,i} = C_{2,i,0} \text{ (bulk liquid)}, \tag{12.130}$$

$$\tau > 0, y \to 1, C_{2,i} = C_{2,i,0} \text{ (bulk liquid)}, \tag{12.131}$$

is

$$C - C_{2,i,0} = \left(C_{2,i}^{1\sigma} - C_{2,i,0}\right) erfc \frac{y}{2\left(D^*_{2,i}\tau\right)^{1/2}}. \tag{12.132}$$

Using the temperature gradient

$$\frac{\partial C}{\partial \tau} = -\frac{C_{2,i}^{1\sigma} - C_{2,i,0}}{\left(\pi D_{2,i}^* \tau\right)^{1/2}} \exp\left[-y^2/\left(4D_{2,i}^*\tau\right)\right] \qquad (12.133)$$

at $y = 0$ we compute the heat flux at the interface

$$\left(\rho w\right)_{2,i}^{1\sigma} = \rho_2 D_{2,i}^* \left.\frac{\partial C_{2,i}}{\partial y}\right|_{y=0} = \rho_2\left(C_{2,i}^{d\sigma} - C_{2,i,0}\right)\sqrt{\frac{2D_{2,i}^*}{\pi\tau}} = \frac{B}{\sqrt{\tau}}\rho_2\left(C_{2,i}^{1\sigma} - C_{2,i,0}\right)$$

$$= \beta_2 \rho_2 \left(C_{2,i}^{1\sigma} - C_{2,i,0}\right). \qquad (12.134)$$

The averaged heat flux over the time period $\Delta\tau$ is

$$\left(\rho w\right)_{2,i}^{1\sigma} = \frac{2B}{\sqrt{\Delta\tau}}\rho_2\left(C_{2,i}^{1\sigma} - C_{2,i,0}\right). \qquad (12.135)$$

12.3.3 Liquid side mass transfer due to turbulence diffusion

Let $\Delta\tau_2$ be the average time in which a turbulent eddy stays in the neighborhood of the free surface before jumping apart, sometimes called the renewal period. During this time the average heat flux from the interface to the eddies by heat conduction is

$$\left(\rho w\right)_{2,i}^{1\sigma} = \frac{2B}{\sqrt{\Delta\tau}}\rho_2\left(C_{2,i}^{1\sigma} - C_{2,i,0}\right) = \beta_2 \rho_2\left(C_{2,i}^{1\sigma} - C_{2,i,0}\right). \qquad (12.136)$$

If the volume-averaged pulsation velocity is V_2' and the length scale of large turbulent eddies in the film is $\ell_{e,2}$ the time in which the eddy stays at the interface is

$$\Delta\tau_2 = f\left(\ell_{e,2}, V_2'\right). \qquad (12.137)$$

Thus, the task to model turbulent mass diffusion from the interface to the liquid is reduced to the task to model

(i) pulsation velocity V_2' and

(ii) turbulent length scale $\ell_{e,2}$.

Usually β_2 is written as a function of the dimensionless turbulent *Reynolds* number

$$Re_2' = \rho_2 V_2' \ell_{e,2} / \eta_2, \qquad (12.138)$$

and *Prandtl* number Pr_2,

$$\beta_2 = \sqrt{\frac{2 D_{2,i}^*}{\pi \Delta \tau_2}} = \sqrt{\frac{2}{\pi}} Sc_2^{-1/2} \sqrt{\frac{\eta_2}{\rho_2 \Delta \tau_2}} = \sqrt{\frac{2}{\pi}} Sc_2^{-1/2} Re_2'^{-1/2} \sqrt{\frac{V_2' \ell_{e,2}}{\Delta \tau_2}}, \qquad (12.139)$$

and is a subject of modeling work and verification with experiments. Once we know β_2 we can compute some effective conductivity by the equation

$$\sqrt{\frac{2 D_{2,i,eff}^*}{\pi \Delta \tau_2}} \rho_2 \left(C_{2,i}^{1\sigma} - C_{2,i,0} \right) = \beta_2 \rho_2 \left(C_{2,i}^{1\sigma} - C_{2,i,0} \right) \qquad (12.140)$$

or

$$D_{2i,eff}^* = \frac{\pi}{2} \Delta \tau_2 \beta_2^2, \qquad (12.141)$$

and use Eq. (12.141) for both laminar and turbulent heat transfer.

12.3.3.1 High Reynolds number

One of the possible ways for computation of the renewal period $\Delta \tau_2$ for high turbulent *Reynolds* numbers

$$Re_2' > 500 \qquad (12.142)$$

is the use of the hypothesis by *Kolmogoroff* for isotropic turbulence

$$\Delta \tau_2 \approx c_1 \left(v_2 \ell_{e2} / V_2'^3 \right)^{1/2}, \qquad (12.143)$$

where

$$\ell_{e2} \approx c_1 \delta_{2F}, \qquad (12.144)$$

where c_1 is a constant. In this region the turbulence energy is concentrated in microscopic eddies (mechanical energy dissipating motion). Substituting $\Delta\tau_2$ from Eq. (12.143) into (12.139) results in

$$\beta_2 = \sqrt{\frac{2}{\pi}} Sc_2^{-1/2} Re_2^{t-1/4} V_2'. \qquad (12.145)$$

The qualitative relationship $\beta_2 Sc_2^{1/2} \approx Re_2^{t-1/4}$ was originally proposed by *Banerjee* et al. (1968) and experimentally confirmed by *Banerjee* et al. (1990). *Lamont* and *Scot*, see in *Lamont* and *Yuen* (1982), describe successfully heat transfer from the film surface to the flowing turbulent film for high Reynolds numbers using Eq. (12.145) with a constant 0.25 instead of $\sqrt{2/\pi}$. Recently *Hobbhahn* (1989) obtained experimental data for condensation on a free surface, which are successfully described by Eq. (12.145) modified as follows

$$\beta_2 = 0.07 Sc_2^{-0.6} Re_2^{t-1/5} V_2' \qquad (12.146)$$

where the discrepancy between the assumption made by the authors

$$\ell_{e2} \approx \delta_{2F} \qquad (12.147)$$

and

$$V_2' \approx |V_2| + \sqrt{\rho_1/\rho_2}\,|V_1 - V_2| \qquad (12.148)$$

and the reality are compensated by the constant 0.07.

12.3.3.2 Low Reynolds number

For low turbulence *Reynolds* number

$$Re_2^t < 500 \qquad (12.149)$$

the turbulence energy is concentrated in macroscopic eddies. The choice of the liquid film thickness, as a length scale of turbulence

$$\ell_{e,2} \approx \delta_{2F} \qquad (12.150)$$

is reasonable. Therefore the renewal period is

$$\Delta\tau_2 \approx \delta_{2F}/V_2' \qquad (12.151)$$

and Eq. (12.139) reduces to

$$\beta_2 = \sqrt{\frac{2}{\pi}} Pr_2^{-1/2} Re_2^{t-1/2} V_2'. \tag{12.152}$$

For low *Reynolds* numbers *Fortescue* and *Pearson* (1967) recommended instead of $\sqrt{2/\pi}$ in Eq. (12.152) to use

$$\text{const} = 0.7(1 + 0.44 / \Delta \tau_2^*), \tag{12.153}$$

where

$$\Delta \tau_2^* = \Delta \tau_2 V_2' / \ell_{e,2} \approx 1, \tag{12.154}$$

which is > 0.85, as recommended by *Brumfield* et al. (1975).

Theofanous et al. (1975) showed that most of their experimental data for condensation on channels with free surface can be described by using Eqs. (12.141) and (12.152) with the above discussed corrections introduced by *Lamont* and *Scot*, and *Fortescue* and *Person*, respectively.

12.3.3.3 Time scales for pulsation velocity

a) Time scale of turbulence pulsation velocity based on average liquid velocity.

A very rough estimate of the pulsation velocity is

$$V' = c_1 V_2 \approx (0.1 \text{ to } 0.3) V_2, \tag{12.155}$$

With this approach and

$$\ell_{e2} \approx c_2 \delta_{2F} \tag{12.156}$$

the two equations for β_2 read

$$Re_{2F} < c_1 500, \tag{12.157}$$

$$\beta_2 = c_2^{-1/4} c_1^{3/4} \sqrt{\frac{2}{\pi}} Sc_2^{-1/2} Re_{2F}^{-1/4} V_2 = \text{const } Sc_2^{-1/2} Re_{2F}^{-1/4} V_2. \tag{12.158}$$

Comparing this equation with the *McEligot* equation we see that

$$\text{const} \approx 0.021 \text{ to } 0.037. \tag{12.159}$$

For

$$Re_{2F} \geq c_1 500 \tag{12.160}$$

$$\beta_2 = c_1^{1/2}\sqrt{\frac{2}{\pi}} Pr_2^{-1/2} Re_{2F}^{-1/2} V_2. \tag{12.161}$$

b) The scale of turbulence pulsation velocity based on friction velocity

Improvement of the above theory requires a close look at the reasons for the existence of turbulence in the liquid film. This is either the wall shear stress, τ_{2w}, or the shear stress acting on the gas–liquid interface, τ_{12}, or both simultaneously. The shear stress at the wall for channel flow is

$$\tau_{2w} = \frac{1}{4}\lambda_{2w}\frac{1}{2}\rho_2 V_2^2, \tag{12.162}$$

where

$$\lambda_{2w} = \lambda_{friction}\left(\frac{\rho_2 V_2 D_{h2w}}{\eta_2}, k_w/D_{h2w}\right). \tag{12.163}$$

Here

$$D_{h2w} = 4\alpha_2 F/Per_{2w}. \tag{12.164}$$

is the hydraulic diameter of the channel for the liquid, and k_w/D_{h2w} is the relative roughness. The shear stress of the gas–liquid interface in channel flow is

$$\tau_{12} = \frac{1}{4}\lambda_{12}\frac{1}{2}\rho_1 \Delta V_{12}^2, \tag{12.165}$$

where

$$\lambda_{w2} = \lambda_{friction}\left(\frac{\rho_1 V_{12} D_{h1}}{\eta_2}, \frac{\delta_{2F}/4}{D_{h1}}\right). \tag{12.166}$$

Here

$$D_{h1} = \frac{4\alpha_1 F}{Per_{1w} + Per_{12}}, \tag{12.167}$$

is the hydraulic diameter of the "gas channel" and $(\delta_{2F}/4)/D_{h1}$ is the relative roughness of the "gas channel" taken to be a function of the waviness of the film.

For vertical plane walls the average shear stress at the wall along Δz is

$$\tau_{w2} = c_{w2}\frac{1}{2}\rho_2 V_2^2 \qquad (12.168)$$

where for laminar flow

$$Re_{2z} < (1 \text{ to } 3)10^5 \qquad (12.169)$$

The averaged *steady state* drag coefficient in accordance with *Prandtl* and *Blasius* is

$$c_{w2} = 1.372/Re_{2z}^{1/2} \qquad (12.170)$$

and

$$Re_{2z} = V_2 \Delta z / v_2 \qquad (12.171)$$

and for turbulent flow

$$c_{w2} = 0.072/Re_{2z}^{1/5}, \qquad (12.172)$$

see *Albring* (1970). Similarly the gas side averaged shear stress is computed using

$$Re_{1z} = \Delta V_{12} \Delta z / v_2. \qquad (12.173)$$

Thus the effective shear stress in the film is

$$\tau_{2\mathit{eff}} = \frac{Per_{2w}\tau_{2w} + Per_{12}\tau_{12}}{Per_{2w} + Per_{12}}. \qquad (12.174)$$

Now we can estimate the time scale of the turbulence in the shear flow

$$\Delta \tau_2 \approx \left(v_2 \delta_{2F}/V_2^{\prime 3}\right)^{1/2}, \qquad (12.175)$$

using the dynamic friction velocity

$$V_{2\mathit{eff}}^* = \sqrt{\tau_{2\mathit{eff}}/\rho_2}. \qquad (12.176)$$

Assuming that the pulsation velocity is of the order of magnitude of the friction velocity

$$V_2' \approx \text{const } V_{2\text{eff}}^*, \qquad (12.177)$$

where the

$$\text{const} \approx 2.9, \qquad (12.178)$$

we obtain for

$$Re_{2F}^* > \text{const } 500, \qquad (12.179)$$

$$\beta_2 = \text{const}^{3/4} \sqrt{\frac{2}{\pi}} Pr_2^{-1/2} Re_{2F}^{*-1/4} V_{2\text{eff}}^*, \qquad (12.180)$$

and for

$$Re_2 < \text{const } 500, \qquad (12.181)$$

$$\beta_2 = \text{const}^{1/2} \sqrt{\frac{2}{\pi}} Pr_2^{-1/2} Re_{2F}^{*-1/2} V_{2\text{eff}}^*, \qquad (12.182)$$

where

$$Re_{2F}^* = \frac{\rho_2 V_{2\text{eff}}^* \delta_{2F}}{\eta_2}. \qquad (12.183)$$

More careful modeling of the turbulent length scale for the derivation of Eq. (12.182) was done by *Kim* and *Bankoff* (1983). The authors used the assumption

$$V_2' \approx V_{2\text{eff}}^*, \qquad (12.184)$$

and modified Eq. (12.182) as follows

$$\beta_2 = 0.061 Sc_2^{-1/2} Re_2'^{0.12} V_{2\text{eff}}^*, \qquad (12.185)$$

where

$$\ell_{e2} = \left[\frac{\sigma}{(\rho_2 - \rho_1)g} \right]^{1/2} 3.03 \times 10^{-8} Re_1^{*1.85} Re_2^{*0.006} Sc_2^{-0.23}, \qquad (12.186)$$

for $3000 < Re_1^* < 18000$ and $800 < Re_2^* < 5000$, $\lambda_{12} = 0.0524 + 0.92 \times 10^{-5} \, Re_2^*$ valid for $Re_2^* > 340$, and $\tau_{12} \gg \tau_{2w}$. Note the special definition of the *Reynolds* numbers as mass flow per unit width of the film

$$Re_1^* = \frac{\alpha_1 \rho_1 V_1}{\eta_1} \frac{F}{Per_{12}}, \qquad (12.187)$$

$$Re_2^* = \frac{\alpha_2 \rho_2 V_2}{\eta_2} \frac{F}{Per_{12}}. \qquad (12.188)$$

Assuming that the pulsation velocity is of order of the magnitude of the friction velocity as before but the time scale of the turbulence is

$$\Delta \tau_2 \approx (v_2/\varepsilon_2)^{1/2} = v_2/V_{2eff}^{*2}, \qquad (12.189)$$

results for

$$Re_{2F}^* > \text{const } 500, \qquad (12.190)$$

in

$$\beta_2 = \sqrt{\frac{2}{\pi}} Pr_2^{-1/2} V_{2eff}^*. \qquad (12.191)$$

This equation is recommended by *Jensen* and *Yuen* (1982) with a constant 0.14 instead $\sqrt{2/\pi}$ for $\tau_{12} \gg \tau_{2w}$. *Hughes* and *Duffey* reproduced an excellent agreement with experimental data for steam condensation in horizontal liquid films by using Eq. (12.180) and the assumption

$$\tau_{2eff} = (\tau_{12} + \tau_{2w})/2. \qquad (12.192)$$

Nevertheless one should bear in mind that Eq. (12.174) is more general.

An example of detailed modeling of the turbulence structure in the film during film condensation from stagnant steam on vertically cooled surfaces is given by *Mitrovich*, see *Rohsenow* and *Choi* (1961). Using the analogy between heat and mass transfer we rewrite their results to

$$(\rho w)_{2,i}^{1\sigma} = \beta_{21,i} \rho_2 \left(C_{2,i}^{1\sigma} - C_{2,i} \right) \qquad (12.193)$$

where

$$\beta_{2,i}\delta_{2F}/D_{2,i}^* = 1.05 Re_{2F}^{-0.33}\left[1+C^{1.9}Re_{2F}^{1.267}Sc_2^{1.11}\right]^{0.526}, \qquad (12.194)$$

$$C = 8.8 \cdot 10^{-3}/(1 + 2.29 \cdot 10^{-5} Ka_2^{0.269}), \qquad (12.195)$$

$$Ka_2 = \rho_2\sigma^3/\left(g\eta_2^4\right). \qquad (12.196)$$

Summarizing the results discussed above we can say the following:

(a) Gas in the two-phase film flow behaves as a gas in a channel. Therefore the gas side mass transfer can be considered as a mass transfer between gas and the interface taking into account the waves at the liquid surface.

(b) The liquid side mass transfer at the interface is due to molecular and turbulent diffusion. The modeling of the turbulent diffusion can be performed by modeling the time and length scale of the turbulence taking into account that turbulence is produced mainly

(i) at the wall–liquid interface, and
(ii) at the gas–liquid interface.

(c) Gas side mass transfer in a pool flow can be considered as a mass transfer at plane interface.

(d) The liquid side mass transfer from the interface into the bulk liquid is governed by the solution of the transient *Fourier* equation in terms of concentrations where in case of the turbulence the use of effective eddy diffusivity instead of the molecular diffusivity is recommended.

Nomenclature

Latin

A	function
a_{cd}	interfacial area density between the continuum and disperse phase, m²/m³
a_{12}	$=1/\Delta z$ gas–liquid interfacial area density in Cartesian coordinates, $1/m$
B	acceleration function defining the diffusion controlled bubble growth or collapse, m/s²
b_i	coefficient defining the activation of nucleation sites for specific gas specie i, 1/m²
$C_{2,i}$	mass concentration of the specie i inside the liquid, dimensionless
$C_{2,i}^{sat}$	saturation mass concentration of specie i inside the liquid, dimensionless

$C_{c,i}$ — bulk mass concentration far from the interface in the continuum, dimensionless

$C_{c,i,0}$ — bulk mass concentration far from the interface in the continuum at the beginning of the considered process, dimensionless

$C_{c,i}^{d\sigma}$ — interface concentration continuum site, dimensionless

$C_{c,i}^{sat}$ — saturation concentration in the continuum, dimensionless

$C_{d,i}$ — mass concentration of the specie i inside the dispersed phase, kg/kg

$\bar{C}_{d,i} := \dfrac{C_{d,i}(r,\tau) - C_{d,i,0}}{C_{d,i}^{c\sigma} - C_{d,i,0}}$, mass concentration of the specie i inside the dispersed phase, dimensionless

$C_{di}^{c\sigma}$ — mass concentration of the specie i at the droplet surface, kg/kg

$C_{d,i,0}$ — mass concentration of the specie i inside the droplet at the beginning of the process considered, kg/kg

$\bar{C}_{m,d,i}$ — volume-averaged mass concentration of the droplet, kg/kg

c_{cw}^{d} — drag coefficient, dimensionless

c_{dif} — geometry constant, dimensionless

c_t — constant, dimensionless

D_1 — bubble diameter, m

D_{1d} — bubble departure diameter, m

$D_{1d,w_2=1}$ — bubble departure diameter for 1m/s liquid velocity, m

D_{1o} — initial size of micro-bubbles, m

$D_{c,i}^{*}$ — diffusion coefficient for specie i in the continuous liquid c, m²/s

D_d — diameter dispersed phase, m

D_{d0} — initial diameter dispersed phase, m

$D_{d,i}^{*}$ — diffusion coefficient for specie i in the dispersed liquid d, m²/s

D_h — hydraulic diameter, m

D_{hu} — hydraulic diameter in x-direction, m

D_{hv} — hydraulic diameter in y-direction, m

D_{hw} — hydraulic diameter in z-direction, m

D_{h1} — hydraulic diameter for the gas, *m*

D_{h2} — hydraulic diameter for the liquid, *m*

D_{h12} — hydraulic diameter for computation of the gas friction pressure loss component in a gas–liquid stratified flow, *m*

e — aspect ratio width/height, dimensionless

F — channel cross section, *m*

f	function
f_{1w}	bubble departure frequency, 1/s
f_c	function originating after time averaging
g	gravitational acceleration, m/s²
Δh	evaporation enthalpy, J/kg
$J_{c,i}$	memory function
k	specific turbulent kinetic energy, m²/s²
k_w	wall roughness, m
l_{ec}	characteristic size of the large eddies in the liquid, m
m_d	mass of the dispersed particle, kg
m_{d0}	initial mass of the dispersed particle, kg
m_{gas}	inert gas mass inside the bubble, kg
m_i	mass of the specie i, kg
n_1''	active nucleation site density for bubble generation, 1/m²
$\dot{n}_1'''$	generated bubbles per unit time and unit mixture volume, 1/(m³s)
n_d	number of dispersed particles per unit mixture volume, 1/m³
$\dot{n}_d''$	number of the micro-bubbles striking the wall per unit time and unit surface, 1/(m²s)
$\dot{n}_d'''$	number of the micro-bubbles transferred in the turbulent boundary layer per unit flow volume and unit time, 1/(m³s)
$P_{n_{10}}$	probability that a micro-bubble with size within ΔD_d is found in unit volume of the liquid, 1/m⁴
p	pressure, Pa
p'	*rms*-values of pressure fluctuation, Pa
p_0	initial pressure, Pa
p_{1i}	partial pressure of specie i in the bubble, Pa
p_c	pressure in the continuum, Pa
p_{di}	partial pressure of specie i in the dispersed phase, Pa
p_{gas}	partial pressure of the non-condensing gases, Pa
p_{H_2O}	partial pressure of water or water steam, Pa
p_τ	$:= w\, dp/dz$, Pa/s
Δp	pressure difference, Pa
Pe_{cd}	$:= D_d |\Delta V_{cd}|/D_{ci}^* = Re_d Sc_c$, bubble diffusion *Peclet* number, dimensionless
Pe_d^*	$:= \left(D_d |\Delta V_{cd}|/D_{di}^*\right) \eta_c / (\eta_c + \eta_d)$, droplet diffusion *Peclet* number, dimensionless

Per_w	perimeter of the rectangular channel, m
Per_{1w}	wetted perimeters for the gas, m
Per_{2w}	wetted perimeters for the liquid, m
R_1	bubble radius, m
R_d	radius of the dispersed phase, m
R_{d0}	initial radius of the dispersed phase, m
$R_{d,cr}$	critical radius of the dispersed phase, m
R_{gas}	gas constant,
Re_d	$:= D_d \rho_c |\Delta V_{cd}|/\eta_c$, Reynolds number, dimensionless
Re_{cw}	$:= D_h w_c / \nu_c$, Reynolds number of the continuum in pipe, dimensionless
Re_1	$= \rho_1 w_1 D_{h1} / \eta_1$, gas Reynolds number, dimensionless
Re_2	$= \rho_2 w_2 D_{h2} / \eta_2$, liquid Reynolds number, dimensionless
Re_1^*	$= \dfrac{\alpha_1 \rho_1 V_1}{\eta_1} \dfrac{F}{Per_{12}}$, modified gas Reynolds number, dimensionless
Re_2^*	$= \dfrac{\alpha_2 \rho_2 V_2}{\eta_2} \dfrac{F}{Per_{12}}$, modified liquid Reynolds number, dimensionless
Re_{2F}	$= \dfrac{\rho_2 V_2 \delta_{2F}}{\eta_2}$ film Reynolds number, dimensionless
Re_{1z}	$= V_1 \Delta z / \nu_1$, gas Reynolds number, dimensionless
Re_2'	$= \rho_2 V_2' \ell_{e2} / \eta_2$, liquid turbulent Reynolds number, dimensionless
Re_{2F}^*	$= \dfrac{\rho_2 V_{2eff}^* \delta_{2F}}{\eta_2}$, liquid turbulent Reynolds number based on the friction velocity, dimensionless
r	radius, m
Sc_c	$:= \eta_c / (\rho_c D_{c,i}^*)$, Schmidt number, dimensionless
Sh_c	$:= \dfrac{\beta_{cd,i} D_d}{D_{ci}^*} = \dfrac{(\rho w)_{i,dc}}{\rho_c (C_{ci}^{d\sigma} - C_{ci})} \dfrac{D_d}{D_{ci}^*}$, Sherwood number, dimensionless
Sh_{cw}	$:= \dfrac{(\rho w)_{c,i}^{d\sigma} D_h}{\rho_c (C_{c,i}^{d\sigma} - C_{c,i,0}) D_{c,i}^*}$, Sherwood number for gas release or absorption in turbulent flow in pipes, dimensionless
$Sh_{d,i}$	$:= \dfrac{\beta_{cd,i} D_d}{D_{c,i}^*}$ Sherwood number, dimensionless
T	temperature, K
T_{20}	initial liquid temperature, K

T_c	continuum temperature, K
ΔT_c	temperature difference in the continuum, K
u	velocity in x-direction, m/s
u'	fluctuation velocity in x-direction, m/s
$V^*_{2\mathit{eff}}$	$:= \sqrt{\tau_{2\mathit{eff}}/\rho_2}$, dynamic friction velocity, m/s
ΔV_{cd}	$:= V_c - V_d$, velocity difference, m/s
v	velocity in y-direction, m/s
v'	fluctuation velocity in y-direction, m/s
w	velocity in z-direction, m/s
w'	fluctuation velocity in z-direction, m/s
w^*_c	friction velocity of the continuum, m/s
Δw_{cd}	velocity difference, m/s
Δx_{eff}	some effective length scale, m
Y_{di}	molar concentration of specie i inside the dispersed phase, dimensionless
z	z-coordinate, m

Greek

α_{1o}	initial gas volume concentration, dimensionless
α_c	volume concentration of the continuum phase, dimensionless
α_d	volume concentration of the dispersed phase, dimensionless
$\beta_{cd,i}$	mass transfer coefficient for specie i from the continuum to the dispersed phase, m/s
$\beta_{dc,i}$	mass transfer coefficient for specie i from the dispersed phase to the continuum phase, m/s
δ_{2F}	film thickness, m
ε	dissipation of the turbulent kinetic energy, m²/s³
ζ	irreversible friction coefficient, dimensionless
η	dynamic viscosity, kg/(ms)
$\Delta\theta$	angular increment, rad
λ	thermal conductivity, W/(mK)
λ_{RT}	$:= \sqrt{\dfrac{\sigma_2}{g\Delta\rho_{21}}}$, *Rayleigh-Taylor* wavelength, m
λ_{fr}	friction coefficient, dimensionless
λ_{w1}	friction coefficient for the gas–wall contact, dimensionless
$\mu_{1,nucl}$	generated gas mass due to the production of bubbles with departure diameter per unit time and unit mixture volume, kg/(m³s)

$\mu_{c,i}^{d\sigma}$ time-averaged mass transfer from the surface to the continuum per unit time and unit mixture volume, kg/(m³s)

$\mu_{d,i}^{c\sigma}$ time-averaged mass transfer from the surface to the droplet per unit time and unit mixture volume, kg/(m³s)

ρ density, kg/m³

$(\rho w)_{c,i}^{d\sigma}$ mass flow rate of specie i into the continuum from the interface with the dispersed phase, kg/(m²s)

$(\rho w)_{d,i}^{c\sigma}$ mass flow rate of specie i into the dispersed phase from the interface with the continuum, kg/(m²s)

σ cavitation number, dimensionless

σ_{cd} surface tension between continuum c and dispersed phase, N/m

τ time, s

τ' time, s

$\Delta\tau$ time interval, s

$\Delta\tau_2$ time for which the eddy stays at the interface, renewal period, s

$\Delta\tau_{1d}$ bubble departure time, s

$\Delta\tau_d$ $:= D_d/|\Delta w_{cd}|$

$\Delta\tau_c^*$ $:= D_d^2/(6 D_{c,i}^* Sh_c)$, time constant, dimensionless

$\Delta\tau_2^*$ $:= \Delta\tau_2 V_2'/\ell_{e,2}$, time scale, dimensionless

$\Delta\tau_t$ characteristic time constant, s

τ_{1w} $:= \lambda_{1w}\frac{1}{2}\rho_1 V_1^2$, shear stress, N/m²

τ_{2w} liquid–wall shear stress, N/m²

τ_{12} gas–liquid interface shear stress, N/m²

Φ_{2o}^2 two phase friction multiplier, dimensionless

ω_c effective mass transfer coefficient at the continuum site of the interface, kg/(sm³)

ω_d effective mass transfer coefficient at the dispersed site of the interface, kg/(sm³)

Subscripts

c   continuum
d   disperse
1   gas
2   liquid
2F   liquid film

References

Abramson B, Sirignano WA (1989) Droplet vaporization model for spray combustion calculations, Int. Heat Mass Transfer, vol 32 pp 1605-1618

Acrivos A, Taylor TD (1962) Heat and mass transfer from single spheres in stokes flow, The Physics of Fluids, vol 5 no 4 pp 387-394

Acrivos A, Goddard J (1965) Asymptotic expansion for laminar forced- convection heat and mass transfer, Part 1, J. of Fluid Mech., vol 23 p 273

Aksel'Rud GA (1953) Zh. fiz. Khim. vol 27 p 1445

Albring W (1970) Angewandte Stroemungslehre, Verlag Theodor Steinkopf, Dresden, 4. Auflage

Al-Diwani HK and Rose JW (1973) Free convection film condensation of steam in presence of non condensing gases, Int. J. Heat Mass Transfer, vol 16 p 1959

Avdeev AA (1986) Growth and condensation velocity of steam bubbles in turbulent flow, Teploenergetika, in Russian, vol 1 pp 53-55

Banerjee S (1990) Turbulence structure and transport mechanisms at interfaces, Proc. Ninth Int. Heat Transfer Conference, Jerusalem, Israel, vol 1 pp 395-418

Bankoff SS (1980) Some Condensation Studies Pertinent to LWR Safety, Int. J. Multiphase Flow, vol 6 pp 51-67

Batchelor FK (1953) The theory of homogeneous turbulence, The University Press, Cambridge, England

Berman LD (1961) Soprotivlenie na granize razdela fas pri plenochnoi kondensazii para nizkogo davleniya, Tr. Vses. N-i, i Konstrukt In-t Khim. Mashinost, vol 36 p 66

Billet ML and Holl JW (1979) Scale effects on various types of limited cavitation, in Parkin BR and Morgan WB eds. (1979) Int. Symposium on Cavitation Inception ASME Winter Annual Meeting, New Orleans

Boussinsq M (1905) Calcul du pourvoir retroidissant des courant fluids, J. Math. Pures Appl., vol 1 p 285

Brauer et al H (1976) Chem. Ing. Tech., vol 48 pp 737-741

Brennen CE (1995) Cavitation and Bubble Dynamics, Oxford University Press, New York, Oxford

Brumfield LK, Houze KN, and Theofanous TG (1975) Turbulent mass transfer at free, Gas-Liquid Interfaces, with Applications to Film Flows. Int. J. Heat Mass Transfer, vol 18 pp 1077-1081

Bunker RS and Carey VP (1986) Modeling of turbulent condensation heat transfer in the boiling water reactor primary containment, Nucl. Eng. Des., vol 91 pp 297-304

Calderbank PH and Moo-Young MB (1962) Chem. Eng. Sc., vol 16 p 39

Calderbank PH (Oct. 1967) Gas absorption from bubbles, Chemical Engineering, pp CE209-CE233

Carlsow HS, Jaeger JC (1959) Condution of heat in solids, Oxford

Celata GP, Cumo M, D'Annibale F, Farello GE (1991) Direct contact condensation of steam on droplets, Int. J. Multiphase Flow, vol 17 no 2 pp 191-211

Cha YS and Henry RE (February 1981) Bubble growth during decompression of liquids, Transaction of the ASME, Journal of Heat Transfer, vol 103 pp 56-60

Chiang CH, Sirignano WA (Nov. 1990) Numerical analysis of convecting and interacting vaporizing fuel droplets with variable properties, Presented at the 28th AIAA Aerospace Sciences Mtg, Reno

Churchill RV (1958) Operational mathematics, McGraw Hill, New York, pp 132-134

Churchill SW, Brier JC (1955) Convective heat transfer from a gas stream at high temperature to a circular cylinder normal to the flow, Chem. Engng. Progr. Simp. Ser. 51, vol 17 pp 57-65
Clift R, Ggarce JR, Weber ME (1978) Bubbles, drops, and particles, Academic Press, New York 1978
Condie KG et al. (August 5-8, 1984) Comparison of heat and mass transfer correlation with forced convective non equilibrium post-CHF experimental data, Proc. of 22nd Nat. Heat Transfer Conf. & Exhibition, Niagara Falls, New York, in Dhir VK, Schrock VE (ed) Basic Aspects of Two-Phase Flow and Heat Transfer, pp 57-65
Daily JW and Johnson JR (November 1956) Turbulence and boundary-layer effect on cavitation inception from gas nuclei, Transactions of the ASME, vol 78 pp 1695-1705
Deitsch ME, Philiphoff GA (1981) Two-phase flow gas dynamics, Moscu, Energoisdat, in Russian
Eddington RI and Kenning DBR (1978) Comparison of gas and vapor bubble nucleation on brass surface in water, Int. J. Heat Mass Transfer vol 21 pp 855-862
Epstein PS and Plesset MS (1950) On the stability of gas bubbles in liquid-gas solution, J. Chem. Phys. Vol 18 pp 1505-1509
Fedorovich ED, Rohsenow WM (1968) The effect of vapor subcooling on film condensation of metals, Int. J. of Heat Mass Transfer, vol 12 pp 1525-1529
Feng Z-G and Michaelides EE (1986) Unsteady heat transfer from a spherical particle at finite Peclet numbers, ASME Journal of Fluids Engineering, vol 118 pp 96-102
Fieder J, Russel KC, Lothe J, Pound GM (1966) Homogeneous nucleation and growth of droplets in vapours. Advances in Physics, vol 15 pp 111-178
Fortescue GE and Pearson JRA (1967), On gas absorption into a turbulent liquid, Chem. Engng. Sci., vol 22 pp 1163-1176
Ford JD, Lekic A (1973) Rate of growth of drops during condensation. Int. J. Heat Mass Transfer, vol 16 pp 61-64
Fourier J (1822) Theory analytique de la chaleur
Frenkel FI (1973) Selected works in gasdynamics, Moscu, Nauka, in Russian
Friedlander SK (1957) A. I. Ch. Eng. J., vol 3 p 43
Friedlander SK (1961) A. I. Ch. Eng. J, vol 7 p 347
Froessling N (1938) Beitr. Geophys. vol 32 p 170
Furth R (1941) Proc. Cambr. Philos. Soc., vol 37 p 252
Gates EM and Bacon J (1978) Determination of the cavitation nuclei distribution by holography, J. Ship Res. vol 22 no 1 pp 29-31
Gibbs JW (1878) Thermodynamische Studien, Leipzig 1982, Amer. J. Sci. and Arts, vol XVI, pp 454-455
Gnielinski V (1975) Berechnung mittlerer Waerme- und Stoffuebertragungskoeffizienten an laminar und turbulent ueberstroemenden Einzelkoerpern mit Hilfe einer einheitlichen Gleichung, Forsch. Ing. Wes, vol 41 no 5 pp 145-153
Hadamard J (1911) Dokl. Akad. Nauk, SSSR, vol 152 p 1734
Hammitt FG (1980) Cavitation and multiphase flow phenomena, McGraw-Hill Inc.
Hausen H (1958) Darstellung des Waermeueberganges in Roehren durch verallgemeinerte Potenzgleichungen, Verfahrenstechnik, vol 9 no 4/5 pp 75-79
Hertz H (1882) Wied. Ann., vol 17 p 193
Hobbhahn WK (Oct. 10-13, 1989) Modeling of condensation in light water reactor safety, Proc. of the Fourth International Topical Meeting on Nuclear Reactor Thermal-Hydraulics, Mueller U, Rehme K, Rust K, Braun G (eds) Karlsruhe, vol 2 pp 1047-1053
Huang TT (1984) The effect of turbulence simulation on cavitation inception of axisymmetric headforms, in Parkin BR and Morgan WB eds. (1984) Int. Symposium on Cavitation Inception ASME Winter Annual Meeting, New Orleans

Hughes ED, Paulsen MP, Agee LJ (Sept. 1981) A drift-flux model of two-phase flow for RETRAN, Nuclear Technology, vol 54 pp 410-420

Hunt DL (1970) The effect of delayed bubble growth on the depressurization of vessels Containing high temperature water. UKAEA Report AHSB(S) R 189

Isenberg J, Sideman S (June 1970) Direct contact heat transfer with change of phase: Bubble condensation in immiscible liquids, Int. J. Heat Mass Transfer, vol 13 pp 997-1011

Jakob M, Linke W (1933) Der Waermeübergang von einer waagerechten Platte an sidendes Wasser, Forsch. Ing. Wes., vol 4 pp 75-81

Jensen RJ and Yuen MC (1982) Interphase transport in horizontal stratified concurrent flow, U.S. Nuclear Regulatory Commission Report NUREG/CR-2334

Katz J (1978) Determination of solid nuclei and bubble distribution in water by holography, Calif. Inst. of Techn., Eng. and Appl. Sci. Div. Rep. No. 183-3

Keller AP (1979) Cavitation inception measurements and flow visualization on asymetric bodies at two different free stream turbulence levels and test procedures, Int. Symposium on Cavitation Inception, ASME Winter Annual Meeting, New York

Kendouch AA (1976) Theoretical and experimental investigations into the problem of transient two-phase flow and its application to reactor safety, Ph.D.Thesis, Department of Thermodynamics and Fluid Mechanics, University of Strathclyde, U.C.

Kim HJ and Bankoff SG (Nov. 1983) Local heat transfer coefficients for condensation in stratified countercurrent steam - water flows, Trans. ASME, vol 105 pp 706-712

Kim MH, Corradini ML (1990) Modeling of condensation heat transfer in a reactor containment, Nucl. Eng. and Design, vol 118 pp 193-212

Knapp RT, Levy J, Brown FB and O'Neill JP (1948) The hydrodynamics laboratory at the California Institute of Technology, Trans. ASME vol 70 p 437

Knowles J B (1985) A mathematical model of vapor film destabilization, Report AEEW-R-1933

Knudsen M (1915) Ann. Physic, vol 47 p 697

Kolev NI (2002, 2004) Multiphase Flow Dynamics 2, Mechanical and Thermal Interactions, Springer

Kremeen RW, McGraw JT and Parkin BR (1955) Mechanism of cavitation inception and the released scale effects problem, Trans. Am. Soc. Mech. Engs vol 77 p 533

Kuiper G (1979) Some experiments with distinguished types of cavitation on ship properties, in Morgan WB and Parkin BR eds. (1979) Int. Symposium on Cavitation Inception ASME Winter Annual Meeting, New Orleans

Labunzov DA (1974) State of the art of the nuclide boiling mechanism of liquids, Heat Transfer and Physical Hydrodynamics, Moskva, Nauka, in Russian, pp 98 – 115

Labunzov DA, Krjukov AP (1977) Processes of intensive flushing, Thermal Engineering, in Russian, vol 24 no 4 pp 8-11

Lamont JC and Scott DS (1970) An eddy cell model of mass transfer into the surface of a turbulent liquid, AIChE Journal, vol 16 no 4 pp 513-519

Lamont JC and Yuen MC (1982) Interface transport in horizontal stratified concurrent flow, U. S. Nuclear Regulatory Commission Report NUREG/CR-2334

Langmuir I (1913) Physik. Z., vol 14 p 1273

Langmuir I (1927) Jones HA and Mackay GMJ, Physic., rev vol 30 p 201

Lienhard JH A heat transfer textbook, Prentice-Hall, Inc., Engelwood Cliffts, New Jersey 07632

Lee K, Ryley DJ (Nov. 1968) The evaporation of water droplets in superheated steam, J. of Heat Transfer, vol 90

Levich VG (1962) Physicochemical Hydrodynamics, Prentice-Hall, Englewood Cliffs, NJ

Lochiel AC (1963) Ph. D. Thesis, University of Edinburgh

Ludvig A (1975) Untersuchungen zur spontaneous Kondensation von Wasserdampf bei stationaerer Ueberschalllstroemung unter Beruecksichtigung des Realgasverhaltens. Dissertation, Universitaet Karlsruhe (TH)

Mason BJ (1951) Spontaneous condensation of water vapor in expansion chamber experiments. Proc. Phys. Soc. London, Serie B, vol 64 pp 773-779

Mason BJ (1957) The Physics of Clouds. Clarendon Press, Oxford

Michaelides EE (March 2003) Hydrodynamic force and heat/mass transfer from particles, bubbles and drops – The Freeman Scholar Lecture, ASME Journal of Fluids Engineering, vol 125 pp 209-238

Mills AF The condensation of steam at low pressure, Techn. Report Series No. 6, Issue 39. Space Sciences Laboratory, University of California, Berkeley

Mills AF, Seban RA (1967) The condensation coefficient of water, J. of Heat Transfer, vol 10 pp 1815-1827

Malnes D, Solberg K (May 1973) A fundamental solution to the critical two-phase flow problems, applicable to loss of coolant accident analysis. SD-119, Kjeller Inst., Norway

Nabavian K, Bromley LA (1963) Condensation coefficient of water; Chem. Eng. Sc., vol 18 pp 651-660

Nigmatulin RI (1978) Basics of the mechanics of the heterogeneous fluids, Moskva, Nauka, in Russian

O'Hern TJ, Katz J and Acosta AJ (1988) Holographic measurements of cavitation nuclei in the see, Proc. ASME Cavitation and Multiphase Flow Forum, pp 39-42

Parkin and Kermeen RW (1963) The roles of convective air diffusion and liquid tensile stresses during cavitation inception, Proc. IAHR Symposium of Cavitation and Hydraulic Machinery, Sendai, Japan

Peterson FB et al. (1975) Comparative measurements of bubble and particulate spectra by three optical measurements methods, Proc. 14th Int. Towing Conf.

Petukhov BS, Popov VN (1963) Theoretical calculation of heat exchange and friction resistance in turbulent flow in tubes of an incompressible fluid with variable physical properties. High Temperature, vol 1 pp 69-83

Pohlhausen E (1921) Der Waermeaustausch zwischen festen Koerpern und Fluessigkeiten mit kkleiner Reibung und kleiner Waermeleitung, Z. angew. Math. Mech., vol 1 no 2 pp 115-121

Ranz W, Marschal W Jr (1952) Evaporation from drops, Ch. Eng. Progress, vol 48 pp 141-146

Rohsenow WM, Choi H (1961) Heat, mass and momentum transfer, Prentice - Hall Publishers, New Jersey

Rosenberg B (1950) Report no 727, Washington DC: The David Taylor Model Basin)

Rouse H (January-February 1953) Cavitation in the mixing zone of a submerged jet, La Houille Blanche

Saha P (1980) Int. J. Heat and Mass Transfer, vol 23 p 481

Samson RE, Springer GS (1969) Condensation on and evaporation from droplets by a moment method, J. Fluid Mech., vol 36 pp 577-584

Siddique M, Golay MW (May 1994) Theoretical modeling of forced convection condensation of steam in a vertical tube in presence of non condensable gas, Nuclear Technology, vol 106

Simoneau RJ (1981) Depressurization and two-phase flow of water containing high levels of dissolved nitrogen gas, NASA Technical Paper 1839, USA

Skripov WP, Sinizyn EN, Pavlov PA, Ermakov GW, Muratov GN, Bulanov NB, Bajdakov WG (1980) Thermophysical properties of liquids in meta-stable state, Moscu, Atomisdat, in Russia

Slattery JC 1990 Interfacial transport phenomena, Springer Verlag
Soo SL (1969) Fluid dynamics of multiphase systems, Massachusetts, Woltham
Spalding DB (1953) The combustion of liquid fuels, Proc. 4th Symp. (Int.) on Combustion, Williams & Wilkins, Baltimore MD, pp 847-864
Tanaka M (Feb. 1980) Heat transfer of a spray droplet in a nuclear reactor containment. Nuclear Technology, vol 47 p 268
Takadi T and Maeda S (1961) Vhem. Engng, Tokyo, vol 25 p 254
Taylor GI (1935) Proc. Roy. Soc. A vol 151 p 429
Taylor GI (1936) The mean value of the fluctuation in pressure and pressure gradient in a turbulent flow, Proceedings of the Cambridge Philosophical Society, pp 380-384
Theofanous TG, Houze RN and Brumfield LK (1975) Turbulent mass transfer at free gas-liquid interfaces with applications to open-channel, bubble and jet flows, Int. J. Heat Mass Transfer, vol 19 pp 613-624
Uchida U, Oyama A, Togo Y (1964) Evolution of post - incident cooling system of light water reactors, Proc. 3th Int. Conf. Peaceful Uses of Atomic Energy, International Atomic Energy Agency, Vienna, Austria, vol 13 p 93
Van Vingaarden L (1967) On the growth of small cavitation bubbles by convective diffusion, Int. J. Heat Matt Transfer, vol 10 pp 127-134
VDI-Waermeatlas (1984) 4. Auflage VDI-Verlag
Volmer M (1939) Kinetik der Phasenbildung, Dresden und Leipzig, Verlag von Theodor Steinkopff
Ward DM, Trass O and Johnson AI (1962) Can. J. Chem. Eng., vol 40 p 164
Wilson JF (1965) Primary separation of steam from water by natural separation. US/EURATOM Report ACNP-65002
Wolf L et al. (1982-1984) HDR Sicherheitsprogramm. (Investigations of RPV internals during blow down) Kernforschungszentrum Karlsruhe. Reports in German
Zeldovich JB (1942) To the theory of origination of the new phase, cavitation, Journal of Experimental and Theoretical Physics, in Russian, vol 12 no 11/12 pp 525-538

13 Thermodynamic and transport properties of diesel fuel

This Chapter provides a review of the existing data on thermodynamic and transport properties of diesel fuel from 20 references. From the collected data a set of approximations is generated enabling the use of the available information in computer code models. The emphasis is on the strict consistency of the thermodynamic representations. Even being an approximation of the reality the generated correlations are mathematically consistent to each other. The generated saturation line is consistent with the Clausius-Clapeyron equation, with the definition of the latent heat of evaporation and the other vapor properties. The collection of analytical equations approximating the collected data set can be applied for the liquid diesel fuel being stable or meta-stable, as well for the diesel fuel vapor being stable or meta-stable.

13.1 Introduction

Modeling of processes in modern combustion motors requires sophisticated mathematical models that culminate finally in very complex computer-code-models. Multiphase flow analyses require a complete set of thermodynamic, caloric and transport properties that are *inherently consistent*. Inherently consistent means that even being approximations, the thermodynamic relationships among the properties have to be strictly satisfied. Using for instance density function taken from one reference and the velocity of sound from other source may result in considerable mass losses in sensitive applications by using compressible fluid dynamic tools. Similar is the case by trying to simulate evaporation and condensation with densities and enthalpies at the saturation lines that do not satisfy e.g. *Clausius-Clapayron* relation. Non consistent enthalpy functions of temperature and pressure used for instance for phase transition, that do not take into account exactly the dependence of the latent heat of evaporation on temperature, may lead also to considerable energy losses during the analysis. The purpose of this work is to review the openly available sources of information and to attempt to derive from them a consistent set of thermodynamic, caloric and transport functions. Even being forced to accept the uncertainty of the available data we will explicitly document them. The so obtained functions are then recommended for use in computer codes for consistent multi-phase dynamic analysis.

The generated analytical approximations are:

For liquid and gas

$\rho = \rho(p,T)$ Density as a function of pressure and temperature, kg/m³

$(\partial \rho / \partial p)_T = f(p,T)$ Derivative of the density with respect to pressure at constant temperature, kg/(m³Pa)

$(\partial \rho / \partial T)_p = f(p,T)$ Derivative of the density with respect to temperature at constant pressure, kg/(m³K)

The volumetric thermal expansion coefficient, the isothermal coefficient of compressibility, isothermal bulk modulus

$$\beta = -(\partial \rho / \partial T)_p / \rho, \quad k = (\partial \rho / \partial p)_T / \rho = 1/B, \quad B = 1/k$$

are then easily computed.

$h = h(p,T)$ Specific enthalpy as a function of temperature and pressure, J/kg

$(\partial h / \partial p)_T = f(p,T)$ Derivative of the specific enthalpy with respect to pressure at constant temperature, J/(kgPa)

$(\partial h / \partial T)_p = c_p(p,T)$ Derivative of the specific enthalpy with respect to temperature at constant pressure – specific thermal capacity at constant pressure J/(kgK)

$s = s(p,T)$ Specific entropy as a function of temperature and pressure, J/(kgK)

$(\partial s / \partial p)_T = f(p,T)$ Derivative of the specific entropy with respect to pressure at constant temperature, J/(kgKPa)

$(\partial s / \partial T)_p = f(p,T)$ Derivative of the specific entropy with respect to temperature at constant pressure, J/(kgK²)

$a = a(p,T)$ Velocity of sound, m/s

$\lambda = \lambda(p,T)$ Thermal conductivity, W/(mK)

$v = v(p,T)$ Cinematic viscosity, m²/s

In addition approximation of the surface tension at the liquid–gas interface as a function of temperature is given

$\sigma = \sigma(T)$ — Surface tension at the liquid–gas interface as a function of temperature, N/m

Approximation for the saturation line is provided in two forms:

$T' = f(p)$ — Saturation temperature as a function of the pressure, K
$p' = f(T)$ — Saturation pressure as a function of the temperature, Pa
$dT'/dp = f(T)$ — Derivative of the saturation temperature with respect to pressure – the *Clausius-Clapayron* relation, K/Pa
$\Delta h = h'' - h' = f(T)$ — Latent heat of evaporation as a function of the temperature, J/kg

The properties at the saturation line for liquid, designated with ', and for vapor, designated with ", are computed from the p-T functions using the corresponding p'-T or T'-p couples of dependent variables.

13.2 Constituents of diesel fuel

Diesel fuel is a complex mixture consisting among others hydrocarbons having different properties like boiling points, densities etc. Table 13.1 summarizes some of the most important constituents taken from *Wenck* and *Schneider* [17] (1993). Therefore we have for the mass concentrations of the components groups

$$C_{par} = 0.456, \qquad (13.1)$$

$$C_{naph} = 0.256, \qquad (13.2)$$

and for computational purposes

$$C_{arom} = 1 - C_{par} - C_{naph}. \qquad (13.3)$$

Table 13.1. Constituents of diesel

Groups	Mass %
Paraffin	45.6
Naphthalene	25.6
- monocyclic	17.4
- dicyclic	6.3
- tricyclic	1.9
Aromates	28.6
- alkylbenzole	9.6

- indane/tetralie	5.6
- indene	1.3
- monoaromats	16.5
- naphthaline	0.1
- alkylnaphthaline	6.9
- acenaphthene/diphenyle	2.3
- acenaphthene/fluorene	1.6
- diaromats	10.9
- triaromats	0.5

Table 13.2 illustrates that there are also small amount of other constituents like sulfur, ash etc. Note that the data for the properties of the diesel fuel available in the literature spread mainly because of the variety of diesel fuel constituents being different for different geographical origination sources. That is why a practical approach for the approximation of state equation is to define a *model fluid* that represents approximately the diesel fuel. Therefore we will represent the variety of diesel fuels by a single-model of a diesel fuel.

Table 13.2. Properties for diesel fuel

	[12] p. 282	[3] p. 915 light diesel	[3] p. 915 heavy diesel	[2] gas oil	[12] n-Heptane	[12] n-Octane	[17] diesel
Formula		$C_nH_{1.8n}(l)$	$C_nH_{1.7n}(l)$		$C_7H_{16}(l)$	$C_8H_{18}(l)$	
% mass composition							
Carbon	86.5	86.9	87.6	87	84	84.2	
Hydrogen	13.2	13.1	12.4	13	16	15.8	
Sulfur	0.3	0	0	0			max 0.2
Ash							max 0.01
Water in mg/kg							200
Molecular weight	148.6 [12] p.133	≈ 170	≈ 200		100	114	
Density, kg/m³	840 (15.6°C, 1atm)	840-880 (0°C, 1atm)	820-950 (0°C, 1atm)	820-860	683.8 (20°C, 1atm)	702.5 (20°C, 1atm)	820-850 (15°C, 1atm)
Cetane number	52				56		47.4-63.9
Initial boiling point, °C	180				98.4	125.67	≈ 160, p. 72
10% vol., °C	230						
50% vol., °C	270						
65% vol., °C							until 250
85% vol., °C							until 350
90% vol., °C	320						
95% vol., °C							until 370

Property							
Final boiling point, °C	340						
Flush point *, °C	70			36.85			min 55
Auto ignition, °C				206.85			
Freezing point					-90.61	-56.80	
Cloud point, °C	-12 to -3						
Coefficient of cubic thermal expansion, K^{-1}	0.00067						
Heat of vaporization kJ/kg	192 at mid boiling point	270	230		316.3 at boiling point	302.2 at boiling point	731, p. 72
Liquid specific heat, kJ/(kg K)	2.81 at mid boiling point	2.2	1.9				
Vapor specific heat, kJ/(kg K)		≈ 1.7	≈ 1.7		1.659 (25°C, 0bar)	1.656 (25°C, 0bar)	
Caloric value, higher/lower, MJ/kg	45.7/42.9 at 15°C	44.8/42.5	43.8/41.4	46/			
Cinematic viscosity, m²/s	5×10^{-6} at 15.6°C, 3 at 37.8°C						3.8×10^{-6} at 20°C 2.4 to 2.6×10^{-6} at 40°C, p. 10, 11
Critical point							
T°C					267.01	295.6	
p, bar					27.31	24.83	
Density, kg/m³					232	232	
Surface tension, N/m				0.025-0.03 [2] p.284			
Saturation curve							

* Flush point: minimum temperature for spark ignition near the condense phase.
Siemens reported in [15] density for diesel liquid fuel at 27.7 to 33.9°C to be 833 kg/m³.

13.3 Averaged boiling point at atmospheric pressure

Due to different boiling point of the constituents the diesel fuel appears as not having single boiling point. It has rather a temperature for boiling inception and temperature at which the last liquids molecule evaporates at given pressure. From Table 13.2 we select the temperature at atmospheric pressure at which the boiling starts and ends as follows:

$$T'_{boiling\_incipiation} = 273.15 + 180 = 453.15K, \qquad (13.4)$$

$$T'_{boiling\_completed} = 273.15 + 380 = 653.15K. \qquad (13.5)$$

The so-called averaged boiling temperature sometimes called *mid boiling temperature* at atmospheric pressure is then

$$T'_{mid\ boiling}\left(1\times10^5\ Pa\right) = 273.15 + 280 = 553.15K. \qquad (13.6)$$

Some authors introduced a characteristic number defined as follows

$$K = \frac{\sqrt[3]{1.8 T'_{mid\ boiling}}}{\rho_{2,ref}/1000}. \qquad (13.7)$$

With this selected reference density and averaged boiling temperature at atmospheric pressure we obtain

$$K = \frac{\sqrt[3]{1.8 T'}}{\rho_{2,ref}/1000} = \frac{\sqrt[3]{1.8\times 553.15}}{840/1000} = 12. \qquad (13.8)$$

This number classifies the diesel fuel as a hydrocarbon mixture from the type of those presented by *Wenck* and *Schneider* in p. 22, Fig. 4 [17] (1993). We will finally not use the dependence presented on this figure because of its inconsistency at higher temperatures.

Thus we select as a reference boiling temperature at atmospheric pressure

$$\boxed{T'_{1\ bar} = 273.15 + 180 = 453.15K.} \qquad (13.9)$$

This is of course an *idealization*. Up this moment we will considered diesel-fuel as a "single component" liquid.

13.4 Reference liquid density point

The reference liquid density point is the result of the specific combination of hydrocarbons for each diesel fuel. It is used for approximate estimate of other properties. From Table 13.2 we select the following diesel liquid reference density:

$$\rho_{ref} = \rho(273.15+15K, 1\times 10^5 \, Pa) = 833.69 kg/m^3 \,. \tag{13.10}$$

13.5 Critical temperature, critical pressure

The critical point of a liquid–vapor system is defined as a (T_c, p_c)-point where there is no more difference in the properties between the liquid and the vapor. Critical point for diesel fuel is not known to me. One orientation is the critical point for n-heptan and n-octane. Using the *Reid* et al. [11] (1982) Eq. (2.2.4) p. 20

$$\lg T_c = A + B \lg \frac{\rho_{ref}}{\rho_{H_2O,ref}} + C \lg T'_{1 \, bar}, \tag{13.11}$$

the constants for the hydrocarbons from Table 13.3, and weighting the resulting temperatures by using the mass concentrations we obtain

$$T_c = 658.4K \,. \tag{13.12}$$

Table 13.3. Constants for computation of the critical temperature from Table 2.3 *Reid* et al. [11] (1982) p.23

hydrocarbons	A	B	C
paraffins	1.359397d0	0.436843d0	0.562244d0
naphthalenes	0.658122d0	- 0.071646d0	0.811961d0
aromatics	1.057019d0	0.227320d0	0.669286d0

There are several methods for computation of the critical pressure discussed by *Reid* at al. in [11] (1982), and by *Philipov* in [8] (1978). Unfortunately for this purpose I need more accurate experimental information than I could obtain up to now. Therefore I select the critical pressure around the values for n-heptan and n-octane

$$p_c = 30\times 10^5 \, Pa \,. \tag{13.13}$$

13.6 Molar weight, gas constant

The molar weight of the diesel fuel varies between about 148 and 200 as seen from Table 13.2. I select for the molar weight

$$M = 170 kg/1-mole.\quad(13.14)$$

The universal gas constant is

$$R_{universal} = 8314 J/(kg-mole\ K).\quad(13.15)$$

Therefore the gas constant for diesel fuel vapor is

$$R = R_{universal}/M = 48.906 J/kg.\quad(13.16)$$

13.7 Saturation line

Wenck and *Schneider* reported in [17] p. 56 (1993) the *Grabner*'s data for diesel fuel in winter from 30 to 100°C given in Table 13.4. These are the only data we have. We smooth the data by the approximation

$$p = 977292.0044d0 - 6313.34077d0\ T + 10.4084d0\ T^2,\quad(13.17)$$

and correct the first point and last three points as indicated in Table 13.4. We will use these four points in a moment.

Table 13.4. Saturation pressure as a function of temperature

T in K	$p'(T)$ in Pa
303.15	19000./19933.
311.15	21000.
313.15	21000.
323.15	25000.
333.15	30000.
343.15	36000.
353.15	45000/45818.
363.15	56000/57240.
373.15	72000/70744.
417.97/551.15	100000.
$T_c = 657.04$	3000000.

The saturation line $p' = p(T)$ has to satisfy the *Clausius-Clapayron* relation

$$\frac{dT}{dp} = T\frac{v''-v'}{\Delta h},\quad(13.18)$$

13.7 Saturation line

which is in fact the strict mathematical expression of the thermodynamic equilibrium. Because the critical pressure of about $30 bar$ is relatively low, the *Clausius-Clapayeron* relation can be simplified by considering the specific volume of the steam as much larger than the specific volume of the liquid. The result is

$$\frac{dT}{dp} \approx \frac{Tv''}{\Delta h} . \tag{13.19}$$

Assuming that the vapor behave as a perfect gas we have

$$\frac{dp}{p} = \frac{1}{R}\frac{\Delta h}{T^2} dT . \tag{13.20}$$

We assume that the latent heat of evaporation is a quadratic function of the temperature

$$\Delta h = a_1 + a_2 T + a_3 T^2 \tag{13.21}$$

and therefore

$$\frac{dp}{p} = \frac{1}{R}\left(\frac{a_1}{T^2} + \frac{a_2}{T} + a_3\right) dT . \tag{13.22}$$

After integration between two pressure-temperature points we obtain

$$R \ln \frac{p}{p_0} = a_1\left(\frac{1}{T_0} - \frac{1}{T}\right) + a_2 \ln \frac{T}{T_0} + a_3 (T - T_0) . \tag{13.23}$$

This is the form appropriate for fitting the *Grabner*'s data. The fit of the data as they are for instance using the first and the last three points gives a very low critical pressure which does not reflect the reality. That is why we use the already selected critical point

$$p = p_c \exp\left\{\frac{1}{R}\left[a_1\left(\frac{1}{T_c} - \frac{1}{T}\right) + a_2 \ln \frac{T}{T_c} + a_3 (T - T_c)\right]\right\} . \tag{13.24}$$

in addition to the last three points. The fitting gives for the constants $a_1 = 1.6e5$, $a_2 = -187.629$, $a_3 = 0.365$. The result is plotted on Fig. 13.1. This equation giving the saturation pressure as a function of the temperature is strictly consistent

a) with the *Clausius-Clapayron* equation,
b) with the definition of the critical point,

c) with the definition of the evaporation enthalpy as a function of temperature, and
d) with the assumption that the vapor is a perfect gas.

We see that the agreement is acceptable for practical use.

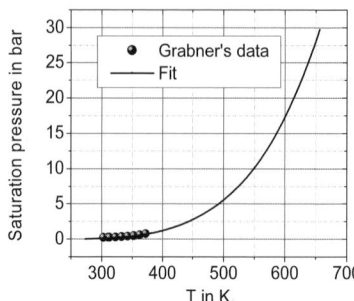 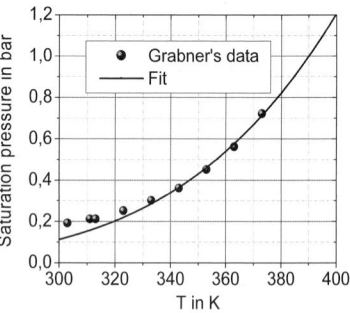

Fig. 13.1. Saturation pressure as a function of the temperature strictly consistent with the *Clausius-Clapayron* equation and with the assumption that the vapor is a perfect gas. Comparison with the *Grabner*'s data

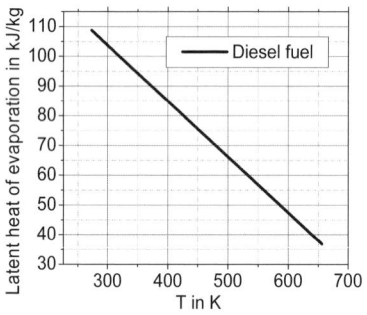

Fig. 13.2. Latent heat of evaporation as a function of temperature

The latent heat of evaporation compatible with this coefficients computed by using Eq. (13.21) is presented in Fig. 13.2.

Note: The points for the latent heat of evaporation given in Table 13.2 for 1 bar pressure are higher. If one fixes this point and generates a line or curve going to zero at critical pressure, the generated saturation line will not satisfy the *Garbner*'s data. That is why we select the opposite way – first to generate the saturation line satisfying the *Garbner*'s data and then to compute the latent heat of vaporization consistent with them.

In many applications there is a need to compute the saturation temperature as a function of the local pressure. For this purpose Eq. (13.24) has to be solved with respect to the temperature by iteration. The function

$$T'(p) = 1/\left[a + b\left(\log_{10} p\right)\right], \tag{13.25}$$

where $a = 0.00638$ and $b = -7.58558\text{e-}4$, is a good representation of this curve and can be used to compute the initial value.

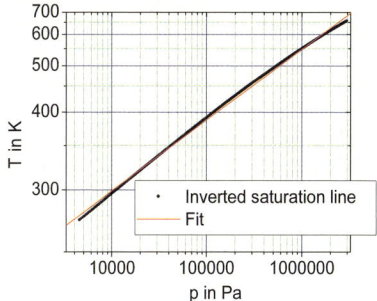

Fig. 13.3. Saturation temperature as a function of pressure for diesel fuel

Iteration method can then designed to improve the so obtained values based on Eq. (13.24)

$$T^{n+1} = a_1 \bigg/ \left[\frac{1}{T_c} - R\ln\frac{p}{p_c} - a_2\ln\frac{T}{T_c} - a_3\left(T - T_c\right)\right]. \tag{13.26}$$

Seven iterations at maximum are then necessary to have accuracy less than 0.001K.

13.8 Latent heat of evaporation

Equation (13.21) gives the latent heat of evaporation for diesel fuel. It is plotted in Fig. 13.4.

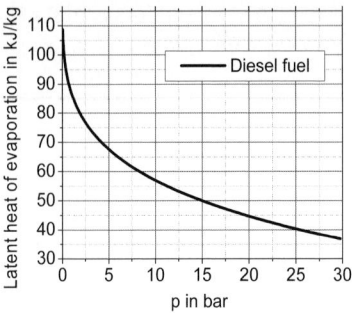

Fig. 13.4. Latent heat of evaporation for diesel fuel as a function of pressure

From Table 13.2 we see that in accordance with *Rose* and *Cooper* [12] (1977) at the mid boiling point 533.15K the reported latent heat of evaporation is $\Delta h_{ref}\left(533.15K, 1bar\right) = \Delta h_{ref}\left(T/T_c = 0.81, 1bar\right) = 192 kJ/kg$. At this temperature Fig. 13.4 gives about 60kJ/kg which, having in mind, is the very approximate data base we have that is acceptable. Note that *Heywood* [3] (1988) reported the values 270kJ/kg and 230kJ/kg for light and heavy diesel, respectively. Note also that n-heptan and n-octane have values of 316.3 and 302.2 kJ/kg, respectively, as given in Table 13.2.

13.9 The liquid density

The Siemens data [15] for the density of summer diesel fuel are well represented by the function

$$\rho = \sum_{i=1}^{3}\left(\sum_{j=1}^{3} a_{ij} T^{j-1}\right) p^{i-1}, \qquad (13.27)$$

where

$$\mathbf{A} = \begin{pmatrix} 828.59744 & 0.63993 & -0.00216 \\ 8.65679\text{e-}07 & -5.93672\text{e-}09 & 1.56678\text{e-}11 \\ -7.59052\text{e-}16 & 8.99915\text{e-}18 & -2.77890\text{e-}20 \end{pmatrix}. \qquad (13.28)$$

The corresponding density derivatives are then

$$\left(\frac{\partial \rho}{\partial p}\right)_T = \sum_{i=2}^{3}\left[(i-1)\left(\sum_{j=1}^{3}a_{ij}T^{j-1}\right)p^{i-2}\right], \qquad (13.29)$$

$$\left(\frac{\partial \rho}{\partial T}\right)_p = \sum_{i=1}^{3}\left[\left(\sum_{j=2}^{3}(j-1)a_{ij}T^{j-2}\right)p^{i-1}\right]. \qquad (13.30)$$

The prediction of the correlation is given in Fig. 13.5.

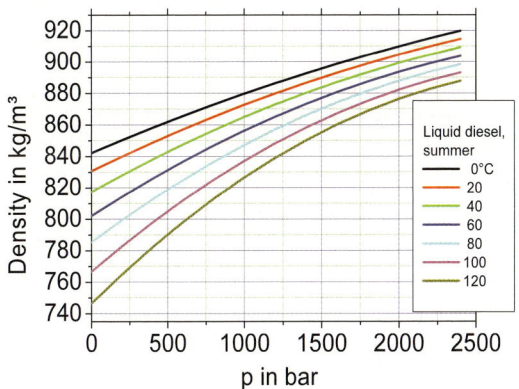

Fig. 13.5. Liquid density as function of pressure for summer diesel fuel. Parameter – temperature

The function can be successfully used for pressures between zero and 2400 bar and temperatures between 20 and 120°C. Because of the good data reproduction the extrapolation is also possible.

The liquid density at the critical point is therefore

$$\rho_c = 327.71 kg/m^3 \qquad (13.31)$$

Next we give additional comparison to available data for the volumetric thermal expansion coefficient and for the isothermal coefficient of compressibility.

13.9.1 The volumetric thermal expansion coefficient

Zenkevich [19] (1968) reported data for the density dependence of heavier diesel fuel as given in Table 13.5. The corresponding volumetric thermal expansion co-efficients defined as

$$\beta = -\frac{1}{\rho}\left(\frac{\partial \rho}{\partial T}\right)_p \quad (13.32)$$

are given in Table 13.6. From Table 13.5 we derive

$$\left(\frac{\partial \rho}{\partial T}\right)_p \approx -0.67, \quad (13.33)$$

see Table 13.6.

Table 13.5. Density at constant pressure for diesel fuel, [19] see [16] p.1265

T,°C, 1atm	ρ_2, kg/m³	ρ_2, kg/m³, Eq. (13.27)
20	878.7	830.62
40	865.4	817.23
60	852.0	802.12
80	838.5	785.28
100	825.1	766.71

Table 13.6. The volumetric thermal expansion coefficient

t,°C, 1atm	β [16] p.1265	β [12] p. 282	β Eqs. (13.30, 13.32)
15.6		0.000670	0.000729
30	0.000763		0.000812
50	0.000780		0.000933
70	0.000799		0.001061
90	0.000805		0.001196

Table 13.7. The derivative $(\partial \rho/\partial T)_p$ corresponding to Table 13.5

t,°C, 1atm	[16] p.1265
15.6	
30	-0.6654
50	-0.6698
70	-0.6754
90	-0.6696

The volumetric thermal expansion coefficients computed by using Eqs. (13.29, 13.32) are presented in Table 13.6 also. They are slightly higher. The density computed using Eq. (13.27) is smaller than the density reported by *Zenkevich* [19] (1968) – see for comparison Table 13.5.

13.9.2 Isothermal coefficient of compressibility

The expression

$$B = -v(\partial p/\partial v)_T \approx \rho a^2 \qquad (13.34)$$

is traditionally called *isothermal bulk modulus*. The data used for verifying the density function for the *isothermal coefficient of compressibility*

$$k = (\partial \rho/\partial p)_T/\rho = 1/\left[-v(\partial p/\partial v)_T\right] = 1/B \qquad (13.35)$$

are presented in Tables 13.8 and 13.9. This data are for gas–oil with reference density close to our reference density.

Table 13.8. Isothermal bulk modulus B in MN/m^2 for gas–oil with density $840 kg/m^3$ at $1 atm$, $15.6\,°C$ Heywood [3] (1988) p. 291

p in MN/m^2, $T\,°C =>$	15.5	30	50
13.8	1500	1400	1300
27.6	1700	1600	1500
41.4	1800	1700	1600

Table 13.9. Isothermal coefficient of compressibility κ in $1/Pa$ corresponding to Table 13.8

p in Pa, $T\,°C =>$	15.5	30	50
13.8×10^6	6.70×10^{-10}	7.14×10^{-10}	7.69×10^{-10}
Eqs. (13.29, 13.35)	5.29×10^{-10}	5.89×10^{-10}	6.88×10^{-10}
27.6×10^6	5.89×10^{-10}	6.25×10^{-10}	6.67×10^{-10}
Eqs. (13.29, 13.35)	5.10×10^{-10}	5.65×10^{-10}	6.56×10^{-10}
41.4×10^6	5.56×10^{-10}	5.89×10^{-10}	6.25×10^{-10}
Eqs. (13.29, 13.35)	4.91×10^{-10}	5.42×10^{-10}	6.26×10^{-10}

Bhatt [1] (1985) reported the values for diesel-fuel liquid velocity of sound at atmospheric pressure presented in Table 13.10. In the last column we compute an approximate estimate of the isothermal coefficient of compressibility for comparison with the data of *Heywood* [3] (1988) – the order of magnitude is confirmed.

The isothermal coefficient of compressibility computed by using Eqs. (13.29) and (13.35) is presented in Table 13.9 also. We see that the agreement is acceptable.

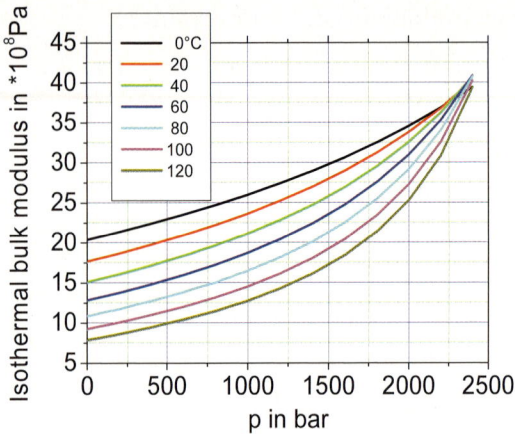

Fig. 13.6. Isothermal bulk modulus as a function of pressure. Parameter – temperature

13.10 Liquid velocity of sound

The Siemens data [15] for the liquid velocity of sound of summer diesel fuel are well represented by the function

$$a = \sum_{i=1}^{3}\left(\sum_{j=1}^{5} b_{ij} p^{j-1}\right) T^{i-1}, \qquad (13.36)$$

where

$$\mathbf{B} = \begin{pmatrix} 2226.4926 & 2.27318\text{e-}6 & 2.75574\text{e-}15 & 3.41172\text{e-}22 & -1.74367\text{e-}30 \\ -2.68172 & 3.79909\text{e-}9 & -8.17983\text{e-}17 & -1.65536\text{e-}24 & 9.50961\text{e-}33 \\ -0.00103 & 1.77949\text{e-}11 & 6.4506\text{e-}20 & 2.19744\text{e-}27 & -1.29278\text{e-}35 \end{pmatrix}. \quad (13.37)$$

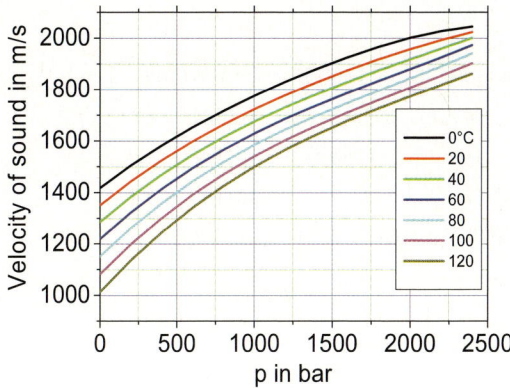

Fig. 13.7. Diesel fuel liquid velocity of sound as a function of pressure. Parameter - temperature

Equation (13.27) gives the results presented in Fig. 13.7. We perform also comparison with the velocity of sound reported by *Bhatt* [1] (1985) – see Table 13.13. The agreement is good.

Table 13.10. Comparison between the measured by *Bhatt* [1] (1985) at 1*bar* and predicted diesel fuel liquid velocity of sound at 1bar

T°C	a_2, m/s, exp.	a_2, m/s, Eq. (13.36)
30	1337	1319
35	1319	1303
40	1298	1286

13.11 The liquid specific heat at constant pressure

The Siemens data [15] for the liquid specific capacity at constant pressure of summer diesel fuel are well represented by the function

$$c_p = \sum_{i=1}^{5}\left(\sum_{j=1}^{3} d_{ij} T^{j-1}\right)\left(\frac{p}{10^5}\right)^{i-1}, \qquad (13.38)$$

where

$$D = \begin{pmatrix} -977.16186\text{d}0 & 14.025100\text{d}0 & -0.01374\text{d}0 \\ 2.22361\text{d-}04 & -1.62143\text{d-}04 & 2.23214\text{d-}09 \\ -1.96181\text{d-}09 & 2.03748\text{d-}07 & -1.78571\text{d-}14 \\ 4.15000\text{d-}14 & -7.54100\text{d-}11 & 4.03897\text{d-}28 \\ -3.48714\text{d-}18 & 1.00688\text{d-}14 & -1.47911\text{d-}31 \end{pmatrix}. \quad (13.39)$$

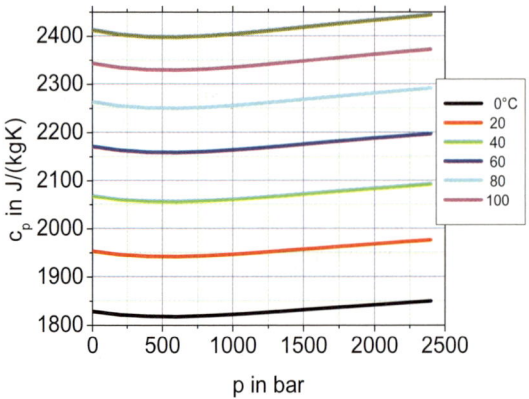

Fig. 13.8. Specific heat at constant pressure as a function of the pressure for diesel liquid. Parameter – temperature

For comparison the *Cragoe* formula from 1929, see *Rose* [12] (1977) p.281, applied for $\rho(15.6°C, 1atm) = 833.33 kg/m^3$ gives

$$c_p = \frac{1684.8 + 3.39(T - 273.15)}{\sqrt{\dfrac{\rho(15.6°C, 1atm)}{1000}}} = 831.25 + 3.714T. \quad (13.40)$$

Wenck and *Schneider* [17] (1993) used the same formula. Thus we will use farther Eq. (13.39).

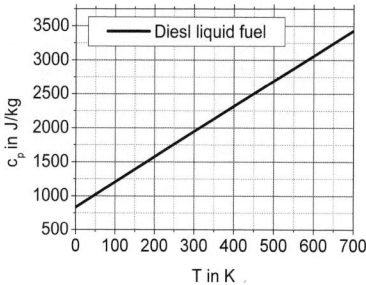

Fig. 13.9. Specific heat at constant pressure for diesel liquid

An example for the behavior of equation (13.40) for 1bar is presented in Fig. 13.9.

Having in mind the definition of the velocity of sound for single phase fluid

$$a = \left[\left(\frac{\partial \rho}{\partial p} \right)_T - \left(\frac{\partial \rho}{\partial T} \right)_p \frac{\rho \left(\frac{\partial h}{\partial p} \right)_T - 1}{\rho c_p} \right]^{-1/2}, \qquad (13.41)$$

we realize that knowing the velocity of sound and the density function with its derivatives, the derivative of the specific enthalpy with respect to the pressure at constant temperature can be computed

$$\left(\frac{\partial h}{\partial p} \right)_T = \frac{1}{\rho} \left\{ 1 + \left[\left(\frac{\partial \rho}{\partial p} \right)_T - \frac{1}{a^2} \right] \frac{\rho c_p}{\left(\frac{\partial \rho}{\partial T} \right)_p} \right\}. \qquad (13.42)$$

The so obtained result is approximated by the following function,

$$\left(\frac{\partial h}{\partial p} \right)_T = \sum_{i=1}^{3} \left(\sum_{j=1}^{3} c_{ij} T^{j-1} \right) p^{i-1}, \qquad (13.43)$$

where

$$C = \begin{pmatrix} 0.00404 & -1.54245\text{E-}5 & 2.20238\text{E-}8 \\ -7.34229\text{E-}11 & 4.84276\text{E-}13 & -8.79805\text{E-}16 \\ 2.23591\text{E-}19 & -1.60598\text{E-}21 & 3.17966\text{E-}24 \end{pmatrix}. \qquad (13.44)$$

Equation (13.43) reproduces the results given in Fig. 13.10.

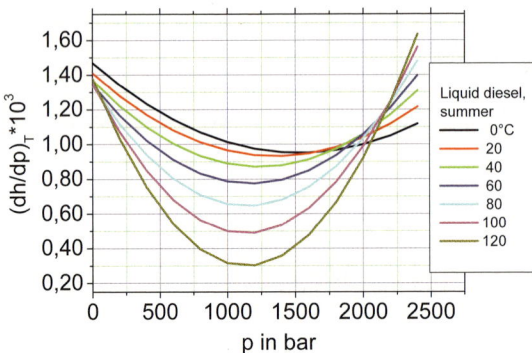

Fig. 13.10. Derivative of the specific enthalpy with respect to the pressure at constant temperature. Parameter – temperature

13.12 Specific liquid enthalpy

The differential form of the dependence of the specific enthalpy on temperature and pressure is

$$dh = c_p dT + \left(\frac{\partial h}{\partial p}\right)_T dp. \qquad (13.45)$$

The derivatives for the liquid in the above equation are already known. We will integrate Eq. (13.45) taking into account that the final result do not depend on the integration path. First we keep the temperature $T_{ref} = (273.15+15)K$ constant and integrate from $p_{ref} = 10^5\,Pa$ to p to obtain

$$h^* = \int_{p_{ref}}^{p} \left(\frac{\partial h}{\partial p}\right)_{T=T_{ref}} dp = \int_{p_{ref}}^{p} \sum_{i=1}^{3}\left(\sum_{j=1}^{3} c_{ij} T_{ref}^{j-1}\right) p^{i-1} dp = \sum_{i=1}^{3} \frac{1}{i}\left(\sum_{j=1}^{3} c_{ij} T_{ref}^{j-1}\right)\left(p^i - p_{ref}^i\right)$$

$$\approx -142.3652 + 0.00142p - 3.46467\text{E}-12 p^2$$

$$+ 8.27876\text{E}-21 p^3 \approx 10088.55051 + 0.00102 p \, . \tag{13.46}$$

The function under the integral is small but not negligible especially at very high pressures.

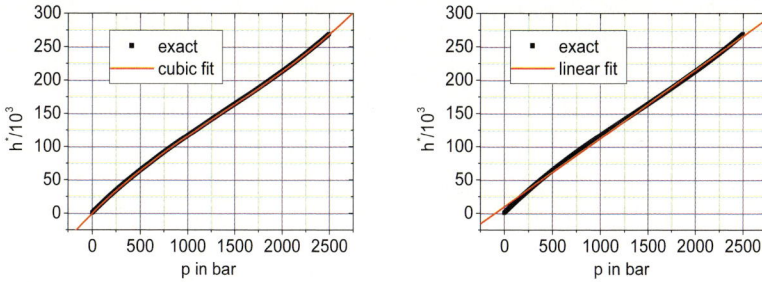

Fig. 13.11. Function under the integral in Eq. (13.46)

Then we keep the pressure p constant and integrate from T_{ref} to T. The result is

$$h = f_0 + c_1 \left(T - T_{ref} \right) + \frac{1}{2} c_2 \left(T^2 - T_{ref}^2 \right) + h^*(p) \, . \tag{13.47}$$

The constant $f_0 = 28538.07825$ can be optionally added to obtain zero enthalpy at the reference pressure and 0°C. The results of Eq. (13.47) are presented in Fig. 13.12.

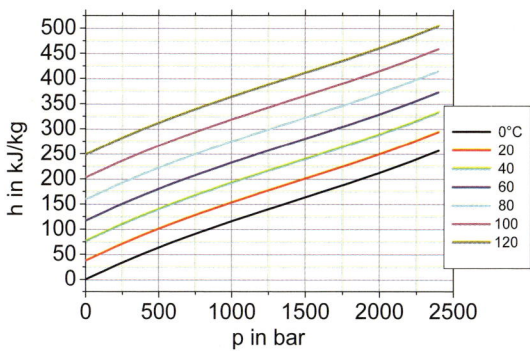

Fig. 13.12. Specific enthalpy as a function of pressure. Parameter – temperature

13.13 Specific liquid entropy

The differential form of the dependence of the specific entropy on temperature and pressure is

$$ds = \frac{c_p}{T}dT + \frac{\rho\left(\frac{\partial h}{\partial p}\right)_T - 1}{\rho T}dp.\qquad(13.48)$$

From this expression we easily see the definitions of the derivatives

$$\left(\frac{\partial s}{\partial T}\right)_p = \frac{c_p}{T},\qquad(13.49)$$

$$\left(\frac{\partial s}{\partial p}\right)_T = \frac{\rho\left(\frac{\partial h}{\partial p}\right)_T - 1}{\rho T}.\qquad(13.50)$$

Again we will integrate Eq. (13.48) taking into account that the final result do not depend on the integration path. First we keep the temperature $T_{ref} = (273.15+15)K$ constant and integrate from $p_{ref} = 10^5\,Pa$ to p to obtain

$$s^* = \int_{p_{ref}}^{p}\left(\frac{\partial s}{\partial p}\right)_{T=T_{ref}}dp.\qquad(13.51)$$

The function under the integral can be replaced with very good accuracy as seen in Fig. 13.13 by the following quadratic polynomial

$$f_s = s_1 + s_2 p + s_3 p^2.\qquad(13.52)$$

where s_1 = 7.80789e-07, s_2 = -2.18515d-14, s_3 = 8.34843d-23.

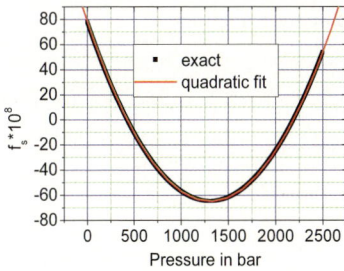

Fig. 13.13. Function under the integral in Eq. (13.50)

Therefore we have

$$s^* = s_1\left(p - p_{ref}\right) + \frac{1}{2}s_2\left(p^2 - p_{ref}^2\right) + \frac{1}{3}s_3\left(p^3 - p_{ref}^3\right). \tag{13.53}$$

Then we keep the pressure p constant and integrate from T_{ref} to T. The result is

$$s = s_4 + c_1 \ln\frac{T}{T_{ref}} + c_2\left(T - T_{ref}\right) + s^*(p). \tag{13.54}$$

An arbitrary constant $s_4 = 5869.20459410$ is introduced in order to have at $T = 1\,K$ and $p = 10^5 Pa$ zero entropy. This allows operating with positive entropies during practical analyses. Fig. 13.14 demonstrates the behavior of Eq. (13.54) for the reference temperature.

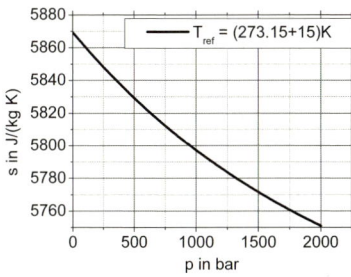

Fig. 13.14. Specific liquid diesel entropy as a function of pressure for the reference temperature of 15°C

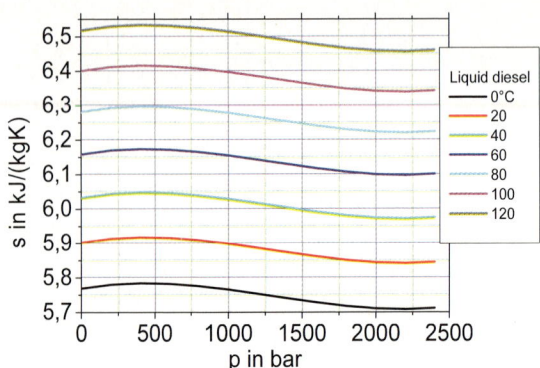

Fig. 13.15. Specific liquid diesel entropy as a function of pressure. Parameter – Temperature

Figure 13.15 presents the pressure dependence of the entropy for different temperatures. The slight increase of the entropy for low pressure is a result of inaccurate approximation. This feature has to be improved in the future.

13.14 Liquid surface tension

Surface tension for gas–oil at atmospheric conditions is reported by *Heywood* [3] (1988) p. 284 to be

$$\sigma_{ref} = 0.025 \text{ to } 0.03 \, N/m \, . \tag{13.55}$$

Using the *Othmer* equation, see *Yaws* and *Chung* [18] (1991), we extend this information up to the critical temperature

$$\sigma = \sigma_{ref} \left(\frac{T_c - T}{T_c - T_{ref}} \right)^{11/9} . \tag{13.56}$$

13.15 Thermal conductivity of liquid diesel fuel

The Siemens data [15] for the thermal conductivity of summer diesel fuel are well represented by the function

$$\lambda = \sum_{i=1}^{3}\left(\sum_{j=1}^{3} a_{ij} T^{j-1}\right) p^{i-1}, \qquad (13.57)$$

where

$$\mathbf{A} = \begin{pmatrix} 0.13924 & 3.78253\text{e-}05 & -2.89732\text{e-}07 \\ 6.27425\text{e-}11 & 6.08052\text{e-}13 & 3.64777\text{e-}16 \\ -1.38756\text{e-}19 & -2.57608\text{E-}22 & -2.70893\text{e-}24 \end{pmatrix}. \qquad (13.58)$$

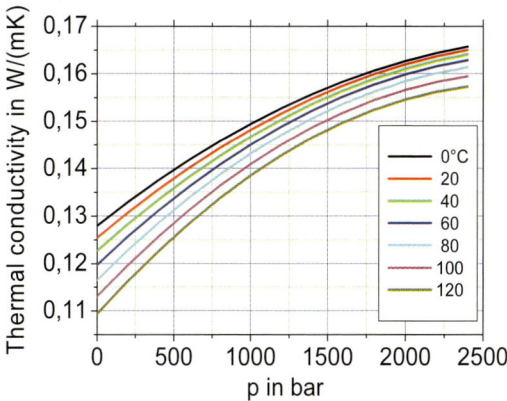

Fig. 13.16. Thermal conductivity of liquid diesel fuel as a function of pressure. Parameter – temperature

The thermal conductivity of a diesel liquid fuel is given by *Wenck* and *Schneider* [17] (1993) p. 69 as

$$\lambda = 0.17\left[1 - 0.00054\left(T - 273.15\right)\right] \Big/ \frac{\rho_{ref}}{1000} = 0.164 - 7.7112 \times 10^{-5} T . \quad (13.59)$$

This equation is very close to the *Cragoe* formula from 1929, see *Rose* and *Cooper* [12] (1977) p. 281, applied for $\rho(15.6°C, 1 atm) = 840 kg/m^3$,

$$\lambda = 0.1601 - 7.5343 \times 10^{-5} T . \qquad (13.60)$$

The comparison with the data reported by *Zenkevich* [19] (1968) and by *Rose* and *Cooper* [12] (1977) presented in Table 13.11 demonstrate the usefulness of Eq. (13.59). The prediction of Eq. (13.57) is also given in Table 13.11. The data for atmospheric pressure are between those reported by *Zenkevich* [19] (1968) and *Rose* and *Cooper* [12] (1977).

Table 13.11. Thermal conductivity of liquid diesel fuel measured by *Zenkevich* [19] (1968) and reported in *Vargaftik* et al. [16] (1996) p.1265, data reported by *Rose* and *Cooper* [12] (1977), prediction of Eq. (13.57).

t, °C	ρ, kg/m³	λ, W/(mK) [19] see [16] p. 1265	λ, W/(mK) see [12] p. 281	Eq. (13.59)	Eq. (13.57)
20	878.7	0.1169	0.1380	0.1414	0.1255
40	865.4	0.1146	0.1365	0.1399	0.1227
60	852.0	0.1122	0.1350	0.1383	0.1197
80	838.5	0.1099	0.1335	0.1368	0.1165
100	825.1	0.1076	0.1199	0.1352	0.1130

13.16 Cinematic viscosity of liquid diesel fuel

The Siemens data [15] for the liquid cinematic viscosity of summer diesel fuel are well represented by the function

$$\log_{10}\left(10^6 \nu\right) = 8.67271 - 0.04287T + 5.31710 \times 10^{-5} T^2$$

$$+ \left(0.00538 - 2.78208 \times 10^{-5} T + 3.74529 \times 10^{-8} T^2\right) 10^{-5} p. \qquad (13.61)$$

The prediction of Eq. (13.61) is presented in Fig. 13.17.

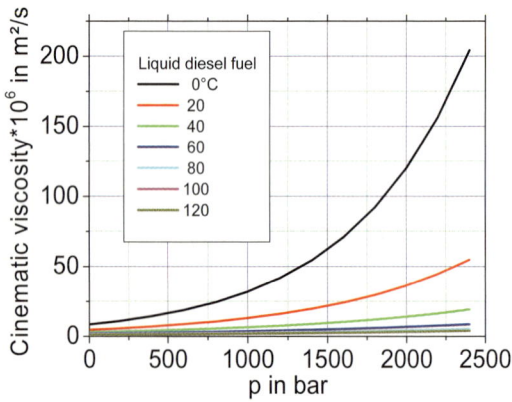

Fig. 13.17. Cinematic viscosity as a function of pressure. Parameter – temperature

For diesel liquid fuel at atmospheric pressure we have the experimental data summarized in Table 13.12. Table 13.12 also contains the prediction of Eq. (13.61). We see that the prediction gives lower values to those by *Vargaftik* and close to the other authors. Note that the consistence of the diesel oil depends very much on the origin of crude oil. Therefore it is advisable to have own measurements for the particular applications.

Table 13.12. Cinematic viscosity of liquid diesel fuel $v \times 10^6$ m^2/s at atmospheric pressure, [19] see *Vargaftik* et al. [16] (1996) p.1265, *Rose* and *Cooper* [12] (1977), *Wenck* and *Schneider* [17] (1993) p. 10, 11, *Zoebl* and *Kruschik* [20] (1978)

t,°C	[16]	[12]	[17]	[20]	Eq. (13.61)
15.6		5			5.34
20	8.94		2.8	1.79-2.97	4.73
37.8		3			3.05
40	4.80		2.5		2.90
60	3.04				1.96
80	2.14				1.46
100	1.62				1.20

13.17 Density as a function of temperature and pressure for diesel fuel vapor

We consider the diesel vapor as a perfect gas. The decision to consider the diesel vapor as a perfect gas results in the following well known relations

$$\rho = p/(RT), \qquad (13.62)$$

and

$$\left(\frac{\partial \rho}{\partial T}\right)_p = -\frac{p}{RT^2} = -\rho/T, \qquad (13.63)$$

$$\left(\frac{\partial \rho}{\partial p}\right)_T = \frac{1}{RT}. \qquad (13.64)$$

13.18 Specific capacity at constant pressure for diesel vapor

The *Cragoe* formulas from 1929, see *Rose* and *Cooper* [12] (1977) p. 281, applied for $\rho(15.6°C, 1atm) = 840 kg/m^3$ gives

$$c_{p,liquid} = 903.35894 + 4.0357143T \qquad (13.65)$$

$$c_{p,gas} = c_{p,liquid} - 448.80952 = 454.55 + 4.0357143T . \qquad (13.66)$$

Wenck and *Schneider* [17] (1993) p. 67 reported an expression for the specific heat of the vapors of mineral oil products depending on their reference density and temperature. In our case this results in

$$c_p = 4186 \left[0.109 + 0.00028(T - 273.15) \right] \left(4 - \frac{\rho_{reff}}{1000} \right) = 430.14 + 3.704T . \qquad (13.67)$$

Comparing Eq. (13.67) with (13.66) we see close similarity having stronger dependence on temperature in Eq. (13.66). For 25°C Eq. (13.67) gives 1534.5 J/(kgK). For comparison see Table 13.2 where values for diesel fuel vapor are given to be about 1700 J/(kgK), and n-heptan and n-octane values – 1659 J/(kgK).

Heywood [3] (1988), p.133, provided a more accurate expression for the specific heat at constant pressure for diesel fuel vapor with molecular weight $M = 148.6$ as a polynomial fit

$$c_p = \left(a_1 + a_2 t + a_3 t^2 + a_4 t^3 + a_5 t^{-2} \right) 4186/148.6 = d_1 + d_2 T + d_3 T^2 + d_4 T^3 + d_5 T^{-2} , \qquad (13.68)$$

where

$$t = T/1000 . \qquad (13.69)$$

The coefficients are given in Table 13.13.

Table 13.13. Diesel vapor specific heat at constant pressure *Heywood* [3] (1988) p.133

a_1	a_2	a_3	a_4	a_5
-9.1063	246.97	-143.74	32.329	0.0518

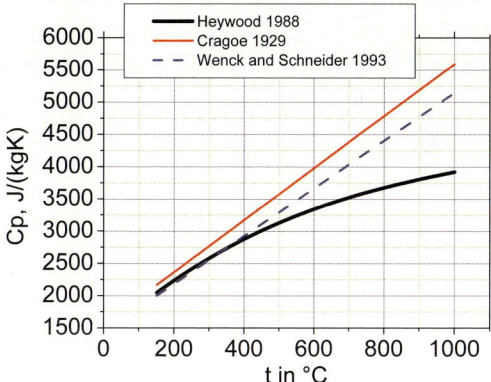

Fig. 13.18. Specific capacity at constant pressure. Comparison of different approximations

In this work we will use Eq. (13.68). Further, for the computation of the specific enthalpies and entropies we will need the integrals

$$h = h_0 + \int_{T_0}^{T} c_p dT = h_0 + \int_{T_0}^{T} \left(d_1 + d_2 T + d_3 T^2 + d_4 T^3 + d_5 T^{-2} \right) dT, \quad (13.70)$$

or

$$h = h_0 + d_1(T - T_0) + \frac{1}{2} d_2 \left(T^2 - T_0^2 \right) + \frac{1}{3} d_3 \left(T^3 - T_0^3 \right)$$

$$+ \frac{1}{4} d_4 \left(T^4 - T_0^4 \right) + d_5 \left(\frac{1}{T_0} - \frac{1}{T} \right), \quad (13.71)$$

and

$$s = s_0 + \int_{T_0}^{T} \frac{c_p}{T} dT = s_0 + \int_{T_0}^{T} \left(\frac{d_1}{T} + d_2 + d_3 T + d_4 T^2 + d_5 T^{-3} \right) dT, \quad (13.72)$$

or

$$s = s_0 + d_1 \ln \frac{T}{T_0} + d_2(T - T_0) + \frac{1}{2} d_3 \left(T^2 - T_0^2 \right) + \frac{1}{3} d_4 \left(T^3 - T_0^3 \right) + \frac{1}{2} d_5 \left(\frac{1}{T_0^2} - \frac{1}{T^2} \right).$$

$$(13.73)$$

13.19 Specific enthalpy for diesel fuel vapor

In accordance with our assumption that the diesel vapor is a perfect gas we have for the caloric equation

$$dh = c_p dT \tag{13.74}$$

where

$$\left(\frac{\partial h}{\partial T}\right)_p = c_p, \tag{13.75}$$

$$\left(\frac{\partial h}{\partial p}\right)_T = 0. \tag{13.76}$$

For given pressure we have from Eqs. (13.24) and (13.26) the corresponding saturation temperature

$$T' = T'(p). \tag{13.77}$$

The saturation enthalpy of the liquid is then computed by the Eq. (13.45)

$$h' = h\left[p, T'(p)\right]. \tag{13.78}$$

The evaporation enthalpy is then computed by using Eq. (13.21)

$$\Delta h = \Delta h\left[T'(p)\right]. \tag{13.79}$$

The saturation enthalpy of the vapor is then

$$h'' = h' + \Delta h = h\left[p, T'(p)\right] + \Delta h\left[T'(p)\right]. \tag{13.80}$$

For temperatures deviating from the saturation temperature at a given pressure we have

$$h(p,T) = h''(p) + \int_{T'(p)}^{T} c_p(T) dT$$

$$= h''(p) + d_1\left[T - T'(p)\right] + \frac{1}{2}d_2\left\{T^2 - \left[T'(p)\right]^2\right\} + \frac{1}{3}d_3\left\{T^3 - \left[T'(p)\right]^3\right\}$$

$$+ \frac{1}{4}d_4\left\{T^4 - \left[T'(p)\right]^4\right\} + d_5\left\{\frac{1}{T'(p)} - \frac{1}{T}\right\}. \tag{13.81}$$

This equation is completely consistent

a) with the liquid enthalpy equation,
b) with the definition of the saturation line, and
c) with the definition of the evaporation enthalpy being a function of the local temperature.

The equation can be also used as a extrapolation into the meta-stable state of the vapor.

13.20 Specific entropy for diesel fuel vapor

Again with accordance with the assumption that the diesel fuel vapor is a perfect gas we have for the differential form of the entropy definition

$$ds = c_p \frac{dT}{T} - R \frac{dp}{p}, \tag{13.82}$$

where

$$\left(\frac{\partial s}{\partial T}\right)_p = \frac{c_p}{T}, \tag{13.83}$$

$$\left(\frac{\partial s}{\partial p}\right)_T = -\frac{R}{p}. \tag{13.84}$$

The integrated form of Eq. (13.82) is

$$s = s_0 + \int_{T_0}^{T} \frac{c_p(T)}{T} dT - R \ln \frac{p}{p_0}$$
$$= s_0 + d_1 \ln \frac{T}{T_0} + d_2 (T - T_0) + \frac{1}{2} d_3 (T^2 - T_0^2)$$
$$+ \frac{1}{3} d_4 (T^3 - T_0^3) + \frac{1}{2} d_5 \left(\frac{1}{T_0^2} - \frac{1}{T^2}\right) - R \ln \frac{p}{p_0}. \tag{13.85}$$

An useful approach is to select as a reference conditions

$$p_0 = p \tag{13.86}$$

$$T_0 = T'(p) \tag{13.87}$$

and

$$s_0 = s''(p) = s'(p) + \Delta s\left[T'(p)\right] = s'(p) + \frac{\Delta h\left[T'(p)\right]}{T'(p)}. \tag{13.88}$$

This equation is completely consistent

a) with the liquid entropy equation,
b) with the definition of the saturation line, and
c) with the definition of the evaporation enthalpy being function of the local temperature.

Like the specific enthalpy equation, the specific entropy equation can be also used for extrapolation into the meta-stable state of the vapor.

13.21 Thermal conductivity of diesel fuel vapor

Data for diesel vapor are not known to me. That is why I take the data for n-octane for 1 bar presented by *Vargaftik* et al. [16] (1996). The approximation is

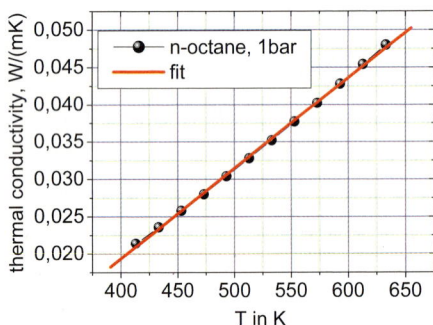

Fig. 13.19. Thermal conductivity of n-octane vapor at 1 bar as a function of temperature. Data taken from *Vargaftik* et al. [16] (1996), p.350

$$\lambda = -0.02912 + 1.21171 \times 10^{-4} T, \tag{13.89}$$

See Fig. 13.19.

13.22 Cinematic viscosity of diesel fuel vapor

Data for viscosity of the diesel fuel vapor are not known to me. Taking an approximation of the data for saturated n-octane vapor from *Vargaftik* et al. [16] (1996), see Appendix 13.1, I obtain

$$v'' = 10^{-9.67068 + 1580.93134(1/T)}, \qquad (13.90)$$

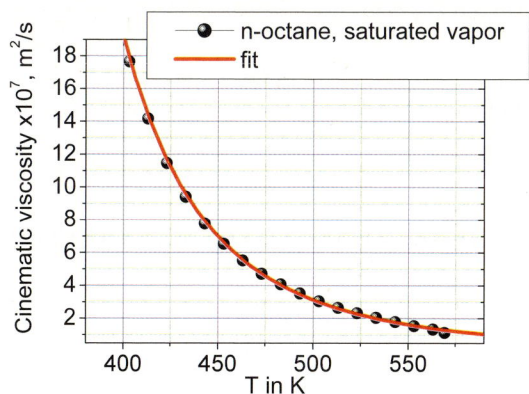

Fig. 13.20. Cinematic viscosity of n-octane vapor as a function of the temperature

See Fig. 13.20.

References

1. Bhatt S (1985) Acoustics Letters, vol 8 no 7 pp 105-108, cited in [17] p. 100
2. Fuel properties: no more available WEB site
3. Heywood JB (1988) Internal combustion engine fundamentals, McGraw-Hill, Inc
4. JSME Data Book (1983): Thermophysical Properties of Fluids
5. Kolev NI (2007, 3d ed.) Multi-Phase Flow Dynamics, Vol. 1 Fundamentals + CD, Springer, Berlin, New York, Tokyo,
6. Kolev NI (2007, 3d ed.) Multi-Phase Flow Dynamics, Vol. 2 Thermal and mechanical interactions, Springer, Berlin, New York, Tokyo
7. Landolt Boernstein Series of Physical Chemistry
8. Philipov P (1978) Similarity of properties of mater, Moscow, Published by the Moscow University in Russia in Russian
9. Prof. Robert Dibble, UC Berkeley, private communication
10. Reid RC, Prausnitz JM and Poling (1987) The Properties of Gases and Liquids, McGraw-Hill
11. Reid RC, Prausnitz JM and Sherwood TK (1971) The properties of gases and liquid, 3th ed, McGraw-Hill Book Company, New York, Russian translation: Chimia, Leningrad (1982)

12. Rose JW and Cooper JR eds. (1977) Technical data on fuel, Seventh edition, The British National Committee of the World Energy Conference, ISBN 0 7073 0129 7
13. Rossini FD et al. (1953) Selected values of physical and thermodynamic properties of hydrocarbons and related compounds, Carnegy Press, Pittsburgh, Pennsylvania
14. Spiers HM, Technical data of fuel, see in Wenck and Schneider [17] (1993) p.72
15. SIEMENS (2003) SIEMENS VDO Automotive, Diesel Systems, AT PT DS CR EIN, Regensburg, Germany, Proprietary
16. Vargaftik NB, Vinogradov YK and Yargin VS (1996) Handbook of physical properties of liquids and gases, Third augmented and revised edition, Begel House
17. Wenck H and Schneider C (November 1993) Collection of chemical and physical data of automotive fuels, German Society for Petroleum and Coal Science and Technology, DGMK-Project 409, Hamburg, ISBN 3-928164-63-5
18. Yaws CL and Chung H (1991) Chem. Eng., vol 98 pp 140-150
19. Zenkevich VB (1968) Izvestiya MVO SSSR, Energetika, no 2 p 8
20. Zoebl H and Kruschik J (1978) Strömung durch Rohre und Ventile, Springer-Verlag, Wien, New York

Appendix 13.1 Dynamic viscosity and density for saturated n-octane vapor

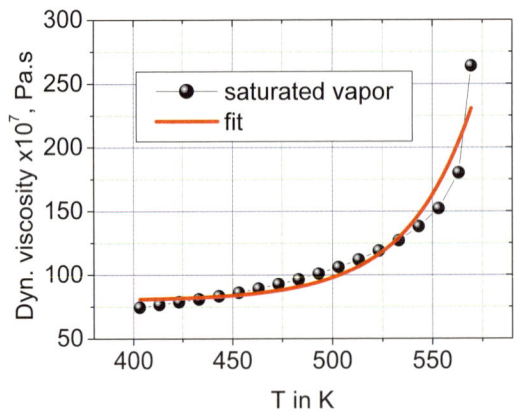

$$\eta'' = 7.98308 \times 10^{-6} + 3.81746 \times 10^{-13} \exp(-T/32.54977)$$

Fig. 13.21. Dynamic viscosity for saturated n-octane vapor as a function of the temperature

Appendix 13.1 Dynamic viscosity and density for saturated n-octane vapor

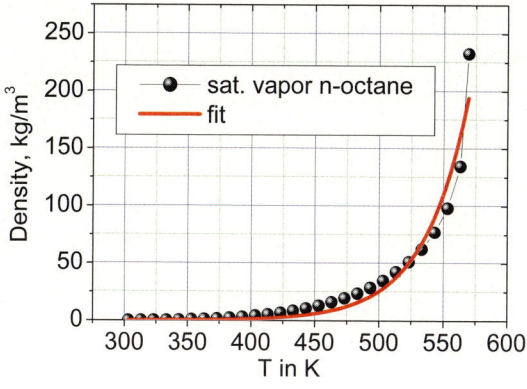

$$\rho'' = 130.69402 + 1.07247 \times 10^{-5} \exp(-T/34.07136)$$

Fig. 13.22. Density for saturated n-octane vapor as a function of the temperature

Index

1/7-th velocity profile, 2

2D-Steady state developed single phase incompressible flow in a circular pipe, 126

Activation of surface crevices, 221
Active nucleation site density, 221
Algebraic models for eddy viscosity, 145
Amsden, 177, 178
Analogy between momentum and heat transfe, 18
Avdeev, 233
Averaged drag coefficient, 2
Averaged volumetric fraction, 175
Axial distribution of the averaged turbulent kinetic energy, 132, 133

Bataille, 145
Bertodano, 127
Blasius, 77
Blasius solution, 6
Blowing and suction, 55
Boiling point, 273
Boundary conditions at the wall, 12
Boundary layer Reynolds number, 5
Boundary layer thickness, 3
Boussinesq, 32, 38
Boussinesq hypothesis, 154, 161
Bubble departure diameter, 61, 222
Bubble departure frequency, 222
Bubble generation dynamics, 156
Bubble growth in wall boundary layer, 238
Bubbles, 216
Buffer layer, 17
Buffer zone, 8
Bunsen absorption coefficient, 188
Buoyancy driven turbulence generation, 56
BWR bundle, 135

Carbon dioxide - Water, 206
Cavitation inception number, 220
Chandesris, 76
Cinematic viscosity of diesel fuel vapor, 301
Clausius-Clapayron, 276
Colebrook and White, 14
Complete rough region, 14
Constituents of diesel fuel, 271
Convection inside the droplet, 249
Critical pressure, 275
Critical temperature, 275

Damping factor, 155
Decay constant for single phase flow, 127
Decay of turbulence in a pipe flow, 128
Deformation of the mean values of the velocities, 76
Deformation of the velocity field, 55
Deformation rate, 32
Density, 295, 302
Deposition, 10
Deposition of the micro-bubbles into the boundary layer, 223
Diesel fuel vapor, 295
Diffusion coefficients, 209
Diffusion controlled bubble growth, 217
Diffusion mass transfer, 224
Dimensionless temperature, 16
Dispersion force, 34, 99
Displacement thickness, 3
Dissolved inert gases, 215
Distribution of the turbulent kinetic energy in the continuous gas, 139
Dittus-Boelter, 79
Drag forces, 93
Droplets deposition, 87
Dynamic friction velocity, 254
Dynamic turbulent viscosity, 33
Dynamic viscosity, 302

Eddy conductivity, 35
Eddy diffusivity, 34
Eddy viscosity, 11
Eddy viscosity in bubbly flow, 145
Effect of the wall boiling on the eddy viscosity, 156
Effective diffusion coefficient, 153
Effective mixing velocity, 147
Effective stagnation pressure difference, 30
Energy conservation in entropy form, 35
Eötvös, 97
Equilibrium solution and dissolution, 211
Euler-Euler description, 109
Euler-Lagrange method, 109
Euler-methods, 125
Existence of micro-bubbles in water, 219

Fanning factor, 4
Field indicator, 174
Film thickness, 247
Film-gas interface in the multi-phase flows, 246
Films, 246
Filtered volume fraction, 175
Filtering, 173, 174
Fine eddy generation, 121
Fine resolution, 46
Flow over plates, 1
Fluctuation, 12
Forced convection without boiling, 83
Fourier equation, 229, 244, 248
Fourier series, 244
Friction coefficient, 1, 4
Friction coefficient of turbulent flow, 233
Friction velocity, 5

Gap Stanton number, 148
Gas constant, 275
Gas flow, 130
Gas side averaged shear stress, 254
Gauss distribution, 174
Gauss function, 40
Gaussian probability distribution, 40
Geometrical film-gas characteristics, 246
Gibbs equation, 39

Hanjalic, 119, 120
Heat transfer, 79
Heat transfer across droplet interface, 242
Heat transfer coefficient, 3, 16

Heat transfer coefficient on the surface of moving solid sphere and water droplets, 224, 240
Heat transfer coefficients, 2
Heat transfer from the interface into the bulk liquid, 248
Heat transfer to fluid in a pipe, 15
Henry's coefficient, 186
Henry's law, 186
Heterogeneous nucleation at walls, 221
Hydraulic smooth wall surface, 6
Hydrocarbons, 271
Hydrogen absorption, 204
Hydrogen water, 203

Inner scale, 42
Interfacial forces, 93
Irreversible dissipated power, 76
Irreversibly dissipated power, 36, 76
Isothermal bulk modulus, 283
Isothermal coefficient of compressibility, 283
Isotropic turbulence, 33, 40, 250
Isotropy, 40, 173

Jukowski, 94

Kataoka, 157
k-eps framework, 44
k-eps models in system computer codes, 125
Kinematics viscosity, 34
Kirillov, 18
Kolmogoroff small scales, 173
Kolmogorov, 43, 77
Kolmogorov-Pandtl expression, 42
Kutateladze, 12

Lagrangian eddy-droplet interaction time, 111
Lagrangian time scale of turbulence, 111
Lahey, 44, 56
Laminar flow over the one site of a plane, 1
Lance, 145
Laplace transformations, 245
Large eddy simulations, 173
Large Scale Simulation, 173
Large scale turbulent motion, 42

Latent heat of evaporation, 279
Laufer, 9
Launder, 46
Legendre transformation, 39
Lift force, 93
Liquid density, 280
liquid specific heat at constant pressure, 285
Liquid surface tension, 292
Liquid velocity of sound, 284
Local algebraic models, 154
Local volume average, 176
Lockhart-Martinelli, 159
Lopez de Bertodano, 120
Lubrication force, 98

Magnus force, 94
Martinelli solution, 16
Martinelli solution for temperature profile, 16
Martinelli-Nelson multiplier, 81
Mass source due to turbulent diffusion, 153
Mass transfer due to turbulence diffusion, 249
Maxwellian distribution, 114
Mei, 96
Micro-bubbles, 219
Mixing length, 8, 34, 158
Model fluid, 276
Modification of the boundary layer share due to modification of the bulk turbulence, 161
Molar weight, 275
Molecular diffusion, 229
Momentum equations, 29
Morton, 97
Multi-group bubble approach, 155

Nikuradse, 7, 9
Nikuradze, 7
Nitrogen absorption, 199
Nitrogen water, 199
Normal Reynolds stresses, 100
Nucleate boiling, 17

One-way coupling, 44, 109
Oxygen absorption, 193
Oxygen in water, 193

Particle-eddy interaction time without collisions, 112
Particle-eddy interaction with collisions, 113
Particle-eddy interaction without collisions, 110
Particle-eddy interactions, 109
Porous body, 46
Porous structure and multiphase flow, 178
Post critical heat transfer, 10
Prandtl mixing length theory, 8
Pseudo-turbulence, 99
Pulsation parallel to the wall, 149
Pulsation through the gap, 147
Pulsation velocity, 250
Pulsations normal to the wall, 146
PWR bundle, 86

Radial fluctuation, 10
Random Dispersion Model, 109
Rate of dissipation of the kinetic energy of isotropic turbulence, 46
Rayleigh, 38
Rectangular rod array, 148
Reference liquid density point, 274
Refractory gas release, 217
Rehme, 10, 22, 146, 147
Reichardt, 10, 155
Reichardt solution, 13
Resolution scale, 174
Responds coefficient for clouds of particles, 112
Response coefficient, 110
Response coefficient for single particle, 110
Reynolds, 32
Reynolds analogy, 16
Reynolds stresses, 32
Reynolds turbulence, 34
Rod bundle, 145
Rodi, 44, 47

Sagaut, 173
Sato, 157
Saturation concentration, 185
Saturation line, 276
Second law of thermodynamics, 39
Sekogushi, 155

Serizawa, 157
Shear stress at the wall, 253
Shear stress in the film, 254
Sherwood number, 216
Simple algebraic models, 150
Single phase flow in rod bundles, 145
Singularities, 81
Smagorinski, 173, 178
Small scale, 42
Small scale turbulent motion, 41
Smoluchowski, 113
Solubility of O_2, N_2, H_2 and CO_2 in water, 185
Solubility of the carbon dioxide into the water, 206
Source terms for k-eps models in porous structures, 75
Sources as derived by Chandesris, 130
Sources for distributed hydraulic resistance coefficients, 130
Specific capacity at constant pressure for diesel vapor, 296
Specific enthalpy for diesel fuel vapor, 298
Specific entropy for diesel fuel vapor, 299
Specific field entropy, 35
Specific liquid enthalpy, 288
Specific liquid entropy, 290
Spin or vortices tensor, 32
Staffman, 95
Statistical theory of turbulence, 235
Steady developed flow, 79
Steady state flow in pipes with circular cross sections, 4
Steady state turbulent kinetic energy and its dissipation, 134
Stokes, 31
Stokes hypothesis, 31, 32
Stokes number, 112
Stokes rate of cubic dilatation, 32
Summer diesel fuel, 294

Taylor, 40
Taylor micro-scale of turbulence, 42
Taylor time micro-scale of turbulence, 41
Thermal conductivity of diesel fuel vapor, 300
Thermal conductivity of liquid diesel fuel, 292

Thermodynamic and transport properties of diesel fuel, 269
Thermodynamic temperature, 35
Thin thermal boundary layer, 229
Three-fluid multi-component model, 125
Tomiyama, 97
Transient diffusion inside the droplet, 243
Transient flow, 80
Transient flow in pipes with circular cross sections, 21
Transient solution and dissolution, 215
Transition region, 14
Triangular arrayed rod bundles, 149
Turbulence dispersion force, 102
Turbulence generated in particle traces, 57
Turbulence generation in the wakes behind the bubbles, 56
Turbulence in the boundary layer, 9
Turbulence model, 44
Turbulent coefficient of thermal conductivity, 35
Turbulent core, 17
Turbulent diffusion, 231
Turbulent flow parallel to plane, 2
Turbulent kinetic energy, 38
Turbulent length scale, 255
Turbulent Prandtl number, 17, 46
Turbulent pulsations, 36
Turbulent Reynolds number, 43
Two group k-eps models, 119
Two phase flow, 150
Two-way coupling, 44, 109

Undisturbed liquid eddy viscosity, 155

van Driest, 155
Van Driest, 9
Velocity distribution, 5
Velocity profile, 6
Viscous dissipation rate, 38
Viscous sub layer, 16
Void diffusion in bundles with sub-channel resolution, 153
Volumetric thermal expansion coefficient, 281
von Karman, 8, 14
von Karman universal velocity profiles, 7